Universitext

Editors

F.W. Gehring
P.R. Halmos

Universitext

Booss/Bleecker: Topology and Analysis
Charlap: Bieberbach Groups and Flat Manifolds
Chern: Complex Manifolds Without Potential Theory
Chorin/Marsden: A Mathematical Introduction to Fluid Mechanics
Cohn: A Classical Invitation to Algebraic Numbers and Class Fields
Curtis: Matrix Groups, 2nd ed.
van Dalen: Logic and Structure
Devlin: Fundamentals of Contemporary Set Theory
Edwards: A Formal Background to Mathematics I a/b
Edwards: A Formal Background to Mathematics II a/b
Endler: Valuation Theory
Frauenthal: Mathematical Modeling in Epidemiology
Gardiner: A First Course in Group Theory
Godbillon: Dynamical Systems on Surfaces
Greub: Multilinear Algebra
Hermes: Introduction to Mathematical Logic
Humi/Miller: Second Order Ordinary Differential Equations
Hurwitz/Kritikos: Lectures on Number Theory
Kelly/Matthews: The Non-Euclidean, The Hyperbolic Plane
Kostrikin: Introduction to Algebra
Luecking/Rubel: Complex Analysis: A Functional Analysis Approach
Lu: Singularity Theory and an Introduction to Catastrophe Theory
Marcus: Number Fields
McCarthy: Introduction to Arithmetical Functions
Mines/Richman/Ruitenburg: A Course in Constructive Algebra
Meyer: Essential Mathematics for Applied Fields
Moise: Introductory Problem Course in Analysis and Topology
Øksendal: Stochastic Differential Equations
Porter/Woods: Extensions of Hausdorff Spaces
Rees: Notes on Geometry
Reisel: Elementary Theory of Metric Spaces
Rey: Introduction to Robust and Quasi-Robust Statistical Methods
Rickart: Natural Function Algebras
Smith: Power Series From a Computational Point of View
Smoryński: Self-Reference and Modal Logic
Stanisić: The Mathematical Theory of Turbulence
Stroock: An Introduction to the Theory of Large Deviations
Sunder: An Invitation to von Neumann Algebras
Tolle: Optimization Methods

Ray Mines Fred Richman
Wim Ruitenburg

A Course in Constructive Algebra

Springer-Verlag
New York Berlin Heidelberg
London Paris Tokyo

Ray Mines
Fred Richman
Department of Mathematical Sciences
New Mexico State University
Las Cruces, NM 88003
U.S.A.

Wim Ruitenburg
Department of Mathematics, Statistics,
and Computer Science
Marquette University
Milwaukee, WI 53233
U.S.A.

AMS Classifications: 03F65, 13-01

Library of Congress Cataloging-in-Publication Data
Mines, Ray.
A course in constructive algebra.
(Universitext)
Bibliography: p.
Includes index.
1. Algebra I. Richman, Fred. II. Ruitenburg, Wim. III. Title. IV. Title: Constructive algebra.
QA155.M53 1988 512 87-26658

Text prepared by the authors using T^3 software output on a LNO3 printer.
Printed and bound by Quinn-Woodbine Inc., Woodbine, New Jersey.
Printed in the United States of America.

9 8 7 6 5 4 3 2 1

ISBN 0-387-96640-4 Springer-Verlag New York Berlin Heidelberg
ISBN 3-540-96640-4 Springer-Verlag Berlin Heidelberg New York

Dedicated to Errett Bishop

Preface

The constructive approach to mathematics has enjoyed a renaissance, caused in large part by the appearance of Errett Bishop's book *Foundations of constructive analysis* in 1967, and by the subtle influences of the proliferation of powerful computers. Bishop demonstrated that pure mathematics can be developed from a constructive point of view while maintaining a continuity with classical terminology and spirit; much more of classical mathematics was preserved than had been thought possible, and no classically false theorems resulted, as had been the case in other constructive schools such as intuitionism and Russian constructivism. The computers created a widespread awareness of the intuitive notion of an effective procedure, and of computation in principle, in addition to stimulating the study of constructive algebra for actual implementation, and from the point of view of recursive function theory.

In analysis, constructive problems arise instantly because we must start with the real numbers, and there is no finite procedure for deciding whether two given real numbers are equal or not (the real numbers are not discrete). The main thrust of constructive mathematics was in the direction of analysis, although several mathematicians, including Kronecker and van der Waerden, made important contributions to constructive algebra. Heyting, working in intuitionistic algebra, concentrated on issues raised by considering algebraic structures over the real numbers, and so developed a handmaiden of analysis rather than a theory of discrete algebraic structures. Paradoxically, it is in algebra where we are most likely to meet up with wildly nonconstructive arguments such as those that establish the existence of maximal ideals, and the existence of more than two automorphisms of the field of complex numbers.

In this book we present the basic notions of modern algebra from a constructive point of view. The more advanced topics have been dictated by our preferences and limitations, and by the availability of constructive treatments in the literature. Although the book is, of necessity, somewhat self-contained, it is not meant as a first introduction to modern algebra; the reader is presumed to have some familiarity with the classical subject.

It is important to keep in mind that constructive algebra is algebra; in fact it is a generalization of algebra in that we do not assume the law of excluded middle, just as group theory is a generalization of abelian group theory in that the commutative law is not assumed. A constructive proof of a theorem is, in particular, a proof of that theorem. Every theorem in this book can be understood as referring to the conventional universe of mathematical discourse, and the proofs are acceptable within that universe (barring mistakes). We do not limit ourselves to a restricted class of 'constructive objects', as recursive function theorists do, nor do we introduce classically false principles, as the intuitionists do.

We wish to express our appreciation to A. Seidenberg, Gabriel Stolzenberg, Larry Hughes, Bill Julian, and Steve Merrin for their suggestions.

Ray Mines
Fred Richman
New Mexico State University

Wim Ruitenburg
Marquette University

Contents

CHAPTER V. PRINCIPAL IDEAL DOMAINS

CHAPTER VI. FIELD THEORY

CHAPTER VII. FACTORING POLYNOMIALS

CHAPTER VIII. COMMUTATIVE NOETHERIAN RINGS

CHAPTER IX. FINITE DIMENSIONAL ALGEBRAS

Chapter I. Sets

1. CONSTRUCTIVE VS. CLASSICAL MATHEMATICS

The classical view of mathematics is essentially descriptive: we try to describe the facts about a static mathematical universe. Thus, for example, we report that every polynomial of odd degree has a root, and that there is a digit that occurs infinitely often in the decimal expansion of π. In opposition to this is the constructive view of mathematics, which focuses attention on the dynamic interaction of the individual with the mathematical universe; in the words of Hao Wang, it is a mathematics of doing, rather than a mathematics of being. The constructive mathematician must show how to construct a root of a polynomial of odd degree, and how to find a digit that occurs infinitely often in the decimal expansion of π.

We picture an idealized mathematician U interacting with the mathematical universe; this is the "you" who finds the δ, and to whom the ϵ is given, when we say "given ϵ you can find δ." The phrases "there exists" and "you can find" mean that U can carry out the indicated constructions. The disjunction of two statements "P_1 or P_2" means that either P_1 is true or P_2 is true, and that U can determine which of these alternatives holds. As "P_1 or P_2" means that there exists i in $\{1,2\}$ such that P_i is true, the meaning of "or" can be derived from the meaning of "there exists", and it is the interpretation of this latter phrase that is fundamental to constructive mathematics.

Classical mathematics can also be encompassed by this picture; the difference lies in what powers we ascribe to U. An omniscient U can decide whether any given mathematical statement is true or false; so U can, for example, survey the decimal expansion of π and determine which digits appear infinitely often. With an omniscient U, our picture is just a more dynamic, anthropomorphic portrayal of classical mathematics.

In constructive mathematics we assume only that U can carry out constructions that are finite in nature. As Errett Bishop put it, "The only way to show that an object exists is to give a finite routine for finding it." In this setting, we are not entitled to say that some digit appears infinitely often in the decimal expansion of π until we are prepared to exhibit such a digit, or at least produce an algorithm that will compute such a digit.

We consider U to be capable of carrying out any finite construction that is specified by an algorithm, but we do not rule out the possibility that U can do other things--even that U might be omniscient. The picture that results when we restrict U to finite constructions is the **computational interpretation** of mathematics. Because any statement that admits a constructive proof is true under the computational interpretation, we say that constructive mathematics has numerical meaning; because any statement that admits a constructive proof is true under the classical interpretation, we say that constructive mathematics is a generalization of classical mathematics.

Constructive mathematics is pure mathematics done algorithmically in order to respect the computational interpretation. The central notion of a finite routine, or algorithm, is taken as primitive. Any attempt to define what an algorithm is ultimately involves an appeal to the notion of existence—for example, we might demand that there exist a step at which a certain computer program produces an output. If the term "exist" is used here in the classical sense, then we have failed to capture the constructive notion of an algorithm; if it is used in the constructive sense, then the definition is circular.

Consider the difference between the constructive and the classical use of the connective "or". In order to prove "P_1 or P_2" constructively, we must construct an algorithm that will either prove P_1 or prove P_2, and by executing that algorithm we (the idealized mathematician) can determine which is true. To prove "P_1 or P_2" classically, it suffices to show that P_1 and P_2 cannot both be false. For example, let P_1 be the statement:

there exist positive integers x, y, z, and n such that

$$x^{n+2} + y^{n+2} = z^{n+2},$$

and let P_2, the famous unproved theorem of Fermat, be the denial of P_1. If P_1 is false then P_2 is true, so "P_1 or P_2" is classically provable; but we don't know, at this time, how to determine which of P_1 or P_2 holds, so we do not yet have a constructive proof of "P_1 or P_2".

A constructive proof of a theorem proves more than a classical one: a constructive proof that a sequence of real numbers converges implies that we can compute the rate of convergence; a constructive proof that a vector space is finite dimensional implies that we can construct a basis; a constructive proof that a polynomial is a product of irreducible polynomials implies that we can construct those irreducible polynomials.

Two statements P and Q may be classically equivalent without being constructively equivalent. Let P be the statement that every subgroup of the additive group $\mathbb{Z}$ of integers is cyclic. This means that we can always produce a generator from the data specifying the subgroup. Let Q be the statement that no subgroup G of $\mathbb{Z}$ can have the property that for each m in G, there is an integer in G that is not a multiple of m. The statements P and Q are immediately equivalent classically, but quite different constructively. The statement Q is true: as 0 is in G, there must be a nonzero integer n in G; as n is in G, there must be an integer properly dividing n in G, and so on until we arrive at a contradiction. But P is unlikely to be true, as we can see by considering the subgroup G generated by the perfect numbers: to construct a generator of G we must construct an odd perfect number or show that all perfect numbers are even.

On the other hand, any two constructively equivalent statements are classically equivalent; indeed any theorem in constructive mathematics is also a theorem in classical mathematics: a constructive proof is a proof.

Suppose we are trying to find a constructive proof of a statement P which is classically true. After many unsuccessful attempts to prove P, we may be inclined to look for a counterexample. But we cannot hope to prove the denial $\neg P$ of P, which is what a *bona fide* counterexample would entail, because $\neg P$ is classically false. As this avenue is closed to us, we need some other alternative to persisting in trying to prove P.

One approach is to fix a formal language in which P is expressible, specify precisely what sequences of words in that language constitute proofs of P, and show that no such proof can be constructed (possibly by giving an unintended interpretation of the formal language, and showing that P is false in that interpretation). Such a program can be illuminating, but it is often doubtful that the formal system adequately reflects the informal mathematics. A more serious objection is that engaging in such independence arguments requires a drastic shift in point of view; a procedure that stays closer to the subject at hand would be preferable. To this end we introduce the idea of an *omniscience principle* and of a *Brouwerian example*.

A rule α that assigns to each positive integer n an element α_n of $\{0,1\}$ is called a **binary sequence.** An **omniscience principle** is a classically true statement of the form "$P(\alpha)$ for all binary sequences α" which is considered not to have a constructive proof. For example, from the classical point of view, for each binary sequence α either

(P) there exists n such that $\alpha_n = 1$,

or (Q) $\alpha_n = 0$ for each positive integer n.

The assertion that either P or Q holds for any binary sequence α is called the **limited principle of omniscience** (**LPO**). As Q is the negation of P, the limited principle of omniscience is a form of the **law of excluded middle:** the assertion that for any statement P, the statement "P or not P" holds. The limited principle of omniscience and, *a fortiori*, the law of excluded middle, is rejected in the constructive approach as no one seriously believes that we can construct an algorithm that, given α, chooses the correct alternative, P or Q.

Another argument against LPO is that if we restrict our idealized mathematician to specific kinds of algorithms, as is done in the Russian school of constructive mathematics, we can show that LPO is false. Fix a computer programming language capable of expressing the usual number theoretic functions and symbol manipulations. It can be shown that there is no computer program that accepts computer programs as inputs, and when applied to a program that computes a binary sequence, returns 1 if the sequence contains a 1, and returns 0 otherwise. So if we require our rules to be given by computer programs, then LPO is demonstrably false. This argues against accepting LPO because any informal algorithm that we use will undoubtedly be programmable, so our theorems will be true in this computer-program intepretation; but we do not restrict ourselves to computer programs lest we rule out the possibility of intepreting our theorems classically: indeed, LPO is classically true.

If a statement P can be shown to imply LPO, then we abandon the search for a constructive proof of P. We do not assert that P is false, as we denied the existence of the computer program in the preceding paragraph;

after all, P may admit a classical proof. Rather, statements like LPO are thought of as *independent* in the sense that neither they nor their denials are valid.

For example, consider the classically true statement P that every subset of the integers is either empty or contains an element. Given a binary sequence α, let $A = \{1\}$ and let $B = \{\alpha_n : n \in \mathbb{N}\}$. Then $A \cap B$ is a subset of the integers. If $A \cap B$ contains an element, then that element must be 1, so, from the definition of B, there exists n such that $\alpha_n = 1$; if $A \cap B$ is empty, then $\alpha_n = 0$ for all n. Thus if P holds, then so does LPO.

A weaker omniscience principle is the **lesser limited principle of omniscience (LLPO)**, which states that for any binary sequence α that contains at most one 1, either $\alpha_n = 0$ for all odd n, or $\alpha_n = 0$ for all even n. This implies that, given any binary sequence, we can tell whether the first occurrence of a 1, if any, occurs at an even or at an odd index n. As in the case of LPO, if we restrict ourselves to rules given by computer programs, then we can refute LLPO. If you think of α as a black box into which you put n and get out α_n, then it is fairly clear that you cannot hope to establish LPO or LLPO. Or consider the sequence α defined by:

$\alpha_{2n} = 1$ if, and only if, there are 100 consecutive 6's in the first n places of the decimal expansion of π.

$\alpha_{2n+1} = 1$ if, and only if, there are 100 consecutive 7's in the first n places of the decimal expansion of π.

As $\pi/4 = 1 - 1/3 + 1/5 - 1/7 + 1/9 - \cdots$, there is an algorithm for computing the terms α_n. But unless we chance upon 100 consecutive 6's or 7's, we are hard put to find an algorithm that will tell us the parity of the index n for which $\alpha_n = 1$ for the first time (if ever).

A **Brouwerian example** E is a construction $E(\alpha)$ based on an arbitrary binary sequence α. We say that a Brouwerian example E **satisfies a condition** C if $E(\alpha)$ satisfies C for each α; we say that E **does not satisfy a condition** C if there is an omniscience principle "$P(\alpha)$ for all α" such that whenever $E(\alpha)$ satisfies C, then $P(\alpha)$ holds. A **Brouwerian counterexample** to a statement of the form "C_1 implies C_2" is a Brouwerian example that satisfies C_1 but does not satisfy C_2.

Our construction $A \cap B$ above is a Brouwerian example of a subset of the integers that neither contains an element nor is empty. We now construct a Brouwerian example of a bounded increasing sequence of real numbers that does not have a least upper bound. For each binary sequence α let $E(\alpha)$ be the sequence β of real numbers such that $\beta_n = \sup_{i=1}^{n}\alpha_i$. Then $E(\alpha)$ is a bounded increasing sequence of real numbers. Let C be the condition, on a sequence of real numbers, that it has a least upper bound. We shall show that E does not satisfy C. Let $P(\alpha)$ be the property that either $\alpha_n = 0$ for all n, or there exists n such that $\alpha_n = 1$, and suppose $E(\alpha)$ satisfies C. If the least upper bound of $E(\alpha)$ is less than 1, then $\alpha_n = 0$ for all n. If the least upper bound of $E(\alpha)$ is positive, then there exists n such that $\alpha_n > 0$, and hence $\alpha_n = 1$. Thus $P(\alpha)$ holds.

EXERCISES

1. Show that LPO implies LLPO.

2. Each subset of $\{0,1\}$ has 0, 1 or 2 elements. Construct a Brouwerian counterexample to that statement.

3. Construct a Brouwerian example of a nonempty set of positive integers that does not contain a smallest element.

4. Construct a Brouwerian example of a subgroup of the additive group of integers that is not cyclic.

5. Construct a Brouwerian example of two binary sequences whose sum contains infinitely many 1's, yet neither of the original sequences does.

6. Call a statement **simply existential** if it is of the form "there exists n such that $\alpha_n = 1$" for some binary sequence α. Show that LLPO is equivalent to

   $$\neg(A \text{ and } B) \text{ if and only if } \neg A \text{ or } \neg B$$

   for each pair of simply existential statements A and B.

7. The **weak limited principle of omniscience** (**WLPO**) is the statement that for each binary sequence α either $\alpha_n = 0$ for all n or it is impossible that $\alpha_n = 0$ for all n. Show that LPO implies WLPO, and that WLPO implies LLPO.

8. Let S be the set of all finite sequences of positive integers. By a **finitary tree** we mean a subset T of S such that

(i) For each $s \in S$, either $s \in T$ or $s \notin T$,

(ii) If $(x_1,\ldots,x_n) \in T$, then $(x_1,\ldots,x_{n-1}) \in T$,

(iii) For each $(x_1,\ldots,x_n) \in T$, there is $m \in \mathbb{N}$ such that if $(x_1,\ldots,x_n,z) \in T$, then $z \leq m$.

An **infinite path** in T is a sequence $\{x_i\}$ of positive integers such that $(x_1,\ldots,x_n) \in T$ for each n. **König's lemma** states that if T is infinite (has arbitrarily large finite subsets), then T has an infinite path. Show that König's lemma implies LLPO.

2. SETS, SUBSETS AND FUNCTIONS

We deal with two sorts of collections of mathematical objects: sets and categories. Our notion of what constitutes a **set** is a rather liberal one.

2.1 DEFINITION. *A set S is defined when we describe how to construct its members from objects that have been, or could have been, constructed prior to S, and describe what it means for two members of S to be equal.*

Following Bishop we regard the **equality relation** on a set as conventional: something to be determined when the set is defined, subject only to the requirement that it be an equivalence relation, that is

reflexive: $a = a$.

symmetric: If $a = b$, then $b = a$.

transitive: If $a = b$ and $b = c$, then $a = c$.

An **n-ary relation** on a set S is a property P that is applicable to n-tuples of elements of S, and is **extensional** in the sense that if $x_i = y_i$, for $i = 1,\ldots,n$, then $P(x_1,\ldots,x_n)$ if and only if $P(y_1,\ldots,y_n)$. Note that equality is a binary relation in this sense. The relation P is **decidable** if for each n-tuple $x_1,\ldots,x_n$ either $P(x_1,\ldots,x_n)$ holds or it doesn't hold.

A unary relation P on S defines a **subset** $A = \{x \in S : P(x)\}$ of S: an element of A is an element of S that satisfies P, and two elements of A are equal if and only if they are equal as elements of S. If A and B are subsets of S, and if every element of A is an element of B, then we say that A is **contained** in B, and write $A \subseteq B$. Two subsets A and B of a set S are **equal** if $A \subseteq B$ and $B \subseteq A$; this is clearly an equivalence relation on subsets of S. We have described how to construct a subset of S, and what it means for two subsets of S to be equal. Thus we have defined the set of all subsets, or the **power set**, of S. A subset of S is **nonempty** if there exists an element in it.

The **union** of two subsets A and B of S is the subset of S defined by $A \cup B = \{x \in S : x \in A \text{ or } x \in B\}$. The 'or' in this definition is interpreted constructively, so given x in $A \cup B$ we can determine which subset x is in (although we may not be able to tell whether it is in both). In terms of existence, $x \in A \cup B$ means that there exists i in $\{1,2\}$ so that if $i = 1$, then $x \in A$, and if $i = 2$, then $x \in B$. The **intersection** of A and B is the subset $A \cap B = \{x \in S : x \in A \text{ and } x \in B\}$.

We regard the relation of **inequality** as conventional and not necessarily the denial of equality; the interpretation of the symbol $a \neq b$ will depend on the context. Every set admits the **denial inequality** defined by $a \neq b$ if $a = b$ is impossible. Some sets admit other more natural inequalities: if a and b are binary sequences, then the good interpretation of $a \neq b$ is that there exists n such that $a_n \neq b_n$. If a set has no specified inequality, we interpret $a \neq b$ as being the denial inequality. We employ the usual terminology involving inequality: to say a and b are **distinct** means $a \neq b$; to say a is **nonzero** means $a \neq 0$.

An inequality on a set may be one or more of the following:

consistent: $a \neq a$ is impossible.

symmetric: if $a \neq b$, then $b \neq a$.

cotransitive: if $a \neq c$, then for any b either $a \neq b$ or $b \neq c$.

tight: if $a \neq b$ is impossible, then $a = b$.

We almost always want an inequality to be symmetric because $a \neq b$ is supposed to embody the idea that a and b are distinct, which should be a symmetric relation. It is also natural to demand consistency, but in practice this property is usually unnecessary. A consistent, symmetric, cotransitive inequality is called an **apartness**; the inequality specified above for the set of binary sequences is a tight apartness, as is the standard inequality on real numbers (see II.3). The denial inequality need not be an apartness, nor need it be tight.

An inequality is said to be **standard** if it can be shown to be equivalent to the denial inequality using the law of excluded middle. A tight consistent inequality is standard because $\neg\neg(a \neq b)$ is equivalent to $\neg(a = b)$. The denial inequality is trivially a standard inequality. With one important exception (local rings), we will be interested only in standard inequalities. It should be noted, however, that the requirement that an inequality be standard has very little constructive content: one cannot even prove that every standard inequality on a one-element set is consistent (a statement that can be refuted using LEM need not be refutable).

A set S with a consistent inequality is **discrete** if for any two elements a and b of S, either $a = b$ or $a \neq b$; if S has no specified inequality, then S is discrete if it is discrete under the denial inequality. The inequality of a discrete set is the denial inequality, and is a tight apartness. However the assertion that a set is discrete does not *a priori* refer to the denial inequality, but rather to whatever inequality comes with S: to say that a set S of binary sequences is discrete means that for all a and b in S, either $a = b$ or there exists n such that $a_n \neq b_n$.

The set $\mathbb{Z}$ of integers is discrete. The set $\mathbb{Q}$ of rational numbers is also discrete: a rational number is a pair of integers m/n with $n \neq 0$, two rational numbers m_1/n_1 and m_2/n_2 being considered equal if $m_1n_2 = m_2n_1$. Another example of a discrete set is the ring $\mathbb{Z}_{12}$ of integers modulo 12: the elements of $\mathbb{Z}_{12}$ are integers, and two elements of $\mathbb{Z}_{12}$ are equal if their difference is divisible by 12. As we can decide whether or not an integer is divisible by 12, the set $\mathbb{Z}_{12}$ is discrete.

If $x \in S$ and A is a subset of S, then we define $x \notin A$ to mean that $x \neq a$ for each $a \in A$; if the inequality on S is the denial inequality, or if S has no specified inequality, then $x \notin A$ if and only if x cannot be in A. The **complement** of A in S is $S\backslash A = \{x \in S : x \notin A\}$. A subset A of S is **proper** if there exists x in S such that $x \notin A$. A subset A of a set S is **detachable** if for each x in S either $x \in A$ or $x \notin A$.

Given sets $S_1, S_2, \ldots, S_n$ we define their **Cartesian product** $S_1 \times S_2 \times \cdots \times S_n$ to consist of the n-tuples $(x_1,x_2,\ldots,x_n)$ with x_i in S_i for each i. Two such n-tuples $(x_1,x_2,\ldots,x_n)$ and $(y_1,y_2,\ldots,y_n)$ are equal if $x_i = y_i$ for each i. Relations may be identified with subsets of Cartesian products: a binary relation on S is a subset of $S \times S$.

If A and B are sets, then a **function** from A to B is a rule that assigns to each element a of A an element $f(a)$ of B, and is **extensional** in the sense that $f(a_1) = f(a_2)$ whenever $a_1 = a_2$. We write $f : A \to B$ to indicate that f is a function from A to B. Two functions f and g from A to B are **equal** if $f(a) = g(a)$ for each a in A. The **identity function** $f : A \to A$ is defined by setting $f(a) = a$ for each a in A.

To construct a function from A to B it suffices to construct a subset S of the Cartesian product $A \times B$ with the properties

(i) for each $a \in A$ there exists $b \in B$ such that $(a,b) \in S$,

(ii) if (a,b_1) and (a,b_2) are elements of S, then $b_1 = b_2$.

In the computational interpretation, the algorithm for the function comes from (i), which specifies the construction of an element b depending on the parameter a. Without (ii), however, the algorithm implicit in (i) need not be extensional. The fact that a subset S of $A \times B$ satisfying (i) and (ii) determines a function f such that $(a,f(a)) \in S$ for each a in A is known as the **axiom of unique choice.**

Let f be a function from A to B. We say that f is

one-to-one if $a_1 = a_2$ whenever $f(a_1) = f(a_2)$,

onto if for each b in B there exists a in A such that $f(a) = b$,

strongly extensional if $a_1 \neq a_2$ whenever $f(a_1) \neq f(a_2)$.

Note that any function between sets with denial inequalities is strongly extensional. If $S \subseteq A$, then the **image** of S under f is the set

$$f(S) = \{b \in B : b = f(a) \text{ for some } a \in A\}.$$

Thus f is onto if and only if $f(A) = B$. If $S \subseteq B$, then the **preimage** of S under f is the set

$$f^{-1}(S) = \{a \in A : f(a) \in S\}.$$

Two sets A and B **have the same cardinality** if there are functions f from A to B, and g from B to A, such that fg is the identity function on B and gf is the identity function on A; we say that the functions f and g are **inverses** of each other, and that each is a **bijection.** If A and B have the same cardinality, we write $\#A = \#B$. The axiom of unique choice implies that a one-to-one onto function is a bijection (Exercise 6). Classically we think of sets of the same cardinality as simply having the same *size*; constructively it is more accurate to think of them as having the same *structure*.

When we refer to the **cardinality** of a set we mean the set itself, ignoring any structure other than equality that it might have. The distinction between referring to a set and referring to its cardinality is primarily one of intent: when we refer to the cardinality of a set, we do not plan on paying attention to any characteristics of the set that are not shared by all sets with the same cardinality. For example, if x is an element of a group, then the set $S = \{1, x, x^2, x^3, \ldots\}$ is the submonoid generated by x, while the cardinality of S is the order of x. It is much the same distinction as that between a fraction and a rational number. A common device for dealing with this kind situation is to introduce *equivalence classes* but, following Bishop (1967), we prefer to deal with the equivalence directly and not introduce cumbersome new entities.

If a set A has the same cardinality as $\{1,\ldots,n\}$ (empty if $n = 0$) for some nonnegative integer n, then we say that A is an **n-element set**, or A has cardinality n, and write $\#A = n$. A **finite** set A is a discrete set which has cardinality n for some nonnegative integer n. Recall that a discrete set must be discrete in its specified inequality, if any, so a set can have finite cardinality without being finite; such sets are somewhat pathological, which is why we give them the longer name.

A set A is **finitely enumerable** if it is empty or there is a function from $\{1,\ldots,n\}$ onto A. Note that a finitely enumerable set is discrete if and only if it is finite. We say that A **has at most n elements** if whenever $a_0,\ldots,a_n \in A$, then there exist $0 \leq i < j \leq n$ such that $a_i = a_j$. A set is **bounded in number**, or **bounded**, if it has at most n elements for some n. A set is **infinite** if it contains arbitrarily large finite subsets.

A set A is **countable** if there is a function from a detachable subset of the positive integers onto A. Thus the empty set is countable, as is the set of odd perfect numbers. Nonempty countable sets are the ranges of functions from the positive integers, so their elements can be enumerated (possibly with repetitions) as $a_1, a_2, \ldots$.

A **sequence** of elements of a set A, or a **sequence in A**, is a function from the nonnegative integers $\mathbb{N}$ to A. We shall also speak of functions from the positive integers as being sequences. A **family of elements** of A, **indexed by** a set I, is a function f from I to A; the image of i in A under f is usually denoted f_i rather than $f(i)$. Thus a sequence is a family indexed by the nonnegative integers $\mathbb{N}$. A **finite family of elements** of A is a family of elements of A indexed by $\{1,\ldots,n\}$ for some positive integer n.

If $\{A_i\}_{i\in I}$ is a family of subsets of S, then its **union** is defined by $\cup_{i\in I}A_i = \{x \in S : \text{there exists } i \in I \text{ such that } x \in A_i\}$, and its **intersection** is defined by $\cap_{i\in I}A_i = \{x \in S : x \in A_i \text{ for all } i \in I\}$.

If S is a set with an inequality, and X is a set, then the set S^X of functions from X to S inherits an inequality from S by setting $f \neq g$ if there exists x in X such that $f(x) \neq g(x)$.

2.2 THEOREM. *Let S be a set with inequality and let X be a set. If the inequality on S is consistent, symmetric, cotransitive, or tight, then so, respectively, is the inequality on S^X.*

PROOF. Consistency and symmetry are clear. Suppose that the inequality on S is tight. If $f_1 \neq f_2$ is impossible, then there cannot exist x such that $f_1(x) \neq f_2(x)$. Thus, given x, it is impossible that $f_1(x) \neq f_2(x)$, whence $f_1(x) = f_2(x)$ for each x, so $f_1 = f_2$. Now suppose the inequality on S is cotransitive. If $f_1 \neq f_3$, then for some x we have $f_1(x) \neq f_3(x)$ so either $f_1(x) \neq f_2(x)$ or $f_2(x) \neq f_3(x)$ whereupon either $f_1 \neq f_2$ or $f_2 \neq f_3$. □

As an example of Theorem 2.2, take S to be the discrete set $\{0,1\}$ and X to be the nonnegative integers. Then S^X is the set of binary sequences. As $\{0,1\}$ is discrete, the inequality on $\{0,1\}$ is a consistent tight apartness, so the inequality on the set of binary sequences is also a consistent tight apartness. However if the set of binary sequences were discrete, then we could establish LPO.

EXERCISES

1. Give an example (not Brouwerian) of a consistent apartness that is not tight.

2. Show that the set of binary sequences is discrete if and only if LPO holds.

3. **A denial inequality that is not an apartness.** Let A be the set of binary sequences. For x and y in A define $x = y$ if there exists N such that $x_n = y_n$ for all $n \geq N$, and let $x \neq y$ be the denial inequality. Show that if this inequality is an apartness, then WLPO holds; show that if it is a tight apartness, then LPO holds.

4. A notty problem. A **difference relation** is a symmetric inequality such that any of the following three conditions holds:

   (i) $x \neq z$ implies $\neg(\neg\, x \neq y$ and $\neg\, y \neq z\,)$
   (ii) $\neg\, x \neq y$ and $\neg\, y \neq z$ implies $\neg\, x \neq z$

(iii) $x \neq z$ and $\neg\, x \neq y$ implies $\neg\neg\, y \neq z$.

Show that these conditions are equivalent, and that an apartness is a difference relation.

5. Define a natural tight apartness on the set of detachable subsets of a set. Show that a subset A of a set S is detachable if and only if it has a **characteristic function**, that is, a function f from S to $\{0,1\}$ such that

$$A = \{s \in S : f(s) = 1\}.$$

6. Show that a function is a bijection if and only if it is one-to-one and onto.

7. Show that a finitely enumerable discrete set is finite. Construct a Brouwerian example of a finitely enumerable set, with a tight apartness, that is not finite. Show that a finitely enumerable set is bounded in number. Construct a Brouwerian example of a set that is bounded in number but is not finitely enumerable.

8. Show that a nonempty set A is countable if and only if there exists a function from $\mathbb{N}$ onto A. Show that a discrete set is countable if and only if it has the same cardinality as a detachable subset of $\mathbb{N}$.

9. Show that a subset A of $\mathbb{N}$ is countable if and only if there is a detachable subset S of $\mathbb{N} \times \mathbb{N}$ such that $A = \pi S$, where π is the projection of $\mathbb{N} \times \mathbb{N}$ on its first component.

10. Show that the set of functions from a discrete bounded set A to $\{0,1\}$ need not be discrete. (Hint: Let A be the range of a binary sequence)

11. Show that if a set S is bounded in number, then any one-to-one function from S to S is onto.

12. Give a Brouwerian example of a subset A of $\mathbb{N}$ such that it is contradictory for A to be finite, but A is not infinite.

13. Let S be a nonempty set with an apartness, and n a positive integer. Show that the following are equivalent.

    (i) There exist elements $x_0,\ldots,x_n$ in S such that $x_i \neq x_j$ for $i \neq j$.

(ii) Given $y_1,\ldots,y_n$ in S, there exists z in S such that $z \neq y_i$ for $i = 1,\ldots,n$.

14. A set with inequality is **Dedekind infinite** if it is isomorphic, as a set with inequality, to a proper subset. Show that any Dedekind-infinite set satisfies (i) of Exercise 13 for each n.

15. Call a set S **ω-bounded** if for each sequence $\{s_i\}$ in S, there exists $m \neq n$ such that $s_m = s_n$. Show that if S is a discrete ω-bounded set, and $\{s_i\}$ is a sequence in S, and m is a positive integer, then there is a finite set I of m positive integers such that $s_i = s_j$ if i and j are in I. Show that if A and B are discrete ω-bounded sets, then so is $A \times B$.

3. CHOICE

The axiom of choice asserts the existence of a function with a certain property, so we might well suspect that its validity would be more dubious in a constructive setting where functions may be interpreted as algorithms. We phrase the axiom of choice as follows:

AXIOM OF CHOICE. *Let A and B be sets, and S a subset of $A\times B$ such that for each a in A there is an element b in B such that $(a,b) \in S$. Then there is a function $f:A \to B$ such that $(a,f(a)) \in S$ for each a in A.*

The axiom of choice can be criticized from the computational point of view on two grounds. The first concerns whether we can find an *algorithm* f (not necessarily extensional) with the required property. We have already come across this issue with the axiom of unique choice, and we take the position that the algorithm is inherent in the interpretation of the phrase "for each a in A there is an element b in B".

A more serious criticism is that, although we can find an algorithm f, we cannot find a *function*. In fact we can construct a Brouwerian counterexample to the axiom of choice.

3.1 EXAMPLE. *Let α be a binary sequence, let $A = \{x,y\}$ with the equality on A defined by setting $x = y$ if and only if there exists n such that $\alpha_n = 1$, and let $B = \{0,1\}$. Consider the subset $S = \{(x,0),(y,1)\}$ of $A\times B$. Suppose $f : A \to B$ satisfies $(a,f(a)) \in S$ for each $a \in A$. If $f(x) = f(y)$, then $\alpha_n = 1$ for some n; if $f(x) \neq f(y)$, then $\alpha_n = 0$ for all n.*

There are two restricted versions of the axiom of choice which are commonly accepted in constructive mathematics. The weaker of these is:

COUNTABLE AXIOM OF CHOICE. *This is the axiom of choice with* A *being the set of positive integers.*

If A is the set of positive integers, then there is no real distinction between an algorithm and a function in the computational interpretation, as each integer has a canonical representation. So this axiom follows from the interpretation of the phrase "for all a there exists b" as entailing the existence of an algorithm for transforming elements of A to elements of B.

Even stronger than the countable axiom of choice is:

AXIOM OF DEPENDENT CHOICES. *Let* A *be a nonempty set and* R *a subset of* A×A *such that for each* a *in* A *there is an element* a' *in* A *with* $(a,a') \in R$. *Then there is a sequence* $a_0,a_1,\ldots$ *of elements of* A *such that* $(a_i,a_{i+1}) \in R$ *for each* i.

The axiom of dependent choices implies the countable axiom of choice as follows. Suppose S is a subset of $\mathbb{N} \times B$ such that for each n in $\mathbb{N}$ there is an element b in B with $(n,b) \in S$. Let A consist of all finite sequences $b_0,b_1,\ldots,b_m$ in B such that $(i,b_i) \in S$ for each i, and let R consist of all pairs (a,a') of elements of A such that deleting the last element of a' gives a. Applying the axiom of dependent choices to R yields a sequence in A whose last elements form the required sequence in B.

The argument for the axiom of dependent choices is much the same as that for the countable axiom of choice. We shall freely employ these axioms, although we will often point out when they are used.

We shall have occasion to refer to the following axiom of choice for which we have no Brouwerian counterexample, yet which we believe is unprovable within the context of constructive mathematics.

WORLD'S SIMPLEST AXIOM OF CHOICE. *Let* A *be a set of two-element sets such that if* $a_1 \in$ A *and* $a_2 \in$ A, *then* $a_1 = a_2$. *Then there is a function* f *from* A *to* $\{x : x \in a$ *for some* $a \in$ A$\}$ *such that* $f(a) \in a$ *for each* $a \in$ A.

EXERCISES

1. Modify (3.1) to show that the axiom of choice implies the law of excluded middle.

2. Show that LLPO, together with dependent choice, implies König's lemma (see Exercise 1.7).

3. Show that the axiom of choice implies the world's simplest axiom of choice.

4. A set P is **projective** if whenever $\pi : A \to B$ is onto, and $f : P \to B$, then there exists $g : P \to A$ such that $\pi g = f$. Show that finite sets are projective. Show that countable discrete sets are projective if and only if the countable axiom of choice holds. Show that if discrete sets are projective, then the world's simplest axiom of choice holds.

4. CATEGORIES.

The collection of binary sequences forms a set because we know what it means for two binary sequences to be equal. Given two groups, or sets, on the other hand, it is generally incorrect to ask if they are equal; the proper question is whether or not they are *isomorphic*, or, more generally, what are the homomorphisms between them.

A **category**, like a set, is a collection of objects. An equality relation on a set constructs, given any two objects a and b in the set, a *proposition* '$a = b$'. To specify a category $\mathcal{C}$, we must show how to construct, given any two objects A and B in $\mathcal{C}$, a *set* $\mathcal{C}(A,B)$. In concrete categories, the objects of $\mathcal{C}$ are mathematical structures of some kind, and the set $\mathcal{C}(A,B)$ is the set of maps from A to B that respect that structure: if $\mathcal{C}$ is the category of sets, then $\mathcal{C}(A,B)$ is the set of functions from A to B; if $\mathcal{C}$ is the category of groups, then $\mathcal{C}(A,B)$ is the set of homomorphisms from A to B.

When we abstract the notion of composition of maps from these concrete situations, we see that a category must have, for any three objects A,B,C in $\mathcal{C}$, a function from $\mathcal{C}(A,B) \times \mathcal{C}(B,C)$ to $\mathcal{C}(A,C)$, called **composition** and denoted by juxtaposition, and an element $1_B \in \mathcal{C}(B,B)$, such that if $f \in \mathcal{C}(C,D)$, $g \in \mathcal{C}(B,C)$, and $h \in \mathcal{C}(A,B)$, then

(i) $1_B h = h$ and $g1_B = g$.

(ii) $(fg)h = f(gh)$

Every set S can be considered a category by setting

$$S(a,b) = \{x \in \{0\} : a = b\}.$$

The reflexive law for equality gives the element 1_B, and the transitive law gives (ii).

The sets and functions introduced in the previous section constitute a category: the **objects** of this category are sets, and the **maps** are the functions between the sets. We may also consider the category whose objects are sets with inequality, and whose maps are strongly extensional functions. The idea of category theory is to forget about the internal structure of the objects and to concentrate on the way the maps combine under composition. For example, a function f from A to B is one-to-one if $a_1 = a_2$ whenever $f(a_1) = f(a_2)$. This definition relies on the internal structure of the sets A and B, that is to say, on the elements of A and B and the equality relations on A and B. The categorical property corresponding to a function f being one-to-one is that if g and h are maps from any set C to A, and $fg = fh$, then $g = h$; that is, f is **left cancellable**. It is routine to show that f is one-to-one if and only if it is left cancellable.

A map f from A to B is onto if for each b in B there exists a in A such that $f(a) = b$. The corresponding categorical property is that f be **right cancellable**, that is, if g and h are maps from B to any set C, and $gf = hf$, then $g = h$. The proof that a function f is right cancellable if and only if it is onto is less routine than the proof of the corresponding result for left cancellable maps.

4.1 THEOREM. *A function is right cancellable in the category of sets if and only if it is onto.*

PROOF. Suppose $f:A \to B$ is onto and $gf = hf$. If $b \in B$, then there exists a in A such that $f(a) = b$. Thus $g(b) = g(f(a)) = h(f(a)) = h(b)$, so $g = h$. Conversely suppose $f:A \to B$ is right cancellable, and let Ω be the set of all subsets of $\{0\}$. Define $g:B \to \Omega$ by $g(b) = \{0\}$ for all b, and define $h:B \to \Omega$ by

$$h(b) = \{x \in \{0\} : b = f(a) \text{ for some } a\}.$$

Thus $h(b)$ is the subset of $\{0\}$ such that $0 \in h(b)$ if and only if there exists a such that $b = f(a)$. Clearly $gf = hf$ is the map that takes every element of A to the subset $\{0\}$. So $g = h$, whence $0 \in h(b)$, which means that $b = f(a)$ for some a. □

An **isomorphism** between two objects A and B of a category $\mathscr{C}$ is an element $f \in \mathscr{C}(A,B)$ such that there is $g \in \mathscr{C}(B,A)$ with $fg = 1_B$ and $gf = 1_A$. The element g is called the **inverse** of f; it is easily shown to be unique. A bijection between sets is an isomorphism in the category of sets. We say that A and B are **isomorphic**, and write $A \cong B$, if there is an isomorphism between A and B.

We will be interested mainly in categories of sets with algebraic structures, in which the maps are the functions that preserve those structures. In this case, the maps are called **homomorphisms**. If a homomorphism is one-to-one, it is called a **monomorphism**; if it is onto, it is called an **epimorphism**. A homomorphism from an object to itself is called an **endomorphism**, and an endomorphism that is an isomorphism is called an **automorphism**.

A **functor** T from a category $\mathscr{A}$ to a category $\mathscr{B}$ is a rule that assigns to each object $A \in \mathscr{A}$ an object $T(A) \in \mathscr{B}$, and to each map $f \in \mathscr{A}(A_1,A_2)$ a map $T(f) \in \mathscr{B}(T(A_1),T(A_2))$ such that

(i) $T : \mathscr{A}(A_1,A_2) \to \mathscr{B}(T(A_1),T(A_2))$ is a function

(ii) $T(fg) = T(f)T(g)$

(iii) $T(1_A) = 1_{T(A)}$.

A functor between two sets, viewed as categories, is simply a function. Note that if f is an isomorphism, then so is $T(f)$.

Using the notion of a functor, we can extend our definition of a family of elements of a set to a family of objects in a category $\mathscr{C}$. Let I be a set. **A family** A **of objects of $\mathscr{C}$ indexed by** I is a functor from I, viewed as a category, to the category $\mathscr{C}$. We often denote such a family by $\{A_i\}_{i\in I}$. If $i = j$, then the map from A_i to A_j is denoted by A_j^i, and is an isomorphism.

An element of the **disjoint union** of a family $\{A_i\}_{i\in I}$ of sets is a pair (i,x) such that $i \in I$ and $x \in A_i$. Two elements (i,x) and (j,y) of the disjoint union are **equal** if $i = j$ and $A_j^i(x) = y$. We identify A_i with the subset $\{(i,x) : x \in A_i\}$ of the disjoint union. Thus after constructing the disjoint union, we may consider the family $\{A_i\}_{i\in I}$ to be a family of elements of the power set of the disjoint union.

Let $\{A_i\}_{i\in I}$ be a family of sets, and let P be a set. Then a map from P to A_i may be identified with a map f from P to the disjoint union of $\{A_i\}_{i\in I}$ such that $f(P) \subseteq A_i$. Let F denote the set of maps f from P to the disjoint union of $\{A_i\}_{i\in I}$ such that $f(P) \subseteq A_i$ for some i in I. By a **family of maps** π_i from P to A_i we mean a family π of elements of F such that $\pi_i(P) \subseteq A_i$ for each i in I.

Let $\{A_i\}_{i\in I}$ be a family of objects in a category $\mathscr{C}$. A **categorical product** of the objects A_i is an object P together with a family of (projection) maps π_i from P to A_i such that for any object S and family of maps f_i from S to A_i, there exists a unique map f from S to P such that $\pi_i f = f_i$ for each i in I. A categorical product is unique up to isomorphism in the sense that if (P',π') is another one, then there exists an isomorphism θ from P to P' such that $\pi_i'\theta = \pi_i$ for each i. If $\mathscr{C}$ is the category of sets, then it is easy to verify that the set of all functions λ from I to the disjoint union of $\{A_i\}_{i\in I}$ such that $\lambda(i) \in A_i$ for each i, and $\pi_i(\lambda)$ defined to be $\lambda(i)$, is a categorical product of the A_i, which we refer to as *the* product and denote by $\Pi_{i\in I} A_i$.

If $I = \{1,\ldots,n\}$, then the product of the sets A_i is the Cartesian product $A_1 \times \cdots \times A_n$. If $A_i = S$ for each i in I, then we write the product, which is the set of functions from I to S, as S^I, or S^n if $I = \{1,\ldots,n\}$.

EXERCISES

1. Show that a function is one-to-one if and only if it is left cancellable.
2. Show that the categorical product of a family of objects is unique up to isomorphism.
3. Show that, in the category of sets, the set of all functions λ from I to the disjoint union of $\{A_i\}_{i\in I}$, such that $\lambda(i) \in A_i$ for each i, is a categorical product of $\{A_i\}_{i\in I}$.
4. Let I be the set of binary sequences, and, for each i in I, let A_i be $\{x \in \{0,1\} : x \geq i_j \text{ for all } j\}$. Show that the natural map from $\Pi_i A_i$ to A_0 is onto if and only if WLPO holds.
5. Consider the category of sets with inequality and strongly extensional functions. Show that the product $\Pi_i A_i$ in this

category is the product in the category of sets equipped with the inequality $\lambda \neq \mu$ if there exists i such that $\lambda(i) \neq \mu(i)$. Generalize Theorem 2.2 to this setting.

6. Let a be an object in a category $\mathcal{A}$. Show how $T(b) = \mathcal{A}(a,b)$ is a functor from $\mathcal{A}$ to the category of sets. Such a functor T is said to be **representable.**

7. If $\mathcal{C}$ is a category, then the **dual category** $\mathcal{C}'$ is defined to have the same objects as $\mathcal{C}$, but $\mathcal{C}'(a,b) = \mathcal{C}(b,a)$. The **coproduct** of a family of objects of $\mathcal{C}$ is the product in the category $\mathcal{C}'$. Describe the coproduct directly. What is the coproduct in the category of sets?

8. **Direct limits.** A direct system is a sequence of objects A_n and maps $f_n : A_n \to A_{n+1}$. An upper bound of a direct system is an object B together with maps $b_n : A_n \to B$ such that $b_{n+1}f_n = b_n$ for each n. A **direct limit** of a direct system is an upper bound L so that, for any upper bound B, there is a unique map $\mu : L \to B$ such that $\mu \ell_n = b_n$ for each n.
   (i) Show that any two direct limits are isomorphic.
   (ii) Show that the direct limit in the category of sets is the disjoint union of the A_n with the equality generated by requiring $a = f_n(a)$ for each $a \in A_n$.
   (iii) Show that the direct limit of discrete sets need not be discrete, but it is discrete if all the maps are one-to-one.

5. PARTIALLY ORDERED SETS AND LATTICES

A **partially ordered set** is a set P with a relation $a \leq b$ satisfying:

(i) $a \leq a$,
(ii) if $a \leq b$ and $b \leq c$, then $a \leq c$,
(iii) if $a \leq b$ and $b \leq a$, then $a = b$.

A **map** between two partially ordered sets P_1 and P_2 is a function f from P_1 to P_2 such that if $a \leq b$, then $f(a) \leq f(b)$. For the most part we will be interested in discrete partially ordered sets; in this case we write $a < b$ for $a \leq b$ and $a \neq b$.

Let a, b and c be elements of a partially ordered set P. We say that c

is the **greatest lower bound**, or **infimum**, of a and b, and write $c = a \wedge b$, if for each $x \in L$ we have $x \leq c$ if and only if $x \leq a$ and $x \leq b$. It is easily seen that such c is unique, if it exists. Similarly $c = a \vee b$ is the **least upper bound**, or **supremum**, of a and b if for each $x \in L$ we have $c \leq x$ if and only if $a \leq x$ and $b \leq x$.

A **lattice** is a partially ordered set in which any two elements have an infimum and a supremum. If S is a set, then the set of all subsets of S, ordered by inclusion, forms a lattice: the supremum of A and B is $A \cup B$ and the infimum is $A \cap B$. The set of positive integers, ordered by setting $a \leq b$ if b is a multiple of a, is a lattice: the supremum of a and b is their least common multiple, the infimum is their greatest common divisor. Note that $a \leq b$ is decidable in a discrete lattice because it is equivalent to $a \wedge b = a$.

If a lattice has a least element, then we denote that element by 0; if the lattice has greatest element, we denote it by 1.

A lattice is **distributive** if it satisfies the identity

$$a \wedge (b \vee c) = (a \wedge b) \vee (a \wedge c).$$

The lattice of all subsets of a set is distributive. The following 5-element discrete lattice is not distributive.

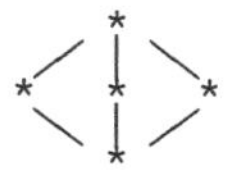

A lattice is **modular** if $a \vee (b \wedge c) = b \wedge (a \vee c)$ whenever $a \leq b$. It is easily seen that any distributive lattice is modular; the 5-element nondistributive lattice shown above is also modular. If G is a finite abelian group, then the set of finite subgroups of G is a modular lattice, which is distributive only if G is cyclic. More generally, the set of submodules of an R-module form a modular lattice. The simplest nonmodular lattice is the following 5-element discrete lattice.

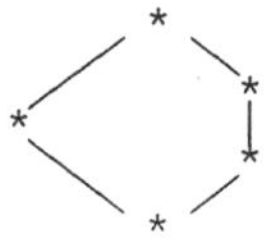

If $a \leq b$ are elements of a partially ordered set P, then we use the interval notation $[a,b]$ to denote the set $\{x \in P : a \leq x \leq b\}$. If P is a lattice, then $[a,b]$ is a lattice with the same suprema and infima as in P.

A key fact about modular lattices is that $[a \wedge d, d]$ and $[a, a \vee d]$ are isomorphic lattices for any elements a and d. We prove this in a slightly disguised form.

5.1 LEMMA. *Let $a \leq b$ and $c \leq d$ be elements of a modular lattice L. Define*

$$f(x) = a \vee (b \wedge x) = b \wedge (a \vee x)$$
$$g(y) = c \vee (d \wedge y) = d \wedge (c \vee y).$$

Then g maps $[f(c), f(d)]$ isomorphically onto $[g(a), g(b)]$ with inverse f.

PROOF. It suffices to show that if $c \leq x \leq d$, then $fgf(x) = f(x)$. We can write $fgf(x)$ as

$$(*) \qquad fgf(x) = b \wedge (a \vee c \vee (d \wedge b \wedge (a \vee x)))$$

or as

$$(**) \qquad fgf(x) = a \vee (b \wedge d \wedge (c \vee a \vee (b \wedge x))).$$

To show that $f(x) \leq fgf(x)$, use (*) and $f(x) = a \vee (b \wedge x)$. To show that $fgf(x) \leq f(x)$, use (**) and $f(x) = b \wedge (a \vee x)$. □

Taking $b = a \vee d$ and $c = a \wedge d$ in (5.1) we see that $[a \wedge d, d]$ and $[a, a \vee d]$ are isomorphic.

A subset C of a partially ordered set P is a **chain** if for each a and b in C, either $a \leq b$ or $b \leq a$; if P itself is a chain, we say that P is **linearly ordered.** A **maximal chain** in a partially ordered set is a chain C such that $C \cup \{a\}$ is a chain only if $a \in C$. The simplest nonmodular lattice above has two maximal finite chains, one of length 2 and one of length 3. For modular lattices this can't happen. We say that two linearly ordered sets C and D are **piecewise isomorphic** if there exist elements $c_1, \ldots c_n$ and $d_1, \ldots, d_n$ such that

(i) $\{x \in C : x \leq c_1\}$ is isomorphic to $\{x \in D : x \leq d_1\}$

(ii) $\{x \in C : x \geq c_n\}$ is isomorphic to $\{x \in D : x \geq d_n\}$

(iii) There is a permutation σ of $\{1, \ldots, n-1\}$ so that $[c_i, c_{i+1}]$ is isomorphic to $[d_{\sigma i}, d_{\sigma i+1}]$ for each $i < n$.

We leave the proof that piecewise isomorphism is a transitive relation as Exercise 4. If C and D are piecewise isomorphic discrete linearly ordered sets, then C and D have the same cardinality (Exercise 5).

5.2 THEOREM (Jordan-Hölder-Dedekind). *If a modular lattice has a maximal finitely enumerable chain X, then each finitely enumerable chain* *is contained in a maximal finitely enumerable chain that is piecewise isomorphic to X.*

PROOF. Let $x_0 \leq x_1 \leq \cdots \leq x_m$ be the maximal chain X; we will refer to m as the **formal length** of X. It is readily verified that $x_0 = 0$ and $x_m = 1$. Let $y_1 \leq \cdots \leq y_n$ be a chain Y. If $x_1 \wedge y_1 = x_1$, then Y is contained in the lattice $[x_1,1]$. By induction on m, the chain Y is contained in a maximal finitely enumerable chain in $[x_1,1]$ that is piecewise isomorphic to $x_1 \leq \cdots \leq x_m$, and therefore in a maximal finitely enumerable chain that is piecewise isomorphic to X. Otherwise $x_1 \wedge y_1 = 0$ and we have the following picture

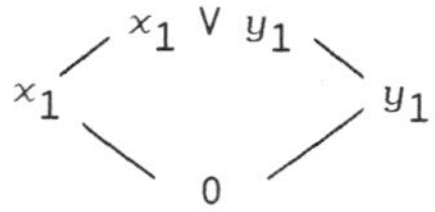

where $[y_1,x_1 \vee y_1]$ is isomorphic to $[0,x_1]$, and $[x_1,x_1 \vee y_1]$ is isomorphic to $[0,y_1]$. By induction on m the chain $x_1 \leq x_1 \vee y_1$ is contained in a maximal finitely enumerable chain in $[x_1,1]$, of formal length $m-1$, consisting of a maximal finitely enumerable chain C in $[x_1,x_1 \vee y_1]$ of formal length ℓ, and a maximal finitely enumerable chain D in $[x_1 \vee y_1, 1]$ of formal length $m-\ell-1$. The chain $\{y_1\} \cup D$ is a maximal chain in $[y_1,1]$ of formal length at most $m-\ell$ so, by induction on m, Y is contained in a maximal chain in $[y_1,1]$ that is piecewise isomorphic to $\{y_1\} \cup D$. Lemma 5.1 shows that the chain C is isomorphic to a maximal chain in $[0,y_1]$. Thus Y is contained in a maximal chain that is piecewise isomorphic to X. □

It follows from (5.2) that if a discrete modular lattice has a maximal finite chain of length n, then any finite chain is contained in a maximal chain of length n.

A partially ordered set P satisfies the **ascending chain condition** if for each sequence $p_1 \leq p_2 \leq p_3 \leq \cdots$ of elements of P, there is n such that $p_n = p_{n+1}$; the **descending chain condition** is defined similarly. Classically, if P satisfies the ascending chain condition, then we can find n such that $p_m = p_n$ for each $m \geq n$. From a constructive point of view, even the two element set $\{0,1\}$ fails to satisfy this form of the ascending chain condition.

We say that an element p of a partially ordered set P has **depth at most**

n if whenever $p = p_0 \leq p_1 \leq p_2 \leq \cdots \leq p_{n+1}$, then $p_i = p_{i+1}$ for some $i \leq n$. If P is discrete, we say that p has **depth at least** n if there exists a chain $p = p_0 < p_1 < \cdots < p_n$. An element has **bounded depth** if it has depth at most n for some n, **finite depth** if it has depth at most n, and at least n, for some n. Similar definitions apply to *height* instead of *depth*.

EXERCISES

1. Show that a lattice is discrete if and only if the relation $a \leq b$ is decidable.
2. Show that a lattice is distributive if and only it satisfies the identity $a \vee (b \wedge c) = (a \vee b) \wedge (a \vee c)$.
3. Let L be a modular lattice containing a maximal chain that is finite (denial inequality). Show that L is discrete.
4. Show that if two linearly ordered sets are piecewise isomorphic to a third, then they are piecewise isomorphic to each other.
5. Show that two piecewise isomorphic discrete linearly ordered sets have the same cardinality.
6. Two intervals A and B of a modular lattice are called **transposes** if they are of the form $[a,a \vee d]$ and $[a \wedge d,d]$ (in either order), **projective** if there is sequence $A = I_1,\ldots,I_n = B$ of intervals such that I_i and I_{i+1} are transposes for $i = 1,\ldots,n-1$. Show that any two maximal finitely enumerable chains in a modular lattice are piecewise projective.
7. Show that a partially ordered set may be considered a category $\mathscr{C}$ in which the set $\mathscr{C}(a,b) = \{x \in \{0\} : a \leq b\}$. What is the categorical description of the infimum of two elements?
8. Suppose that for each binary sequence $a_1 \leq a_2 \leq a_3 \leq \cdots$ we could find m such that $a_n = a_m$ whenever $n \geq m$. Conclude that LPO holds.

6. WELL-FOUNDED SETS AND ORDINALS

Let W be a set with a relation $a < b$. A subset S of W is said to be **hereditary** if $w \in S$ whenever $w' \in S$ for each $w' < w$. The set W (or the relation $a < b$) is **well founded** if each hereditary subset of W equals W. A discrete partially ordered set is well founded if the relation $a < b$ (that is, $a \leq b$ and $a \neq b$) on it is well-founded. An **ordinal,** or a **well-ordered set,** is a discrete, linearly ordered, well-founded set.

Well-founded sets provide the environment for arguments by induction. The prototype well-founded set is the set $\mathbb{N}$ of nonnegative integers, with the usual order.

6.1 THEOREM. *The set $\mathbb{N}$ of nonnegative integers is well founded.*

PROOF. Let S be an hereditary subset of $\mathbb{N}$. Then $0 \in S$ because the hypothesis $w' \in S$ for each $w' < 0$ holds vacuously. From $0 \in S$ we conclude $1 \in S$, from $0 \in S$ and $1 \in S$ we conclude $2 \in S$, etc. □

The set $\mathbb{N}$ of nonnegative integers, viewed as an ordinal, is denoted by ω. We shall see how to construct other well-founded sets from ω. First we observe that any subset of a well-founded set is well founded. More generally we have:

6.2 THEOREM. *Let P and W be sets, each with a relation $a < b$, such that W is well founded. Let φ be a map from P to W such that $\varphi(a) < \varphi(b)$ whenever $a < b$. Then P is well founded.*

PROOF. Let S' be an hereditary subset of P, and let $S = \{w \in W : \varphi^{-1}(w) \subseteq S'\}$. We shall show that S is hereditary, so $S = W$ and therefore $S' = P$. Suppose $v \in S$ whenever $v < w$. If $x \in \varphi^{-1}(w)$ and $y < x$, then $\varphi(y) < w$ so $\varphi(y) \in S$, whence $y \in S'$. As S' is hereditary, this implies that $x \in S'$ for each x in $\varphi^{-1}(w)$, so $w \in S$. Thus S is hereditary. □

In particular, any subset of ω is an ordinal. The range of a binary sequence provides an example of an ordinal for which we may be unable to exhibit a first element. It follows from (6.2) that any subrelation of a well-founded relation is well founded.

We say that a well-founded set, or relation, is **transitive** if $a < b$ and $b < c$ implies $a < c$. An example of a well-founded relation on $\mathbb{N}$ that is not transitive is constructed by taking defining $a < b$ to mean $a + 1 = b$. This relation is well-founded by (6.2). An induction argument with respect to this relation is proof by induction as ordinarily defined; an induction argument with respect to the usual relation $a < b$ is sometimes called proof by complete induction.

A discrete, transitive well-founded set admits a natural partial ordering by defining $a \leq b$ to mean $a < b$ or $a = b$; the only nontrivial thing to check is that $a = b$ if $a \leq b$ and $b \leq a$, but this follows from the fact that $a < a$ is impossible in any well-founded set (Exercise 1). Conversely, the relation $a < b$ on a discrete partially ordered set is transitive.

One way to construct a well-founded set is by adding together previously constructed well-founded sets: we get the ordinal $\omega + \omega$ by placing two copies of the nonnegative integers side by side. More generally, let I be a well-founded set, and let $\{A_i\}_{i\in I}$ be a family of well-founded sets indexed by I. Then the disjoint union $\Sigma_{i\in I} A_i = \{(a,i) : a \in A_i \text{ and } i \in I\}$ admits a relation:

$$(a,i) < (b,j) \text{ if } i < j \text{ or if } i = j \text{ and } a < b$$

If $I = \{1,\ldots,n\}$, with the usual order, we write $A_1 + \cdots + A_n$. Note that if I and each A_i is discrete and transitive (and linearly ordered), then so is $\Sigma_{i\in I} A_i$.

6.3 THEOREM. *If I is a well-founded set, and $\{A_i\}_{i\in I}$ is a family of well-founded sets indexed by I, then $W = \Sigma_{i\in I} A_i$ is a well-founded set.*

PROOF. Suppose S is an hereditary subset of W. For each i in I, let $A_i' = \{a \in A_i : (a,i) \in S\}$, and let $I' = \{i \in I : A_i' = A_i\}$. We shall show that I' is hereditary, so $I' = I$, which says $S = W$. Suppose $i' \in I'$ for each $i' < i$. We shall show that $A_i' = A_i$ by showing that A_i' is hereditary. Suppose $a' \in A_i'$ for each $a' < a$. Then $w' \in S$ for each $w' < (a,i)$, so $(a,i) \in S$ whence $a \in A_i'$. Therefore $A_i' = A_i$ as A_i is well-founded, so $i \in I'$. □

Let $\{A_i\}_{i\in I}$ be a family of well-founded sets indexed by a discrete set I. We say that an element f of $\Pi_{i\in I} A_i$ has **finite support** if there is a finite subset J of I so that for each i in I, either $i \in J$, or $a < f_i$ is impossible for any a in A_i (that is, f_i is a minimal element of A_i). Note that if I is finite, then every element of $\Pi_{i\in I} A_i$ has finite support, while if some element has finite support, then all but finitely many of the A_i have minimal elements. If I is a well-founded set, then the **last-difference** relation on the elements of finite support in the product set $\Pi_{i\in I} A_i$ is defined by $f < g$ if:

(i) there exists $i \in I$ such that $f_i < g_i$, and

(ii) for each $i \in I$, either $f_i = g_i$, or $f_j < g_j$ for some $j \geq i$.

If I, and each A_i, is an ordinal, this may be described as ordering distinct elements according to the last place where they differ (reverse lexicographic order).

6.4 THEOREM. *Let I be an ordinal, and $\{A_i\}_{i \in I}$ a family of well-founded sets indexed by I. Then the set F of elements of finite support in $\Pi_{i \in I} A_i$ is well founded under the last-difference relation.*

PROOF. We first note that the theorem is true when $I = \{1,2\}$; in this case $F = A_1 \times A_2 = \Sigma_{b \in B} A(b)$ where $B = A_2$ and $A(b) = A_1$ for each b in B. So by Theorem 6.3 the last-difference product $A_1 \times A_2$ is well-founded if A_1 and A_2 are.

Adjoining a greatest element ∞ to I and setting $A_\infty = \{0\}$ does not change anything, so we may assume that I has a greatest element ∞. Let F_i be the set of elements of finite support in $\Pi_{j \leq i} A_j$ and let $I' = \{i \in I : F_i$ is well-founded$\}$. We shall show that I' is hereditary, hence equal to I, so $F = F_\infty$ is well founded.

Suppose $k \in I'$ for each $k < i$. Let F_i^* be the elements of finite support in $\Pi_{j<i} A_j$. Then $F_i = F_i^* \times A_i$, so F_i is well founded if F_i^* is. Write $F_i^* = \cup_{k<i} G_k$ where G_k consists of those elements of $\Pi_{j<i} A_j$ of finite support J such that $j \leq k$ for each j in J. The projection from G_k to F_k preserves the relation $x < y$, so G_k is well founded by Theorem 6.2. If S is an hereditary subset of F_i^*, then $S \cap G_k$ is an hereditary subset of G_k for each $k < i$, so $F_i^* = S$. Thus F_i^*, and therefore F_i, is well founded, so $i \in I$. □

If I is an arbitrary set, and A_i is a partially ordered set for each i in I, then the **categorical product** of the A_i is the set ΠA_i equipped with the partial order:

$$f \leq g \text{ if } f_i \leq g_i \text{ for each } i \text{ in } I.$$

If $I = \{1,\ldots,n\}$, and each A_i is discrete and well-founded, then the identity map of the categorical product into the last-difference product preserves order (but not nonorder), so the categorical product is well-founded by (6.2).

If α is an ordinal, and β is a well-founded set, then the well-founded set of functions with finite support from α to β is denoted by β^{α}.

If λ and μ are ordinals, then an **injection** of λ into μ is a function ρ from λ to μ such that if $a < b$ then $\rho a < \rho b$, and if $c < \rho b$, then there is $a \in \lambda$ such that $\rho a = c$. We shall show that there is at most one injection from λ to μ.

6.5 THEOREM. *If λ and μ are ordinals, and ρ and σ are injections of λ into μ, then $\rho = \sigma$.*

PROOF. Let $S = \{a \in \lambda : \rho a = \sigma a\}$, and suppose $a \in S$ for all $a < b$. If $\sigma b < \rho b$, then there is $a \in \lambda$ such that $\rho a = \sigma b < \rho b$, so $a < b$, so $\rho a = \sigma a$; but $\rho a = \sigma b > \sigma a$, a contradiction. Similarly we cannot have $\rho b < \sigma b$. Therefore $\rho b = \sigma b$, so $b \in S$; thus S is hereditary, so $S = \lambda$. □

If there is an injection from the ordinal λ to the ordinal μ we write $\lambda \leq \mu$. Clearly compositions of injections are injections, so this relation is transitive. By (6.5) it follows that if $\lambda \leq \mu$ and $\mu \leq \lambda$, then λ and μ are isomorphic, that is, there is an invertible order preserving function from λ to μ. It is natural to say that two isomorphic ordinals are **equal**.

EXERCISES

1. Show that $a < a$ is impossible in any well-founded set.
2. If $a < b$ is a relation on a set W, define the transitive closure relation $a <^* b$ to mean $a < b$ or there exists $x_1, \ldots, x_n$ such that $a < x_1 < x_2 < \cdots < x_n < b$. Show that $a <^* b$ is well-founded if $a < b$ is well-founded (mimic the proof that ordinary induction on $\mathbb{N}$ implies complete induction on $\mathbb{N}$).
3. A relation $a < b$ is **acyclic** if $a <^* a$ is impossible for any a (see Exercise 2). Show that any acyclic relation on a two-element set is well-founded. Show that any acyclic relation on a set that is bounded in number is well-founded.
4. Show that each discrete, well-founded, partially ordered set satisfies the descending chain condition.
5. Let W be a well-founded set. Let S be a subset of W such that, given w in W, either $w \in S$ or there exists $w' < w$ such that $w \in S$

if $w' \in S$. Show that $S = W$.

6. Show that a discrete partially ordered set that satisfies Exercise 5 for each subset W has the descending chain condition.

7. Show the converse of Exercise 6 holds (this uses dependent choice).

8. Let W be the nonnegative integers with $a < b$ defined to mean $a \leq b$ and $b - a$ is odd. Show that the relation $a < b$ is well-founded but not transitive.

9. Show that the categorical product of partially ordered sets as defined in this section is the categorical product.

10. Show that if every nonempty ordinal has a least element, then LPO holds.

11. A **rank relation** on a discrete partially ordered set W is an ordinal A together with a subset R of $W \times A$ such that

    (i) For each w in W there is a in A such that $(w,a) \in R$.

    (ii) If $v < w$ and $(w,a) \in R$, then there is $b < a$ such that $(v,b) \in R$.

    Prove Theorem 6.4 for I a discrete, well-founded, partially ordered set with a rank relation.

12. A **Grayson ordinal** is a set W with a well-founded relation $a < b$ satisfying:

    (i) If $a < b$ and $b < c$, then $a < c$ (transitivity)

    (ii) If $c < a$ is equivalent to $c < b$ for each c, then $a = b$ (extensionality).

    Define $a \leq b$ in a Grayson ordinal to mean $c < a$ implies $c < b$ for all c. Show that if the relation $a < b$ is decidable, then W is a Grayson ordinal if and only if W is an ordinal. (Hint: prove that $a < b$ or $a = b$ or $b < a$ in a decidable Grayson ordinal)

13. Let α be a binary sequence, and let $S = \{x,y,z\}$ with $x = y$ if α is identically zero. Define a relation $u < v$ on S by setting

    (i) $y < z$

    (ii) $x < y$ if $\alpha_n = 1$ for some n

    (iii) $x < z$ if $x = y$ or $x < y$.

    Show that this turns S into a Brouwerian example of Grayson ordinal with elements such that $x \leq y < z$, but not $x < z$.

NOTES

Bishop's definition of when an object may be said to exist can be found in [Bishop 1967, page 8]. The idea that the notion of an algorithm is primitive has also been advanced by the Russian mathematicians Uspenskii and Semenov (1981):

"*The concept of algorithm like that of set and of natural number is such a fundamental concept that it cannot be explained through other concepts and should be regarded as* [*an*] *undefinable one*."

Attempts to explain constructive existence in classical terms are always somewhat unsatisfactory, but the classical notion of existence is no less mysterious than the constructive one, we are just more familiar with its use. What does it mean for a well-ordering of the real numbers to exist? or a basis of the reals as a rational vector space? or an automorphism of the field of complex numbers that takes e to π? or a noncomputable function? Formal systems that specify the correct use of the phrase "there exists" are available to the constructive mathematician as well as to his classical counterpart.

Our definition of a **set** is a combination of the formulations in [Bridges 1979, page 2] and [Heyting 1971, 3.2.1]. Many authors use the positive term "inhabited" to describe nonempty sets; this avoids possible confusion with the notion of a set that cannot be empty. The notion of a (tight) **apartness** is found in [Heyting 1971, 4.1.1]. The terminology "tight" is due to Scott (1979). Troelstra and van Dalen use the term "pre apartness" to denote what we call an apartness. The part of Theorem 2.2 that states that if the inequality on S is tight, then so is the inequality on $S^{\mathbb{N}}$, is essentially [Bishop 1967, Lemma 5, page 24], which says that the natural inequality on the real numbers is tight.

A standard inequality on the set $\{0\}$ is gotten by setting $0 \neq 0$ if LPO is false. As LPO is refutable in two main branches of constructive mathematics--intuitionism and Russian constructivism--we cannot show that this inequality is consistent. For more on intuitionism and Russian constructivism from the point of view of constructive mathematics see [Bridges-Richman 1987].

Difference relations are symmetric inequalities that satisfy

$$\neg\, x \neq y \text{ and } \neg\, y \neq z \text{ implies } \neg\, x \neq z.$$

They were studied by van Rootselaar (1960) and by Olson (1977).

We could demand that every set come with an inequality, putting inequality on the same footing as equality; it would then be natural to demand that all functions be strongly extensional. With such an approach, whenever we construct a set we must put an inequality on it, and we must check that our functions are strongly extensional. This is cumbersome and easily forgotten, resulting in incomplete constructions and incorrect proofs. An example of the complications: if H is a subgroup of an abelian group G, then the inequality on G/H as a group may differ from the inequality on G/H as a set because the group operation on G/H need not be strongly extensional with respect to the latter inequality (unless the inequality on G is decidable)—see Exercise II.1.6.

Our definition of a **subset** agrees with the informal treatment in [Bishop 1967, page 32]. A categorical definition of a **subset** is found in [Bishop 1967, page 63] where a subset of a set S is defined to be a set A together with a one-to-one map from A to S. The categorical definition is attractive for constructive mathematics, where it is important to keep in mind that an element of a subset, by virtue of belonging to the subset, carries implicitly the additional data that establishes its membership in the subset. The categorical approach allows us to make this additional data explicit. For example, if S is the set of binary sequences, and A is the subset of S consisting of those sequences α such that $\alpha_m = 1$ for some m, then to specify a member of A we must not only construct a sequence α in S, but also an integer m for which $\alpha_m = 1$. Thus an element of A may be viewed as a pair (α,m), two pairs (α,m) and (α',m') being equal if $\alpha = \alpha'$. The map from A to S that takes (α,m) to α is one-to-one but is not really an inclusion map as it forgets the additional datum m. We find the informal approach more natural.

The **axiom of unique choice** allows us to identify a function with its graph; Myhill has called this the **axiom of nonchoice**. It is easy to see that the axiom of unique choice is equivalent to either of the following.

(i) Every one-to-one onto function has an inverse.

(ii) If S is a set, and S^* is the set of one-element subsets of S, then S^* has a choice function: that is, a function f from S^* to S such that $f(x) \in x$ for each x in S^*.

Bishop employs the notion of a nonextensional function or **operation**. In almost all applications, one can consider an operation from a set A to a set B to be a function from A to the set of nonempty subsets of B.

Our definition of a **finite** set differs from that in [Bishop 1967], [Bridges 1979] and [Bishop-Bridges 1985] in that we consider the empty set to be finite. Another, more subtle, difference is that we require that finite sets be discrete with respect to their given inequality. Thus a set S of functions between two discrete sets is finite only if for each f and g in S, either $f = g$ or there exists x such that $f(x) \neq g(x)$.

The functions that we call **onto** are called **surjective** in [Bishop 1967] where the word **onto is** reserved for a map f from A to B that has a cross-section, that is, for which there exists a function g from B to A such that fg is the identity map on B. In [Bishop 1967] a finitely enumerable set is called **subfinite**, and a set is said to have **at most** n **elements** if it can be written in the form $\{x_1,\ldots,x_n\}$. The term 'subfinite' suggests a subset of a finite set to us, while the [Bishop 1967] usage of 'at most' precludes saying that each subset of $\{1,\ldots,n\}$ contains at most n elements.

Greenleaf (1981) examines cardinality of sets, and related questions, from a constructive point of view.

Our definition of a **countable** set differs from that of [Bishop 1967], [Bridges 1979] and [Bishop-Bridges 1985] in that we do not require that a countable set be nonempty (or even that we can decide whether or not it is empty); for discrete sets, it is equivalent to the definition in [Brouwer 1981].

We are unlikely to be able to construct a Brouwerian example of a subset of $\mathbb{N}$ that is not countable. However, any acceptable proof of the theorem T that every subset of $\mathbb{N}$ is countable could probably be converted into a proof that every subset of $\mathbb{N}$ is recursively enumerable, which is false. A well-known variant of T is **Kripke's schema**, which states that for each proposition P there exists a binary sequence α such that P holds if and only if $\alpha_n = 1$ for some n. Kripke's schema has some plausibility within Brouwer's framework of the creating subject, where we imagine the idealized mathematician as effecting a nonpredeterminate sequence of attempts at proving P, and we keep in mind the intuitionistic criterion that to prove $\neg P$ is to show how to convert any proof of P into a contradiction.

Bishop used an italicized *not* to indicate the existence of a Brouwerian counterexample. For example, he would say "there is an inequality that is *not* an apartness" to mean "if every inequality were an apartness, then LPO holds". We shall *not* employ this convention.

The **lesser limited principle of omniscience** (LLPO) was introduced in [Bishop 1973]. Both LPO and LLPO have simple interpretations in terms of real numbers: LPO is equivalent to the statement that, for each real number x, either $x \leq 0$ or $x > 0$; LLPO is equivalent to the statement that, for each real number x, either $x \leq 0$ or $x \geq 0$.

Markov's principle states that if α is a binary sequence, and $\alpha_n = 0$ for all n is impossible, then there exists n such that $\alpha_n = 1$. The idea is that we can construct the number n by successively computing α_1, α_2, α_3, ... until we get a 1. Markov's principle is employed by the Russian constructivist school, which is a constructive form of recursive mathematics; for details see [Bridges-Richman 1987]. An argument against Markov's principle is that we have no prior bound, in any sense, on the length of the computation required to construct n. We regard Markov's principle as an omniscience principle.

The position that "for all a in A there exists b in B" entails the existence of an algorithm (not necessarily extensional) from A to B is suggested by the assertion in [Bishop 1967, page 9] that "A choice function exists in constructive mathematics, because a choice is *implied by the very meaning of existence*." Bishop *defines* a function f from B to A to be surjective if there is an algorithm g from A to B such that $f(g(a)) = a$ for all a in A.

Our Brouwerian counterexample to the axiom of choice, and Exercise 3.1 that the axiom of choice implies the law of excluded middle, is due to Myhill and Goodman (1978). An earlier proof, in a topos-theoretic setting, was given by Diaconescu (1975) who shows that the axiom of choice implies that every subset A of a set B has a **complement** in the sense that there is a subset A' such that $B = A \cup A'$ and $A \cap A' = \phi$.

Fourman and Scedrov (1982) have shown, using topos-theoretic methods, that the world's simplest axiom of choice is not provable in intuitionistic set theory with dependent choice.

The arguments against countable choice, and dependent choices, must be based on more fundamental grounds than we used for the axiom of choice. That is, we must question the interpretation of the phrase "for all a there exists b" as entailing the existence of an algorithm for transforming elements of A to elements of B. One reason for rejecting this interpretation is that by so doing our theorems will also be theorems in other (unintended) models, such as arise in topos theory, which are of interest in classical mathematics (see, for example, [Scedrov 1986]). A more relevant reason, perhaps, is that arguments that depend essentially on this interpretation often have an unsatisfying feel to them which seems to be connected with the gratuitous "completion of an infinity" that occurs when we subsume a potentially infinite number of items of information into a single algorithm.

A **rank function** on a well-founded set W is a map φ from W to an ordinal A such that $\varphi(x) < \varphi(y)$ whenever $x < y$. It seems unlikely that we can always construct a rank function although, classically, every well-founded set admits a unique smallest rank function. Induction on rank is a common technique in the classical theory (see Exercise 6.11).

Chapter II. Basic Algebra

1. GROUPS

A **monoid** is a set G together with a function φ from G×G to G, usually written as $\varphi(a,b) = ab$, and a distinguished element of G, usually denoted by 1, such that for all a, b, c in G

(i) $(ab)c = a(bc)$, (associative law)

(ii) $1a = a1 = a$. (identity)

The function φ is called **multiplication** and the element 1 the **identity.** The associative law allows us to ignore parentheses in products $a_1a_2\cdots a_n$. The monoid is said to be **abelian,** or **commutative,** if $ab = ba$ for all elements a and b. In an abelian monoid the function φ is often called **addition** and written as $\varphi(a,b) = a + b$; the identity element is then denoted by 0. In this case we speak of an **additive** monoid, as opposed to a **multiplicative** monoid. In a multiplicative monoid we write the n-fold product $aa\cdots a$ as a^n for each positive integer n, and set $a^0 = 1$; in an additive monoid we write the n-fold sum $a + a + \cdots + a$ as na, and set $0a = 0$.

If X is a set, then the set of all functions from X to X forms a monoid: multiplication is composition of functions, and the identity element is the identity function. The set $\mathbb{N}$ of nonnegative integers is a commutative monoid under addition, with identity element 0.

A **homomorphism of monoids** is a function f from a monoid G to a monoid H such that $f(1) = 1$, and $f(ab) = f(a)f(b)$ for all a and b in G. If G is a multiplicative monoid, and $a \in$ G, then the map from $\mathbb{N}$ to G that takes n to a^n is a homomorphism. A homomorphism f is **nontrivial** if $\{1\}$ is a proper subset of $im\ f$. A subset H of a monoid G is a **submonoid** if $1 \in H$ and H is closed under multiplication. If S is a subset of the monoid G, then the set consisting of 1 together with all finite products of elements of S is a submonoid of G called the **submonoid generated by** S. The submonoid of $\mathbb{N}$ generated by $\{3,4\}$ is $\mathbb{N}\backslash\{1,2,5\}$.

Let X be a set. Define X^* to be the set of all finite sequences of elements of X, including the empty sequence. The elements of X^* are called **words**. Two words $u \equiv x_1x_2\cdots x_m$ and $v \equiv y_1y_2\cdots y_n$ are **equal** if $m = n$ and $x_i = y_i$ for $i = 1,2,\cdots,m$. If u and v are equal in X^* we will write $u \equiv v$. Define a multiplication on X^* by setting $uv \equiv x_1\cdots x_m y_1\cdots y_n$. This multiplication is associative and the empty word is the identity element, so X^* is a monoid, called the **free monoid on the set** X. If X is a one-element set, then X^* is isomorphic to the additive monoid of nonnegative integers.

If a and b are elements of a monoid, and $ab = 1$, then we say that a is a **left inverse** of b and that b is a **right inverse** of a. If b has a left inverse a and a right inverse c, then $a = a(bc) = (ab)c = c$; in this case we say that a is the **inverse** of b and write $a = b^{-1}$. If b has an inverse we say that b is a **unit**, or that b is **invertible.**

A **group** is a monoid G in which every element is invertible. In an additive group, the inverse of a is denoted by $-a$ rather than a^{-1}. For n a positive integer, we define a^{-n} to be $(a^{-1})^n$; the usual laws of exponents hold. In an additive group this definition takes the form $(-n)a = n(-a)$, and the appropriate associative and distributive laws hold (see the definition of an R-module in Section 3). The prototype abelian group is the group $\mathbb{Z}$ of integers under addition.

The **order** of an element a of a group is the cardinality of the set $\{a^n : n \in \mathbb{N}\}$. The order of a is $n \in \mathbb{N}$ if and only if $a^n = 1$ and $a^m \neq 1$ for $m = 1,\ldots,n-1$. In $\mathbb{Z}$ the element 0 has order 1, as does the identity in any group, and each nonzero element has infinite order. In a discrete group the order of an element a is the cardinality of the set $\{n \in \mathbb{N} : a^m \neq 1 \text{ whenever } 0 < m \leq n\}$ (which contains 0), hence is an ordinal $\beta \leq \omega$.

If G and H are groups, and f is a monoid homomorphism from G to H, then $f(a^{-1}) = f(a)^{-1}$ and $f(a^{-1})f(a) = f(a^{-1}a) = f(1) = 1$. Thus a monoid homomorphism between two groups preserves all the group structures: multiplication, identity, and inverse. If G is a group, and $a \in G$, then the map that takes $x \in G$ to axa^{-1} is easily seen to be an automorphism of G; such an automorphism is called **inner**.

If G and H are abelian groups, then the set $Hom(G,H)$ of homomorphisms from G to H has a natural structure as an abelian group by setting $(f_1 + f_2)(x) = f_1(x) + f_2(x)$. The group $Hom(\mathbb{Z},H)$ is naturally isomorphic to H under the map taking f to $f(1)$. If h is a homomorphism from H to H', then h induces a homomorphism from $Hom(G,H)$ to $Hom(G,H')$ by taking f to hf; that is, $h(f_1 + f_2) = hf_1 + hf_2$. Similarly a homomorphism $g : G' \to G$ induces a homomorphism from $Hom(G,H)$ to $Hom(G',H)$ by taking f to fg. A category $\mathcal{C}$, like the category of abelian groups, such that $\mathcal{C}(G,H)$ is an abelian group for each pair of objects G and H of $\mathcal{C}$, and such that the functions induced from $\mathcal{C}(G,H)$ to $\mathcal{C}(G,H')$ by a map $H \to H'$, and from $\mathcal{C}(G,H)$ to $\mathcal{C}(G',H)$ by a map $G' \to G$, are homomorphisms of groups, is called a **pre-additive category**.

A **permutation** of a set X is a one-to-one map of X onto itself. The set of permutations of X is a group called the **symmetric group** on X. If $x_1,\ldots,x_n$ are distinct elements of a discrete set X, then we denote by $(x_1,\ldots,x_n)$ the permutation π of X such that

$$\pi x_i = x_{i+1} \text{ for } i = 1,\ldots,n-1$$
$$\pi x_n = x_1$$
$$\pi x = x \text{ otherwise.}$$

Such a permutation is called an **n-cycle** with **support** $\{x_1,\ldots,x_n\}$, and two cycles are called **disjoint** if their supports are disjoint. If X is a finite set, then every permutation is a product of disjoint cycles. As $(x_1,\ldots,x_n) = (x_1,x_n)\cdots(x_1,x_3)(x_1,x_2)$, every permutation of a finite set is a product of 2-cycles (not necessarily disjoint). A permutation that can be written as a product of an even number of 2-cycles is said to be **even**, otherwise **odd**. If π is a permutation of a finite set, then we define

$$sgn\ \pi = \begin{cases} 1 \text{ if } \pi \text{ is even} \\ -1 \text{ if } \pi \text{ is odd.} \end{cases}$$

The product of an odd number of 2-cycles is odd (Exercise 7), so $sgn\ \pi_1\pi_2 = (sgn\ \pi_1)(sgn\ \pi_2)$ (Exercise 8).

A **subgroup** of a group is a submonoid that is closed under inverse. If G is a group, then G and $\{1\}$ are subgroups of G; we often denote the subgroup $\{1\}$ by 1. If S is a subset of a group G, then the set

$$\langle S\rangle = \{1\} \cup \{s_1s_2\cdots s_k : s_i \in S \cup S^{-1},\ k \geq 1\}$$

of all finite products of elements that are in S, or are inverses of elements of S, is a subgroup of G called the **subgroup generated by** S. If $\langle S\rangle = G$, then S is called a **set of generators** for G. A group is **finitely generated** if it has a finitely enumerable set of generators, **cyclic** if has a one-element set of generators. The additive group $\mathbb{Q}$ of rational numbers is not finitely generated; in fact, each finitely generated subgroup of $\mathbb{Q}$ is cyclic (it is easy to show that each finitely generated subgroup of $\mathbb{Q}$ is *contained* in a cyclic subgroup).

A subgroup H of G is **normal** if $ghg^{-1} \in H$ for each g in G and h in H. Every subgroup of an abelian group is normal. If f is a homomorphism from G to H, then the **kernel** of f,

$$\mathbf{ker}\ f = \{x \in G : f(x) = 1\} = f^{-1}(1)$$

is easily seen to be a normal subgroup of G. The size of the kernel of f measures how badly f fails to be one-to-one, as the following are equivalent:

$$f(a) = f(b)$$
$$f(ab^{-1}) = f(a)f(b^{-1}) = f(a)f(b)^{-1} = 1$$
$$ab^{-1} \in ker\ f,$$

so f is a monomorphism if and only if $ker\ f = 1$. We will be studying algebraic structures that are abelian groups with additional structure. In these cases, the kernel of a homomorphism f means the kernel of f as a homomorphism of groups; if the group is written additively, as will normally be the case for the more complex structures, $ker\ f = f^{-1}(0)$.

Each normal subgroup H of a group G is the kernel of a homomorphism that is constructed as follows. Let G/H be the set whose elements are precisely the elements of G, but equality is defined by setting $a = b$ if $ab^{-1} \in H$. When it is necessary to distinguish between the equalities on G and on G/H we will write $a = b\ (mod\ H)$ to denote the equality in G/H. Multiplication and inverse remain functions with respect to the equality on G/H, so G/H is a group, called the **quotient group** of G by H. The prototype of a quotient group is obtained by taking G to be the group $\mathbb{Z}$ of integers, and letting H be the subgroup of $\mathbb{Z}$ consisting of all multiples of a fixed integer n; the resulting group G/H is the group $\mathbb{Z}_n$ of integers modulo n.

The fundamental facts relating normal subgroups, quotients, and homomorphisms are contained in the following theorem.

1.1 THEOREM. *Let N be a normal subgroup of a group G, and f a homomorphism from G to a group L with $f(N) = 1$. Then f is a homomorphism from G/N to L. If f is onto, and the kernel of f is N, then f is an isomorphism from G/N to L.*

PROOF. If $a = b \pmod{N}$, then $ab^{-1} \in N$, so $f(a) = f(b)$; therefore f is a function on G/N, which is clearly a homomorphism. Conversely, if $f(a) = f(b)$, then $f(ab^{-1}) = 1$, so $ab^{-1} \in N$ and $a = b \pmod{N}$. Therefore f is a one-to-one map from G/N to L; so if f is onto, then $f: G/N \to L$ has an inverse g. Clearly g is also a homomorphism. □

Let N be a normal subgroup of the group G. A (normal) subgroup of G/N is a (normal) subgroup H of G that is a subset of G/N, that is, if $a \in H$ and $a = b \pmod{N}$, then $b \in H$. It is easily seen that a subgroup H of G is subset of G/N just in case $N \subseteq H$. The difference between a subgroup H of G containing N, and a subgroup H of G/N, is the equality relation on H. We distinguish between H as a subgroup of G, and H as a subgroup of G/N, by writing H/N for the latter. If H is normal subgroup of G, containing N, then $(G/N)/(H/N)$ is isomorphic to G/H; in fact, the elements of both groups are simply the elements of G, and the equalities are the same.

1.2 THEOREM. *Let H and K be subgroups of the group G. If K is normal, then*

(i) *The set $HK = \{hk : h \in H \text{ and } k \in K\}$ is a subgroup,*

(ii) *The subgroup $H \cap K$ is normal in H, and*

(iii) *The quotient groups HK/K and $H/(H \cap K)$ are isomorphic.*

PROOF. Exercise. □

In an additive group, the subgroup HK is written as $H + K$.

If $a \in G$, and H is a subgroup of G, then $Ha = \{ha : h \in H\}$ is called a **right coset** of H, while $aH = \{ah : h \in H\}$ is called a **left coset** of H. The inverse function induces a bijection between left and right cosets of H that takes aH to Ha^{-1}, so we can speak unambiguously of the cardinality of the set of cosets of H in G. This cardinality is called the **index** of H in G and is denoted by $[G:H]$. If H is normal, then $Ha = aH$ for each a in G.

EXERCISES

1. Show that in a finite monoid, if a has a left inverse or a right inverse, then $a^n = 1$ for some positive integer n; so an element can have at most one left or right inverse. Give an example of an element of a monoid that has two distinct left inverses; two

distinct right inverses.

2. Show that a monoid may be identified with a one-object category, and that homomorphisms of monoids are functors between such categories. Which of these categories correspond to groups?

3. Show that the set of units of a monoid is a group.

4. Show that the set S of binary sequences forms an abelian group G under coordinatewise addition modulo 2. Let $a \in S$ and define $b, c \in S$ by $b_n = 1$ if and only if $a_n = 1$ and $a_m = 0$ for all $m < n$, and $c_n = 1$ if and only if $b_{n-1} = 1$. Show that if the subgroup of G generated by b and c is generated by a finite subset of G, then either $a = 0$ or $a \neq 0$.

5. Let G be a multiplicative group with an inequality. The inequality is called **translation invariant** if $x \neq y$ implies $zx \neq zy$ and $xz \neq yz$. Assuming the inequality is translation invariant, show that

   (i) the function taking x to x^{-1} is strongly extensional if and only if the inequality is symmetric.

   (ii) the inequality is cotransitive if and only if multiplication is strongly extensional (the inequality on $G \times G$ is given by $(x_1, x_2) \neq (y_1, y_2)$ if $x_1 \neq y_1$ or $x_2 \neq y_2$).

   Finally, show that if the inequality is consistent, and multiplication is strongly extensional, then the inequality is translation invariant.

6. Considering an inequality on G to be a subset of $G \times G$, show that the union of a family of inequalities on a group G, under which the group operation is strongly extensional, is again such an inequality. Show that there is a unique inequality on G/N that makes Theorem 1.1 true in the category of groups with inequality, and strongly extensional homomorphisms. Show that if G is the set of binary sequences under coordinatewise addition modulo 2, and $N = \{x \in G : \text{there is } m \text{ such that } x_n = 0 \text{ for all } n \geq m\}$, then $x \neq 0$ in G/N if and only if $x_n = 1$ infinitely often and LPO.

7. Let $x_1, \ldots, x_m, y_1, \ldots, y_n$ be distinct elements of a finite set X, let G be the symmetric group on X, and let $1 \leq i < j \leq m$. Verify the following two equalities in G.

(i) $(x_i,x_j)(x_1,\ldots,x_m) = (x_1,\ldots,x_{i-1},x_j,\ldots,x_m)(x_i,\ldots,x_{j-1})$

(ii) $(x_1,y_1)(x_1,\ldots,x_m)(y_1,\ldots,y_n) = (y_1,\ldots,y_n,x_1,\ldots,x_m)$.

For π in G, we can write π in an essentially unique way as a product of disjoint cycles whose supports exhaust X. Let N_π be the number of cycles in such a product. Use (i) and (ii) to show that if τ is a 2-cycle, then $N_{\tau\pi} = N_\pi \pm 1$. Conclude that an even permutation cannot be written as a product of an odd number of 2-cycles.

8. Show that *sgn* is a homomorphism from the symmetric group on a finite set to the group $\{-1,1\}$ under multiplication.

9. Give a Brouwerian example of a countably generated subgroup of a finite abelian group that is not finitely generated.

10. Show that the set of normal subgroups of a group is a modular lattice.

2. RINGS AND FIELDS

A **ring** is an additive abelian group R which is also a multiplicative monoid, the two structures being related by the **distributive laws:**

$$\text{(i)}\quad a(b + c) = (ab) + (ac),$$

$$\text{(ii)}\quad (a + b)c = (ac) + (bc).$$

A ring is said to be **trivial** if $0 = 1$. If the monoid structure is commutative, then R is a **commutative ring.** A **unit** of R is a unit of the multiplicative monoid of R. A ring is said to have **recognizable units** if its units form a detachable subset.

If A is an abelian group, then the set of endomorphisms $E(A) = Hom(A,A)$ is a ring (multiplication is composition) called the **endomorphism ring** of A. More generally, if $\mathcal{C}$ is a pre-additive category, then $\mathcal{C}(A,A)$ is a ring for each object A in $\mathcal{C}$.

A ring k is a **division ring** if, for each a and b in k,

$$a \neq b \text{ if and only if } a - b \text{ is a unit.}$$

We remind the reader that the interpretation of the symbol $a \neq b$ depends on the context: if k comes with an inequality, then $a \neq b$ refers to that inequality, otherwise $a \neq b$ refers to the denial inequality. An immediate consequence of the definition is that if k is a division ring, then the inequality on k is symmetric, and translation invariant: if $a \neq b$, then $a + c \neq b + c$. Note that the denial inequality is automatically translation invariant because addition is a function.

Of course you could define the inequality $a \neq b$ on any ring to mean that $a - b$ is a unit and, technically, you would have a division ring; thus the general theory of division rings will include the theory of rings. However the idea is to use the symbol $a \neq b$ to represent relations that can reasonably be called inequalities: if you pick a silly inequality and get a silly division ring, don't blame us. As a rule of thumb, you should use a standard inequality. For the most part we will be interested only in discrete division rings, and, to a lesser extent, in division rings with a tight apartness.

The nonzero elements of a division ring are exactly the units; in the discrete case, this characterizes division rings (Exercise 3). A **field** is a commutative division ring. A **Heyting field** is a field with a tight apartness. In a Heyting field, or more generally in a field with a cotransitive inequality, the arithmetic operations are strongly extensional (Exercise 5). The rational numbers $\mathbb{Q}$ form a discrete field; the real numbers (next section) form a Heyting field. The rational quaternions (Exercise 4) form a noncommutative discrete division ring.

A subset of a ring is a **subring** if it is an additive subgroup and a multiplicative submonoid. Let S be a subring of a commutative ring R, and let $a_1,\ldots,a_n$ be elements of R. Then the set of sums of elements of R of the form

$$sa_1^{m(1)}a_2^{m(2)}\cdots a_n^{m(n)},$$

with $s \in S$ and $m(i) \in \mathbb{N}$, is a subring of R denoted by $S[a_1,\ldots,a_n]$ which is contained in each subring of R that contains $S \cup \{a_1,\ldots,a_n\}$. If S and R are fields, then $S(a_1,\ldots,a_n)$ denotes the set of quotients f/g with $f,g \in S[a_1,\ldots,a_n]$ and $g \neq 0$. It is easily seen that $S(a_1,\ldots,a_n)$ is a field which is contained in each subfield of R that contains $S \cup \{a_1,\ldots,a_n\}$.

An **integral domain**, or simply a **domain**, is a ring that admits an inequality preserving isomorphism with a subring of a field; more informally, an integral domain is simply a subring of a field. If R is a subring of a field, then $\{ab^{-1} : a,b \in R \text{ and } b \neq 0\}$ is a field containing R called the **field of quotients** of R. The field of quotients of an integral domain is essentially unique (Exercise 6).

If R is a nontrivial discrete integral domain, then, for each a and b in R,

$$(*) \qquad \text{if } a \neq 0 \text{ and } b \neq 0, \text{ then } ab \neq 0.$$

Conversely, if R is a discrete commutative ring satisfying (*), then we can embed R in a discrete field k by imitating the construction of the rational numbers $\mathbb{Q}$ from the integers $\mathbb{Z}$. Let $k = \{(a,b) \in R \times R : b \neq 0\}$ with $(a,b) = (c,d)$ if $ad = bc$. Define multiplication on k by $(a,b)\cdot(c,d) = (ac,bd)$ and addition by $(a,b) + (c,d) = (ad + bc,bd)$. It is routine to verify that this makes k a ring with additive identity (0,1) and multiplicative identity (1,1). If $(a,b) \neq (0,1)$, then $a \neq 0$ so the element (b,a) is in k and $(a,b)(b,a) = (1,1)$; conversely, if $(a,b)(c,d) = (1,1)$, then $ac = bd \neq 0$, hence $a \neq 0$, so $(b,a) \in k$. Thus k is a field. We embed R into k by taking a to $(a,1)$.

An intrinsic characterization of an arbitrary integral domain is given in Exercise 7. To prove that this characterization is correct, you construct a field of quotients as in the discrete case.

If k is a field, then the subfield of k consisting of all elements of the form $(n\cdot 1)/(m\cdot 1)$ where m and n are integers and $m\cdot 1 \neq 0$ is the smallest subfield contained in k, and is called the **prime field** of k. This field is the field of quotients of the subring $\{n\cdot 1 : n \in \mathbb{Z}\}$ of k. If $a_1,\ldots,a_n \in k$, and k_0 is the prime field of k, then we say that k is **generated by** $a_1,\ldots,a_n$ if $k_0(a_1,\ldots,a_n) = k$.

A function f from a ring R to a ring S is a **ring homomorphism** if it is a homomorphism of the additive groups, and a homomorphism of the multiplicative monoids. The map from the ring of integers $\mathbb{Z}$ to S that takes n to $n\cdot 1$ is a ring homomorphism. An additive subgroup I of a ring R is an **ideal** if for each x in I and each r in R the elements rx and xr are in I. It is easily seen that if f is a ring homomorphism, then $\ker f = f^{-1}(0)$ is an ideal. An ideal I is **proper** if $1 \notin I$. A **left** (resp. **right**) **ideal** I of a ring R is an additive subgroup of R such that for each r in R and x in I the element rx (resp. xr) is in I. A left ideal I of a ring is **nonzero** if I contains a nonzero element. To avoid confusion between left ideals, right ideals and ideals, an ideal is often referred to as a **two-sided ideal**. If I is a two-sided ideal of the ring R, then multiplication is a function on the quotient group R/I, so R/I is a ring.

Let X and Y be subsets of the ring R. Define XY to be the additive subgroup of R generated by $\{xy : x \in X \text{ and } y \in Y\}$. If X, Y, and Z are subsets of R, then $(XY)Z = X(YZ)$. A subset I is an ideal if and only if $RIR = I$, while I is a left (resp. right) ideal if $RI = I$ (resp. $IR = I$).

If S is a subset of a ring R then $(S) = RSR$ is the smallest ideal of R containing S, called the **ideal generated by** S. If S is the finite family $\{s_1,\ldots,s_n\}$, then the ideal generated by S is denoted by $(s_1,\ldots,s_n)$. The **left ideal generated by** S is RS, while the **right ideal generated by** S is SR; if S is a one-element set $\{s\}$, then the corresponding left or right ideal is called **principal**, and is denoted Rs or sR.

If I and J are ideals, then IJ and $I \cap J$ are ideals. The set $I \cup J$ need not be an ideal; the ideal generated by $I \cup J$ is $\{i + j : i \in I \text{ and } j \in J\}$ and is denoted by $I + J$. More generally, if $\{I_i\}$ is a family of ideals, then the ideal generated by $\cup_i I_i$ is denoted by $\Sigma_i I_i$. The **quotient** of a left ideal I by a set S is the left ideal $I:S = \{x \in R : xS \subseteq I\}$. The **radical** of an ideal I in a commutative ring is the ideal $\sqrt{I} = \{x \in R : x^n \in I \text{ for some } n\}$.

The fundamental theorem of homomorphisms of rings follows immediately from the corresponding theorem for groups.

2.1 THEOREM. *Let I be an ideal in the ring R. If f is a homomorphism from R to a ring S with $f(I) = 0$, then f is a homomorphism from R/I to S. If f maps onto S and the kernel of f is I, then f is an isomorphism from R/I to S.* □

Let I be an ideal in the ring R. A (left) ideal of R/I is a (left) ideal J of R containing I. If J is an ideal in R containing I, then $R/J \cong (R/I)/(J/I)$.

2.2 THEOREM. *Let R be a ring, S a subring and I an ideal of R. Then $S + I$ is a subring of R containing I as an ideal, $S \cap I$ is an ideal in S, and $(S + I)/I \cong S/(S \cap I)$.*

PROOF. Clearly $S + I$ is a subring and I is an ideal in $S + I$. Define a function f from $S/(S \cap I)$ to $(S + I)/I$ by setting $f(s) = s$. Note that f is indeed a function because if $s_1 = s_2$ in $S/(S \cap I)$, then $s_1 - s_2 \in I$ so $s_1 = s_2$ in $(S + I)/I$. Clearly f is a homomorphism. Now define a function g from $(S + I)/I$ to $S/(S \cap I)$ by setting $g(s+i) = s$. To see that g is a function we note that if $s_1 + i_1 = s_2 + i_2$ in $(S + I)/I$, then $s_1 - s_2 \in I$, so $s_1 = s_2$ in $S/(S \cap I)$. It follows that f is an isomorphism. □

If P is an ideal in a commutative ring R, then we say that P is a **prime ideal** if whenever $xy \in P$, then either $x \in P$ or $y \in P$. If P is a detachable proper ideal of R, then it is easy to see that P is prime if and only if R/P is an integral domain. If p is a prime number, then the ideal (p) in $\mathbb{Z}$ is a detachable proper prime ideal, as is the ideal 0.

2.3 THEOREM. *Let $P_1,\ldots,P_n$ be detachable ideals of a commutative ring R such that P_i is prime for $i \leq n-2$. If I is a finitely generated ideal of R, then either $I \subseteq P_i$ for some i, or there exists $z \in I \setminus \cup_i P_i$.*

PROOF. Let $x_1,\ldots,x_m$ generate I, and let F be the set of finite subsets S of $\{1,2,\ldots,n\}$ such that $\{x_1,\ldots,x_m\} \subseteq \cup_{j\in S} P_j$. We proceed by induction on $\#F$, the number of elements in F. If $\#F = 0$, then $x_j \in I \setminus \cup_i P_i$ for some j. Otherwise choose S in F minimizing $\#S$. If $\#S \leq 1$, then $I \subseteq P_i$ for some i. Otherwise $\#S \geq 2$ and for each $i \in S$, there exists $a_i \in \{x_1,\ldots,x_m\}$ such that $a_i \in P_i \setminus \cup_{S\setminus\{i\}} P_j$. If $\#S = 2$, then set $x_{m+1} = \Sigma_{i\in S} a_i \in I \setminus \cup_{i\in S} P_i$. If $\#S > 2$, then P_i is prime for some $i \in S$, so $x_{m+1} = a_i + \Pi_{j\in S\setminus\{i\}} a_j \in I \setminus \cup_{i\in S} P_i$. In either case we can enlarge $\{x_1,\ldots,x_m\}$, without enlarging I, so that $S \notin F$, and we are done by induction on $\#F$. □

2.4 THEOREM. *Let $I_1,\ldots,I_n$ be finitely generated ideals of a commutative ring R, and P a prime ideal of R such that the product $I_1\cdots I_n \subseteq P$. Then $I_i \subseteq P$ for some i.*

PROOF. By induction on n it suffices to consider the case $n = 2$. Let $I_1 = (a_1,\ldots,a_s)$ and $I_2 = (b_1,\ldots,b_t)$. As $a_i b_j \in P$, for each i either $a_i \in P$, or $b_j \in P$ for all j. □

A **denial field** is a commutative ring that is a field under the denial inequality, and such that 0 is a prime ideal. Any discrete field is a denial field. A **maximal ideal** in a commutative ring R is an ideal M such that R/M is a denial field; so an ideal M of R is maximal if and only if it is a prime ideal, and $x \in R\setminus M$ if and only if $rx - 1 \in M$ for some $r \in R$. A detachable maximal ideal is an ideal M such that R/M is a discrete field.

The **characteristic** of a ring k is the order of the element 1 in the additive group of k. It is a standard convention to say that k has

characteristic 0 if the order of 1 is infinite. Thus the characteristic of a discrete ring k is the least positive integer n such that $n \cdot 1 = 0$, if such a positive integer exists, and is 0 if no such positive integer exists. The field of rational numbers has characteristic 0, and the field $\mathbb{Z}_p = \mathbb{Z}/(p)$ has characteristic p. The characteristic of a discrete field need not be in $\mathbb{N}$ as the following Brouwerian example shows.

2.5 EXAMPLE. Let a be a binary sequence with at most one 1, and define p_n by

$$p_n = \begin{cases} 0 \text{ if } a_n = 0 \\ \text{the } n^{\text{th}} \text{ prime if } a_n = 1 \end{cases}$$

Let P be the ideal of the integers $\mathbb{Z}$ generated by the numbers p_n, and let $R = \mathbb{Z}/P$. Then R is a discrete integral domain; let k be its field of quotients. The characteristic of k is not in $\mathbb{N}$. □

EXERCISES

1. Show that in any ring the following identities hold.

   (i) $a0 = 0a = 0$,

   (ii) $a(-b) = (-a)(b) = -ab$.

2. Use the rings $\mathbb{Z}$ and $\mathbb{Q}$ to construct a Brouwerian example of a discrete integral domain that does not have recognizable units.

3. Show that a discrete ring is a division ring if and only if the nonzero elements form a multiplicative group.

4. **The rational quaternions.** Let the elements of (a,b,c,d) in $R = \mathbb{Q}^4$ be written as formal sums $a + bi + cj + dk$, where elements of $\mathbb{Q}$ commute with i, j, and k, and $i^2 = j^2 = -1$, and $k = ij = -ji$. Note that

   $$(a + bi + cj + dk)(a - bi - cj - dk) = a^2 + b^2 + c^2 + d^2$$

   and show that R is a noncommutative discrete division ring.

5. Show that in a cotransitive field the operations of addition, subtraction, multiplication and division (restricted to units) are strongly extensional (see Exercise 1.5).

6. Let R be a subring of a field K. Show that

   $$k = \{ab^{-1} : a,b \in R \text{ and } b \neq 0\}$$

is a field containing R. Show that if R is a subring of another field K', and the inequalities on K and K' agree on R, then k is isomorphic to k'. Show that the inequality on R is consistent, cotransitive, tight or discrete if and only if the inequality on k is.

7. Show that a commutative ring is an integral domain if and only if the following conditions hold:

   (i) $1 \neq 0$,

   (ii) $a \neq b$ if and only if $a - b \neq 0$,

   (iii) if $a \neq 0$ and $ab = 0$, then $b = 0$,

   (iv) $a \neq 0$ and $b \neq 0$ if and only if $ab \neq 0$.

   Show that a necessary and sufficient condition for a commutative ring with the denial inequality to be an integral domain is that $a \neq 0$ if and only if a is cancellable ($ab = 0$ implies $b = 0$).

8. Show that the following hold for ideals in a commutative ring.

   (i) $IJ \subseteq I \cap J$

   (ii) $IJ \subseteq K$ if and only if $I \subseteq K:J$

   (iii) If $I \subseteq J$ then $K:J \subseteq K:I$

   (iv) $(\cap_i I_i):J = \cap_i (I_i:J)$

   (v) $I:\Sigma_i J_i = \cap_i (I:J_i)$

   (vi) $I:JK = (I:J):K$

9. Show that the following hold for ideals in a commutative ring.

   (i) $\sqrt{IJ} = \sqrt{I \cap J} = \sqrt{I} \cap \sqrt{J}$.

   (ii) If $I^n \subseteq J$ for some n, then $\sqrt{I} \subseteq \sqrt{J}$.

   (iii) $\sqrt{I + J} = \sqrt{\sqrt{I} + \sqrt{J}}$.

   (iv) $\sqrt{\sqrt{I}} = \sqrt{I}$.

10. Let $\varphi : R \to R'$ be a map of commutative rings, and let I and J be ideals of R'. Show that $\varphi^{-1}(I) \cap \varphi^{-1}(J) = \varphi^{-1}(I \cap J)$, that $\varphi^{-1}(I)\varphi^{-1}(J) \subseteq \varphi^{-1}(IJ)$, and that $\sqrt{\varphi^{-1}(I)} = \varphi^{-1}(\sqrt{I})$. Show that if φ is onto, then $\varphi^{-1}(I:J) = \varphi^{-1}I:\varphi^{-1}J$.

11. Show that $(12) \cup (45)$ is not an ideal in the ring $\mathbb{Z}$ of integers. Show that $(12) + (45)$ and $(12):(45)$ are principal ideals.

12. In (2.3) we don't have to know *which* $n-2$ ideals are prime. Prove (2.3) under the weaker hypothesis that if $a_i b_i \in P_i$ for $i = 1,\ldots,n$, then for at least $n-2$ indices i, either $a_i \in P_i$ or

$b_i \in P_i$.

13. Change (2.3) so that none of the P_i are assumed prime, and the conclusion is that $I \subseteq P_i$ for some i, or there exist three distinct indices j such that P_j is nonprime (in a suitably strong sense). Can you prove that the three P_j are distinct (and not just their indices)?

14. Let B and C be two detachable subgroups of a group G. Show that if A is a finitely-generated subgroup of G, then either $A \subseteq B$ or $A \subseteq C$ or there exists $x \in A \setminus (B \cup C)$. Give a counterexample (not Brouwerian) to show that the result is false for three subgroups.

15. Show that a nontrivial discrete ring R is a division ring if and only if each finitely generated ideal is either R or 0. Give a Brouwerian example of an ideal in $\mathbb{Q}$ that is neither $\mathbb{Q}$ nor 0.

16. Show that an ideal of a commutative ring R is a proper prime ideal if and only if it is the kernel of a homomorphism of R into a denial field.

17. Show that a finite integral domain is a field. Show that a detachable ideal I in the ring $\mathbb{Z}$ is maximal if and only if $I = (p)$ for some prime number p.

3. REAL NUMBERS

The prototype Heyting field is the field $\mathbb{R}$ of real numbers. The set $\mathbb{Q}^{\mathbb{N}}$ of sequences of rational numbers forms a commutative ring under coordinatewise addition and multiplication. A sequence $\{q_n\}$ of rational numbers is a **Cauchy sequence** if for each positive $\epsilon \in \mathbb{Q}$, there exists $N \in \mathbb{N}$ such that

$$|q_n - q_m| \leq \epsilon \text{ whenever } m,n \geq N.$$

It is readily seen that the set C of Cauchy sequences of rational numbers forms a subring of $\mathbb{Q}^{\mathbb{N}}$.

A sequence $\{q_n\}$ of rational numbers **converges to 0** if for each positive $\epsilon \in \mathbb{Q}$, there exists $N \in \mathbb{N}$ such that $|q_n| \leq \epsilon$ for all $n \geq N$. The set I of rational sequences that converge to 0 is easily seen to be an ideal in the ring C of Cauchy sequences. The set $\mathbb{R}$ of **real numbers** is the quotient ring C/I. By associating with each $q \in \mathbb{Q}$ the sequence whose elements are all equal to q, we get a natural embedding of $\mathbb{Q}$ in $\mathbb{R}$.

The set $\mathbb{R}$ of real numbers admits a natural order. Define $a \in \mathbb{R}$ to be **positive,** if there exist positive $\epsilon \in \mathbb{Q}$ and $N \in \mathbb{N}$, such that $a_n \geq \epsilon$ whenever $n \geq N$. It is easily checked that this definition respects the equality on $\mathbb{R} = C/I$, and that the set of positive real numbers is closed under addition and multiplication. We write $a < b$, or $b > a$, if $b - a$ is positive; in particular, $a > 0$ means a is positive.

3.1 THEOREM. *The following conditions on a real number a are equivalent.*

(i) *There exist positive $\epsilon \in \mathbb{Q}$ and $N \in \mathbb{N}$ such that $|a_n| \geq \epsilon$ for all $n \geq N$.*

(ii) $a < 0$ or $a > 0$.

(iii) *a is invertible.*

PROOF. Suppose (i) holds. We may assume that $|a_N - a_n| < \epsilon/2$ whenever $n \geq N$. If $a_N \geq \epsilon$, then $a_n > \epsilon/2$ whenever $n \geq N$, so $a > 0$; if $a_N \leq -\epsilon$ then $a < 0$.

Now suppose (ii) holds. If $a > 0$, then there are positive $\epsilon \in \mathbb{Q}$ and $N \in \mathbb{N}$ such that $a_n > \epsilon$ whenever $n \geq N$. We may assume that $a_n > \epsilon$ for *all* $n \in \mathbb{N}$. Then the sequence $\{1/a_n\}$ is Cauchy and is the inverse of a in $\mathbb{R}$.

Finally suppose (iii) holds. Then there exists a Cauchy sequence $\{b_n\}$ such that $a_nb_n - 1$ converges to 0. Choose positive $\epsilon \in \mathbb{Q}$ such that $|b_n| < 1/\epsilon$ for all n. Then $|a_n|$ is eventually greater than ϵ. □

We get an inequality on $\mathbb{R}$ by defining $a \neq b$ to mean that $b - a$ is invertible, so (3.1) says that $a \neq b$ if and only if $a < b$ or $b < a$. The cotransitivity of $a < b$ is the constructive substitute for the classical trichotomy.

3.2 THEOREM (cotransitivity). *Let a, b and c be real numbers. If $a < c$, then either $a < b$ or $b < c$.*

PROOF. Choose $m \in \mathbb{N}$ and $\epsilon > 0$ so that

$$a_m < c_m - 6\epsilon$$
$$|a_n - a_m| < \epsilon$$
$$|c_n - c_m| < \epsilon$$
$$|b_n - b_m| < \epsilon$$

whenever $n \geq m$. Either $b_m < c_m - 3\epsilon$, in which case $b_n < c_n - \epsilon$ for all $n \geq m$, so $b < c$, or $b_m > a_m + 3\epsilon$, in which case $b_n > a_n + \epsilon$ for all $n \geq m$, so $b > a$. □

We write $a \lesssim b$ if $a < b + \epsilon$ for all $\epsilon > 0$. This relation is clearly transitive and reflexive. To show that it gives a partial order, we need the following.

3.3 THEOREM. *If $a \lesssim b$ and $b \lesssim a$, then $a = b$.*

PROOF. Suppose $\epsilon \in \mathbb{Q}$ is positive. As $a \lesssim b$, we can find $N \in \mathbb{N}$ so that $a_n - b_n < \epsilon$ for all $n \geq N$. As $b \lesssim a$, we can find such N so that also $b_n - a_n < \epsilon$ for all $n \geq N$. But this says that $|a_n - b_n| < \epsilon$ for all $n \geq N$, so we have shown that $a_n - b_n$ converges to 0, which means that $a = b$. □

3.4 COROLLARY. *$\mathbb{R}$ is a Heyting field.*

PROOF. If $a + b \neq 0$, then either $a + b > 0$ or $a + b < 0$, and we may assume that $a + b > 0$. Then either $a > 0$ or $a < a + b$, by (3.2); in the former case $a \neq 0$, in the latter case $0 < b$ so $b \neq 0$. Thus the inequality on $\mathbb{R}$ is cotransitive.

To show that the inequality on $\mathbb{R}$ is tight, suppose $a \neq 0$ is impossible. For each $\epsilon > 0$, either $a > 0$ or $a < \epsilon$ by (3.2); the former is impossible, so $a < \epsilon$. Thus $a \lesssim 0$. Similarly $a \gtrsim 0$, so $a = 0$ by (3.3). □

Not only is $\mathbb{R}$ a partially ordered set, it is a lattice. If a and b are real numbers, then $c_n = max(a_n,b_n)$ defines a real number c that is the supremum of a and b, written $c = sup(a,b)$. The infimum of a and b is $-sup(-a,-b)$. The absolute value of a real number a may be defined as $|a| = sup(a,-a)$.

The field $\mathbb{C}$ of **complex numbers** is obtained from the vector space $\mathbb{R}^2$ by setting $(a,b)(c,d) = (ac - bd, ad + bd)$. It is readily verified that $\mathbb{C}$ is a Heyting field with multiplicative identity (1,0). Setting $i = (0,1)$ we see that $i^2 = -1$ and $\mathbb{C} = \mathbb{R} + \mathbb{R}i$.

A **metric space** is a set S together with a function d, called a **metric**, from S × S to $\mathbb{R}$ such that

(i) $d(x,y) = d(y,x) \gtrsim 0$,

(ii) $d(x,y) = 0$ if and only if $x = y$,

(iii) $d(x,z) \lesssim d(x,y) + d(y,z)$.

The real numbers form a metric space under the metric $d(x,y) = |x - y|$. A Cauchy sequence in a metric space is a sequence $\{x_n\}$ in S such that for each positive $\epsilon \in \mathbb{Q}$, there exists $N \in \mathbb{N}$ such that

$$d(x_n,x_m) \leq \epsilon \text{ whenever } m,n \geq N.$$

A sequence in $\{x_n\}$ in S **converges** to $y \in S$ if for each positive $\epsilon \in \mathbb{Q}$ there exists $N \in \mathbb{N}$ such that $d(x_n,y) \leq \epsilon$ whenever $n \geq N$. If $\{x_n\}$ converges to y, we say that y is the **limit** of $\{x_n\}$. It is easy to verify that each convergent sequence is a Cauchy sequence. If, conversely, each Cauchy sequence converges to some element of S, we say that S is **complete**. The space $\mathbb{R}$ is complete.

By imitating the construction of $\mathbb{R}$ from $\mathbb{Q}$, we can embed any metric space S in its **completion** $\hat{S}$, whose elements are Cauchy sequences in S, with $d(a,b)$ equal to the limit of $d(a_n,b_n)$, and $a = b$ defined by $d(a,b) = 0$. The space S is **dense** in $\hat{S}$, that is, for each positive ϵ, and $s \in \hat{S}$, there is $a \in S$ such that $d(a,s) < \epsilon$. The space $\hat{S}$ is complete.

EXERCISES

1. Show that $a > b$ is impossible if and only if $a \leq b$.

2. Show that the following are equivalent.
   (i) For all $a \in \mathbb{R}$, either $a > 0$ or $a \leq 0$.
   (ii) LPO.

3. Show that $\mathbb{R}$ is a distributive lattice under $\leq$.

4. Show that $|a| \geq 0$, and that $|a| > 0$ if and only if $a \neq 0$. Show that $|a + b| \leq |a| + |b|$.

5. Show that the following are equivalent.
   (i) LLPO,
   (ii) For all $a \in \mathbb{R}$, either $a \leq 0$ or $a \geq 0$.
   (iii) For all $a,b \in \mathbb{R}$, if $ab = 0$, then $a = 0$ or $b = 0$.
   (iv) If $a,b \in \mathbb{R}$, then there is $c \in \mathbb{R}$ such that $a = cb$ or $b = ca$.
   (v) For all $a,b \in \mathbb{R}$, if $sup(a,b) = 1$, then $a = 1$ or $b = 1$.

6. Show that Markov's principle is equivalent to $\mathbb{R}$ being a denial field.

7. **The field of p-adic numbers.** If p is a prime, the **p-adic metric on $\mathbb{Q}$** is defined, for $x_1 \neq x_2$, by setting $d(x_1,x_2) = p^n$ such that $p^n(x_1 - x_2)$ can be written with numerator and denominator not divisible by p. Show that d is a metric, and that the completion

of $\mathbb{Q}$ with respect to this metric is a Heyting field.

8. Show that each metric space is dense in its completion, and that its completion is complete.

4. MODULES

If R is a ring, then a **left R-module** is an additive abelian group M together with a function μ from $R \times M$ to M, written $\mu(r,a) = ra$ and called **scalar multiplication**, satisfying

$$\begin{aligned} &\text{(i)} & r(a + b) &= ra + rb \\ &\text{(ii)} & (r + s)a &= ra + sa \\ &\text{(iii)} & (rs)a &= r(sa) \\ &\text{(iv)} & 1 \cdot a &= a \end{aligned}$$

for all r, s in R and a, b in M. **Right R-modules** are defined similarly, except the scalar multiplication is on the right: the only real difference is in (iii) which will read $a(rs) = (ar)s$, so for right modules, multiplying by rs is the same as first multiplying by r and then multiplying by s, while for left modules multiplying by rs is the same as first multiplying by s and then multiplying by r. Any abelian group is a $\mathbb{Z}$-module. The set R^n of n-tuples of elements of R is an R-module under coordinatewise addition and scalar multiplication. If R is a division ring, then an R-module is called a **vector space** over R.

Let M and N be left R-modules. A group homomorphism f from M to N is an **R-module homomorphism** if $f(ra) = rf(a)$ for all r in R and a in M. The **kernel** of f is ***ker*** $f = \{a \in M : f(a) = 0\}$, and the **image** of f is ***im*** $f = \{f(a) : a \in M\}$. The category of R-modules is easily seen to be pre-additive.

The set $E(M)$ of endomorphisms of an abelian group M forms a ring, where multiplication is composition of functions. If R is a ring and f is a ring homomorphism from R to $E(M)$, then M can be endowed with a left R-module structure by setting $rm = f(r)m$ for r in R and m in M. Conversely if M is an R-module, then we can define a ring homomorphism φ from R to $E(M)$ by setting $\varphi(r)(m) = rm$. The homomorphism φ is called a **representation** of R as a ring of endomorphisms of M. Representations and modules are two ways of looking at the same thing. If the kernel of the representation is 0 then the representation is said to be **faithful.** An R-module M is **faithful** if $r = 0$ whenever $rm = 0$ for each m in M.

A subgroup N of an R-module M is an **R-submodule** if $ra \in N$ for each a in N and r in R. The submodule of M generated by a subset X is the additive subgroup generated $\{rx : r \in R \text{ and } x \in X\}$. The quotient group M/N is an R-module because scalar multiplication is a function on it. Theorems 1.1 and 1.2 hold for R-modules if we replace the term 'group' by 'module' throughout, and consider all submodules to be normal (which, as subgroups, they are).

If R and S are subrings of a ring A, then A is, among other things, a left R-module and a right S-module. This kind of situation occurs often enough to deserve a name. Let R and S be rings and let M be a left R-module and a right S-module. Then M is an **R-S-bimodule** if $(ra)s = r(as)$ for all each $r \in R$, $a \in M$, and $s \in S$. Thus the ring R is an R-R-bimodule, and every left R-module is an R-$\mathbb{Z}$-bimodule. If R is a commutative ring, then there is no distinction between left modules and right modules, and each is an R-R-bimodule.

Ideals in R can be described in terms of the module structures on R: a left ideal is a submodule of the left module R, a right ideal is a submodule of the right module R, and a two-sided ideal is a submodule of the R-R-bimodule R.

Let M be an R-module and $\{A_i\}_{i \in I}$ a family of submodules of M. The submodule of M generated by $\cup_{i \in I} A_i$ is denoted by $\Sigma_{i \in I} A_i$. The A_i are said to be **independent** if whenever $i(1),\ldots,i(n)$ and j are elements of I, and $x \in A_j \cap (A_{i(1)} + \cdots + A_{i(n)})$, then either $x = 0$ or $i(m) = j$ for some m. We say that M is the (internal) **direct sum** of the submodules A_i, and write $M = \oplus_{i \in I} A_i$, if $M = \Sigma_{i \in I} A_i$ and the A_i are independent. If $I = \{1,\ldots,n\}$, then we write $M = A_1 \oplus \cdots \oplus A_n$. If $I = \{1,2\}$, then $M = A_1 \oplus A_2$ if and only if $M = A_1 + A_2$ and $A_1 \cap A_2 = 0$. As an example let $M = \mathbb{Z}/(6)$, let $A_1 = \{0,2,4\}$ and let $A_2 = \{0,3\}$.

The product of a family of R-modules $\{A_i\}_{i \in I}$ admits a natural R-module structure under which it is the categorical product, or the **direct product**, of the modules A_i. If f and g are functions in the product, define $f + g$ by $(f + g)(i) = f(i) + g(i)$, and rf by $(rf)(i) = rf(i)$. The direct product is denoted by $\Pi_{i \in I} A_i$.

Let I be a discrete set and $\{A_i\}_{i \in I}$ a family of R-modules. We can construct a categorical coproduct (external direct sum) as follows. An element $f \in \Pi_{i \in I} A_i$ has **finite support** if there is a finite subset J of I so that $f(i) = 0$ if $i \in I \setminus J$. For I a discrete set, the external direct sum of the family $\{A_i\}_{i\in I}$ is the set of elements in the direct product that have finite support. It is easy to see that if f and g have finite support, then so do $f + g$ and rf, so the external direct sum is a submodule of the direct product. Note that if I is finite, then the direct product and the external direct sum are the same. If we identify the module A_i with the submodule $\{f : f(j) = 0 \text{ for } j \neq i\}$, then we see that the external direct sum is a direct sum.

The above construction runs into trouble if the index set I is not discrete, because if A_i is discrete and $\{f : f(j) = 0 \text{ for } j \neq i\}$ contains a nonzero element, then $\{i\}$ is detachable from I. To construct an **external direct sum** when the index set I is not necessarily discrete, let F be the set of finite sequences of elements of the disjoint union of the A_i. Let the equality on F be generated by

(i) $(a_1,\ldots,a_n) = (a_{\sigma(1)},\ldots,a_{\sigma(n)})$ if σ is a permutation of $\{1,\ldots,n\}$.

(ii) $(a_1,\ldots,a_{n-1},a_n) = (a_1,\ldots,a_{n-1})$ if $a_n = 0$.

(iii) $(a_1,\ldots,a_{n-1},a_n) = (a_1,\ldots,a_{n-1}+a_n)$ if a_{n-1} and a_n are in the same A_i.

More precisely, call two sequences in F **adjacent** if some permutation of one is gotten by applying (ii) or (iii) to some permutation of the other. Then σ and τ are **equal** in F if there is a chain

$$\sigma = s_1, s_2, \ldots, s_m = \tau$$

of sequences in F such that s_i is adjacent to s_{i+1} for $i = 1,\ldots,m-1$.

We want to identify $\{(a) \in F : a \in A_i\}$ with A_i. To do this we must show that if $(a) = (b)$, then $a = b$. This follows from the Church-Rosser property of F:

4.1 LEMMA (Church-Rosser property). *Suppose σ and τ are equal in F, and let $\ell(s)$ denote the length of the sequence s. Then then there is a chain $\sigma = s_1, s_2, \ldots, s_m = \tau$ of sequences in F such that s_i is adjacent to s_{i+1} for $i = 1,\ldots,m-1$, and for $i = 2,\ldots,m-1$, if $\ell(s_{i-1}) < \ell(s_i)$, then $\ell(s_i) < \ell(s_{i+1})$.*

PROOF. Let $\sigma = s_1, s_2, \ldots, s_m = \tau$ be a chain of sequences in F such that s_i is adjacent to s_{i+1} for $i = 1,\ldots,m-1$. We will proceed by induction on $N = \Sigma_i \ell(s_i)$, showing that if $\ell(s_{i-1}) < \ell(s_i)$ and $\ell(s_i) > \ell(s_{i+1})$ for some i, then we can reduce N.

Suppose we get from s_i to s_{i+1} by deleting a zero z. If z appears in s_{i-1}, then we can replace s_i by s_{i-1} with z deleted. Otherwise s_{i-1} comes from s_i by deleting z, in which case we can omit s_i. The same argument applies if we get from s_i to s_{i-1} by deleting a zero.

Suppose we get from s_i to s_{i+1}, and from s_i to s_{i-1}, by applying (iii). Depending on how many distinct positions in s_i are involved, we can represent the various cases by

s_{i-1}	s_i	s_{i+1}
$(a+b)$	(a,b)	$(a+b)$
$(a+b,c)$	(a,b,c)	$(a,b+c)$
$(a+b,c,d)$	(a,b,c,d)	$(a,b,c+d)$

In the first case we can omit s_i and s_{i+1}. In the second we can replace s_i by $(a+b+c)$, and in the third we can replace s_i by $(a+b,c+d)$. □

Two sequences in F are added by concatenating them, scalar multiplication is done coordinatewise, and the empty sequence serves as an identity. Because of the equality relation, we can safely, and unambiguously, write an element $(a_1,a_2,\ldots,a_n)$ of F as a formal sum $a_1 + a_2 + \cdots + a_n$

If the module A_i is identified with $\{(a) : a \in A_i\}$, then F is an internal direct sum of the A_i; the verification of this rash claim is left as Exercise 5. If I is discrete, and $(a_1,\ldots,a_n) \in F$, then we may assume that $a_m \in A_{i(m)}$ with $i(m) \neq i(m')$ if $m \neq m'$, and we can identify F with the set of elements of $\Pi_{i\in I}A_i$ of finite support, as above. If each A_i is a fixed module M, and I is an arbitrary index set, we denote the external direct sum by $M^{(I)}$.

The following theorem says that direct sums are categorical coproducts.

4.2 THEOREM. *If* $M = \oplus_{i\in I}A_i$, *and* $f_i : A_i \to N$ *is a family of homomorphisms, then there is a unique homomorphism* f *from* M *to* N *such that* $f = f_i$ *on* A_i *for each* $i \in I$.

PROOF. If $x \in M = \Sigma_{i\in I}A_i$, then $x = \Sigma_{m=1}^{n} a_{i(m)}$, with $a_{i(m)} \in A_{i(m)}$. Therefore $f(x)$ must equal $\Sigma_{m=1}^{n} f_{i(m)}(a_{i(m)})$, so f is unique. If we define $f(x)$ to be $\Sigma_{m=1}^{n} f_{i(m)}(a_{i(m)})$, we must show that $f(x)$ is well defined; it suffices to show that if $x = 0$, then $f(x) = 0$. Suppose $x = \Sigma_{m=1}^{n} a_{i(m)} = 0$. As the A_i are independent, either $a_{i(m)} = 0$ for each m, or there exist $m \neq m'$ such that $i(m) = i(m')$. In the latter case, we can add $a_{i(m)}$ and

$a_{i(m')}$ within $A_{i(m)}$, so $f(x) = 0$ by induction on n. It is readily seen that f is a homomorphism. □

An R-module F is **free** on a family of elements $\{x_i\}_{i \in I}$ of F if for each function f mapping I into an R-module M, there is a unique R-module homomorphism f^* from F to M such that $f^*(x_i) = f(i)$. We say that $\{x_i\}_{i \in I}$ is a **basis** for F. The uniqueness of f^* implies that free modules on $\{x_i\}_{i \in I}$ and $\{y_i\}_{i \in I}$ are isomorphic under an isomorphism taking x_i to y_i, so any two free modules whose bases have the same index set are essentially the same. If $F = \oplus_{i \in I} Rx_i$, and the map from R to Rx_i that takes r to rx_i is an isomorphism for each i in I, then (4.2) shows that F is free on $\{x_i\}_{i \in I}$. If R is a nontrivial ring, then $x_i \neq 0$ for each $i \in I$, so if $x_i = x_j$, then $i = j$; so the basis elements x_i are in one-to-one correspondence with the elements i of I. Thus if we were to restrict ourselves to nontrivial rings, we could define a basis to be a *set* rather than a family. If R is a trivial ring, then any family of elements of any R-module M is a basis for M.

Let I be a discrete set, and for each $i \in I$ let $\delta_i \in R^{(I)}$ be such that $\delta_i(i) = 1$, and $\delta_i(j) = 0$ for $j \neq i$. Then $R^{(I)}$ is free on $\{\delta_i\}_{i \in I}$. Similarly, suppose I is an arbitrary index set, and $\{R_i\}_{i \in I}$ is a family such that each R_i is a copy of R. If for each $i \in I$ we let x_i be the sequence of length one whose term is the identity element of R_i, then $R^{(I)}$ is free on $\{x_i\}_{i \in I}$. By abuse of language we say that $R^{(I)}$ is the **free module on** I.

If $I = \{1,\ldots,n\}$, then we write R^n instead of $R^{(I)}$. A module M has a basis of n elements if and only if it is isomorphic to R^n, in which case we say that M is a **free module of rank** n. In Section 6 we shall see that, for nontrivial commutative R, the rank of M is an invariant. Exercise 3 contains an example of a nontrivial noncommutative ring R such that the left R-modules R and R^2 are isomorphic.

An R-module M is **finitely generated** if there is a map from R^n onto M for some positive integer n; that is, if there exist elements $x_1,\ldots,x_n$ in M such that each element of M can be written in the form $\Sigma_{i=1}^n r_i x_i$. An R-module M is **cyclic** if there is a map from R onto M; that is, if there exists $x \in M$ such that each element of M is a scalar multiple of x.

4.3 THEOREM. *Let R be a subring of the ring E. If M is a finitely generated (free of rank n) E-module, and E is a finitely generated (free* *of rank m) R-module, then M is a finitely generated (free of rank mn) R-module.*

PROOF. Let $\varphi : R^m \to E$ be an epimorphism (isomorphism) of R-modules, and $\psi : E^n \to M$ be an epimorphism (isomorphism) of E-modules. Then $\varphi^n : R^{mn} \to E^n$ is an epimorphism (isomorphism) of R-modules, and $\psi\varphi^n : R^{mn} \to M$ is an epimorphism (isomorphism) of R-modules. □

An R-module P is said to be **projective** if whenever g maps an R-module A onto an R-module B, and f maps P into B, then there exists a map $h : F \to A$ such that $gh = f$. Finite-rank free modules are projective: if $x_1,\ldots,x_n$ is a basis for P, then there exist $a_1,\ldots,a_n$ in A such that $g(a_i) = f(x_i)$ for each i, and there is a map h such that $h(x_i) = a_i$ for each i; the maps gh and f are equal because they agree on the basis.

Let M be a finitely generated module. If π maps a finite-rank free module F onto M with kernel K, then M is isomorphic to F/K. The following theorem shows what happens when we do this for different F's and π's; the result is similar to the rule for determining when two fractions are equal.

4.4 THEOREM (Schanuel's trick). *Let M be an R-module, P_1 and P_2 projective R-modules, and π_i an map from P_i onto M. If K_i is the kernel of π_i, then $K_1 \oplus P_2$ is isomorphic to $K_2 \oplus P_1$.*

PROOF. As the P_i are projective, we can find maps $\varphi_1 : P_2 \to P_1$ and $\varphi_2 : P_1 \to P_2$, such that $\pi_1\varphi_1 = \pi_2$ and $\pi_2\varphi_2 = \pi_1$. Map $K_1 \oplus P_2$ to $K_2 \oplus P_1$ by taking (k_1,p_2) to (k_2,p_1) where

$$k_2 = p_2 - \varphi_2(k_1 + \varphi_1 p_2)$$
$$p_1 = k_1 + \varphi_1 p_2$$

and map $K_2 \oplus F_1$ to $K_1 \oplus F_2$ by setting

$$k_1 = p_1 - \varphi_1(k_2 + \varphi_2 p_1)$$
$$p_2 = k_1 + \varphi_2 p_1$$

It is readily checked that these maps are inverses of each other. □

An element e of a ring is **idempotent** if $e^2 = e$. A submodule A of M is a (direct) **summand** of M if there exists a submodule B of M, called a **complementary summand** of A, so that $M = A \oplus B$. The submodules 0 and M of M are always summands.

4.5 THEOREM. *Let A be a submodule of an R-module M. Then A is a direct summand of M if and only if there is an idempotent endomorphism e of M such that A = eM. In this case the submodule (1−e)M is a complementary summand of A.*

PROOF. Suppose $M = A \oplus B$. If $x \in M$, then we can write x uniquely as $a + b$ for some a in A and b in B. Define an endomorphism e of M by setting $e(x) = a$. It is easy to see that e is idempotent and that $eM = A$.

Conversely suppose e is an idempotent endomorphism of M, and $A = eM$. Set $B = (1-e)M$. As $x = ex + (1-e)x$ we have $A + B = M$. If $x \in A \cap B$, then $x = (1-e)y$ and $x = ez$. Thus $ex = e(1-e)y = (e-e^2)y = 0$ and $ex = e^2z = ez = x$, so $x = 0$. So $A \cap B = 0$. □

The idempotent e in (4.5) is called the **projection** of M on A (along B).

EXERCISES

1. The **opposite ring** R^{op} of a ring R consists of the additive group R with the product ab in R^{op} defined to be the product ba of R. Show that every left R-module is a right R^{op}-module in a natural way. If R is a commutative ring, the opposite ring is isomorphic to R, so each left R-module is also a right R-module, and we need not distinguish between left and right R-modules.

2. Let R be a ring and M an R-module. Show that there exists a free R-module mapping onto M.

3. Let A and B be vector spaces over a discrete field k, each having a countably infinite basis. Let $V = A \oplus B$ and let R be the ring of endomorphisms of V. Construct $x \in R$ such that $xA = 0$ and $x : B \to V$ is an isomorphism, and $y \in R$ such that $yB = 0$ and $y : A \to V$ is an isomorphism. Show that $Rx \cong Ry \cong R$ as R-modules, and that $R = Rx \oplus Ry$. What is the point of this exercise?

4. Let M be an R-module and $\{A_i\}_{i\in I}$ a family of R-modules. Let $\{f_i\}_{i\in I}$ be a family of R-module homomorphisms such that f_i maps M to A_i. Let π_i be the projection of $\Pi_{i\in I} A_i$ to A_i. Show that there exists a unique R-module homomorphism f taking M to $\Pi_{i\in I} A_i$ such that $\pi_i f = f_i$ for each i in I.

5. Let F be the external direct sum of $\{A_i\}_{i\in I}$, as constructed for an arbitrary index set I. Show that the map from A_i into F given

by taking $a \in A_i$ to the sequence (a) is a monomorphism. Show that if we identify A_i with its image under this monomorphism, then $F = \oplus_{i \in I} A_i$.

6. **Summands need not be summands.** Let a be a binary sequence with at most one 1, and let $S = \{0, s, 2s, t, 2t\}$ with

   $s = t$ and $2s = 2t$ if $a_n = 1$ for an even n,

   $s = 2t$ and $2s = t$ if $a_n = 1$ for an odd n.

   Let $I = \{x,y\}$ with $x = y$ if $a_n = 1$ for some n. Finally let $A_x = \{0,s,2s\}$ and $A_y = \{0,t,2t\}$ with the obvious three-element group structures. Show that this is a Brouwerian example (LLPO) with A_x not a summand of $\oplus_{i \in I} A_i$.

7. Let $\{A_i\}_{i \in I}$ be a family of modules and $f_i : A_i \to A$ a family of isomorphisms. Show that the kernel of the map $f : \oplus_{i \in I} A_i \to A$ induced by the ismorphisms f_i is a complementary summand of each submodule A_i.

8. Show that $A \oplus B$ is projective if and only if A and B are projective. Construct a two-element projective module over the the ring $\mathbb{Z}/(6)$.

9. Show that the free module on a projective set (see Exercise I.3.4) is projective. What is the free module on an empty set?

10. **Free modules need not be projective.** Construct a Brouwerian example of a map α from a rank-2 free module F_1 onto a free module F_2 such that there is no map φ from F_2 to F_1 with $\alpha\varphi$ equal to the identity on F_1. Hint: Let F_2 and F_1 be free k-modules on the sets A and B of Example I.3.1, where k is the ring of integers modulo 2.

11. Show that if free modules with discrete bases are projective, then the world's simplest axiom of choice holds.

12. Let I be a discrete set and φ a nontrivial $\mathbb{Z}$-module map from $\mathbb{Z}^{(I)}$ to $\mathbb{Z}$. Show that $ker\ \varphi$ is a summand of $\mathbb{Z}^{(I)}$ if and only if $im\ \varphi$ is cyclic. Construct a Brouwerian example of a (not necessarily nontrivial) φ such that $ker\ \varphi$ is a summand but $im\ \varphi$ is not cyclic.

5. POLYNOMIAL RINGS

If M is a monoid, and R is a ring, let $\mathbf{R^{(M)}}$ denote the free R-module on the set M. We may think of the elements of $R^{(M)}$ as formal finite sums $r_1m_1 + \cdots + r_nm_n$ with each m_i in M and each r_i in R. Define the product of two elements of $R^{(M)}$ by

$$(\Sigma_{i=1}^{n} r_im_i)(\Sigma_{j=1}^{n'} r'_jm'_j) = \Sigma_{i=1}^{n} \Sigma_{j=1}^{n'} (r_ir'_j)(m_im'_j).$$

The product $m_im'_j$ is the product in the monoid M while $r_ir'_j$ is the product in the ring R. This makes $R^{(M)}$ into a ring, called a **monoid ring**, with identity element 1, the identity of M. If M is a group, then $R^{(M)}$ is called a **group ring**. The map taking r to $r1$ maps R isomorphically onto a subring of $R^{(M)}$, and we shall consider R to be a subset of $R^{(M)}$ via this embedding, that is, the element $r1$ will be denoted by r.

Let M be the free monoid on the one-element set $\{X\}$. Then

$$R^{(M)} = \{r_0 + r_1X + \cdots + r_nX^n : r_i \in R \text{ and } n \in \omega\}.$$

The element X is called an **indeterminate** and the elements of $R^{(M)}$ are called **polynomials**. The monoid ring $R^{(M)}$ is denoted by $R[X]$, and is called the **polynomial ring** in the indeterminate X over R.

The polynomial ring in n indeterminates, $R[X_1,\ldots,X_n]$, is defined inductively to be $R[X_1,\ldots,X_{n-1}][X_n]$. An element of $R[X_1,\ldots,X_n]$ of the form $X_1^{e(1)}\cdots X_n^{e(n)}$ is called a **monomial** of **degree** $\Sigma_{i=1}^{n}e(i)$. If R is discrete, then the **(total) degree** of a nonzero polynomial f in $R[X_1,\ldots,X_n]$ is the maximum of the degrees of the monomials that appear in f with nonzero coefficients. The ring $R[X_1,\ldots,X_n]$ is a free R-module on the monomials; in fact the monomials form a commutative monoid, and $R[X_1,\ldots,X_n]$ is the monoid ring, over R, on that monoid.

If R is commutative, then each polynomial f in $R[X_1,\ldots,X_n]$ defines a function from R^n to R: if $a_1,\ldots,a_n$ are in R, or in any commutative ring containing R, we let $f(a_1,\ldots,a_n)$ be the result of substituting a_i for X_i in the formal expression for f, and interpreting the formal operations as operation in R. We require commutativity because the indeterminates commute with each other, and with the elements of R. As the a_i commute with each other, and with the elements of R, the map that takes f to $f(a_1,\ldots,a_n)$ is a homomorphism of rings.

For $n \in \mathbb{N}$, a polynomial f in $R[X]$ that can be written as $\Sigma_{i=0}^{n-1} r_iX^i$ is said to have **degree at most** $n-1$, written $deg\ f \leq n-1$, or $deg\ f < n$. A polynomial is zero if and only if it has degree at most -1. If $deg\ f \leq d$, and $r_d = 1$, then we say that f is **monic.** Note that these definitions make no reference to an inequality on R. If $r_i \neq 0$ for some $i \geq d$, then we say that f has **degree at least** d, and write $deg\ f \geq d$. If $deg\ f \leq d$ and $deg\ f \geq d$, then we say that f has **degree** d, written $deg\ f = d$, and we call r_d the **leading coefficient** of f. If R is not discrete, then f need not have a degree even if f has a nonzero coefficient.

If f and g are polynomials, then we write $deg\ f \leq deg\ g$ if $deg\ g < n$ implies $deg\ f < n$ for each $n \in \mathbb{N}$; and we write $deg\ f < deg\ g$ if $deg\ g < n + 1$ implies $deg\ f < n$ for each $n \in \mathbb{N}$. Note that if $g = X^n + r_{n-1}X^{n-1} + \cdots + r_0$, then $deg\ f \leq deg\ g$ if and only if $deg\ f \leq n$, even if the ring might be trivial.

5.1 THEOREM. *Let k be a field, and $f,g \in k[X]$. If $deg\ f = m$ and $deg\ g = n$, then $deg\ fg = m + n$.*

PROOF. Let a be the leading coefficient of f, and b the leading coefficient of g. Then ab is the leading coefficient of fg, and $deg\ fg = deg\ f + deg\ g$. □

5.2 THEOREM (Division algorithm). *Let R be a commutative ring, and let $f,g \in R[X]$ be polynomials such that $deg\ f \leq m$ and $deg\ g \leq n \leq m+1$. If a is the coefficient of X^n in g, then there exist $q,r \in R[X]$ such that $a^{m-n+1}f = qg + r$ and $deg\ r \leq n-1$.*

PROOF. We proceed by induction on $m-n$. If $m-n = -1$, choose $q = 0$ and $r = f$. If $m-n \geq 0$, let $f = b_0 + b_1X + \cdots + b_mX^m$, and set $f_1 = af - b_mX^{m-n}g$. Then $deg\ f_1 \leq m-1$, so by induction, $a^{m-n}f_1 = q_1g + r$ for some $q_1,r \in R[X]$ with $deg\ r \leq n-1$. Thus $a^{m-n+1}f = (q_1 + a^{m-n}b_mX^{m-n})g + r$. □

If the divisor g is monic, which, if R is a discrete field, we may assume it to be, the division algorithm has a much tidier form.

5.3 COROLLARY. *Let $f,g \in R[X]$ be polynomials over a commutative ring R, with g monic. Then there is a unique pair $q,r \in R[X]$ such that $f = qg + r$, and $deg\ r < deg\ g$.*

PROOF. By Theorem 5.2 there exist q and r in $R[X]$, with $deg\ r < deg\ g$, such that $f = qg + r$. To prove uniqueness, suppose $f = q_1g + r_1$ with $deg\ r_1 < deg\ g$. Then $(q - q_1)g = r_1 - r$ and $deg(r_1 - r) < deg\ g$, so

$q - q_1 = 0$, because g is monic (induction on n such that $deg(q - q_1) \leq n$), and so $r_1 - r = 0$. □

5.4 COROLLARY (Remainder theorem). *Let $f \in R[X]$ be a polynomial over a commutative ring R, and let $a \in R$. Then there exists a unique $q \in R[X]$ such that $f(X) = q(X)(X-a) + f(a)$.*

PROOF. By Theorem 5.2 there exist unique $q \in R[X]$ and $r \in R$ such that $f(X) = q(X)(X-a) + r$. But then $f(a) = q(a)(a-a) + r = r$. □

We can construct a polynomial of degree at most n over a field taking on prescribed values on $n + 1$ distinct points. The remainder theorem shows that this polynomial is unique, so a nonzero polynomial cannot have $n + 1$ distinct roots. This is one of the few results in the general theory of fields (as opposed to discrete fields, or Heyting fields).

5.5 THEOREM (Unique interpolation). *Let $a_0,\ldots,a_n$ be distinct elements of a field k, and let $v_0,\ldots,v_n$ be any $n + 1$ elements of k. Then there is a unique polynomial $f \in k[X]$ of degree at most n such that $f(a_i) = v_i$ for all i.*

PROOF. We prove existence by induction on n. If $n = 0$ take $f = v_0$. If $n > 0$, then, by induction, there is a polynomial g of degree at most $n - 1$ such that $g(a_i) = (v_i - v_0)/(a_i - a_0)$ for $1 \leq i \leq n$. Take $f(X) = (X - a_0)g(X) + v_0$.

To prove uniqueness it suffices to show that if f is a polynomial of degree at most n, and $f(a_i) = 0$ for all i, then $f = 0$. We proceed by induction on n. If $n = 0$, then f is a constant, so $f = 0$ because $f(a_0) = 0$. Suppose $n > 1$. By the remainder theorem we can write $f(X) = (X-a_n)g(X)$ where $deg\ g \leq n-1$. As $a_j \neq a_n$ for $j < n$, and k is a field, it follows that $g(a_j) = 0$ for $j < n$, so $g = 0$ by induction. Therefore $f = 0$. □

The proof of Theorem 5.5 gives a recursive construction of the coefficients λ_i of the **Newton Interpolation Formula**

$$f = \lambda_0 + \lambda_1(X - a_0) + \lambda_2(X - a_0)(X - a_1) + \cdots$$
$$\cdots + \lambda_n(X - a_0)(X - a_1)\cdots(X - a_{n-1}).$$

For the **Lagrange Interpolation Formula**, see Exercise 5.

We state the Euclidean algorithm for discrete commutative rings with recognizable units, rather than just for discrete fields. The algorithm constructs either the desired common factor or a nonzero nonunit. The model application is to the ring $k[X]/(f)$ where k is a discrete field: the construction of a nonzero nonunit results in a factorization of f.

5.6 THEOREM (Euclidean algorithm). *Let R be a discrete commutative ring with recognizable units, and I a finitely generated ideal of $R[X]$. Then either I is principal or R has a nonzero nonunit.*

PROOF. Either $I = 0$, so I is principal, or there is $n \in \mathbb{N}$ and a nonzero polynomial f in I with $\deg f = n$. We may assume that f is monic (otherwise we have a nonzero nonunit), and we proceed by induction on n. If $n = 0$, then $f = 1$, so $I = k[X] = (f)$. If $n > 0$, then each generator g of I can be written as $g = qf + r$ with $\deg r < n$. Note that $r \in I$. Either each $r = 0$, or some $r \neq 0$. If each $r = 0$, then $I = (f)$. If some $r \neq 0$, then we have a nonzero polynomial in I of degree less than n, and we are done by induction. □

If $c = ab$ in a commutative ring, then we say that a **divides** c; if a divides c we say that a is a **divisor**, or a **factor**, of c.

5.7 COROLLARY. *Let k be a discrete field, and $a,b \in k[X]$. Then there exist $s,t \in k[X]$ such that $sa + tb$ divides both a and b. Hence $sa + tb$ is the greatest common divisor of a and b in the sense that every common divisor of a and b divides $sa + tb$.*

PROOF. Let I be the ideal of $k[X]$ generated by a and b. As k is a discrete field, Theorem 5.6 says that I is principal; that is, there exist s and t such that $sa + tb$ divides a and b. □

We say that two polynomials a and b over a commutative ring are **strongly relatively prime** if there exist polynomials s and t such that $sa + tb = 1$. Thus (5.7) implies that if two polynomials over a discrete field have no common factors of positive degree, then they are strongly relatively prime. It is easy to see that if a and b are strongly relatively prime, and a and c are strongly relatively prime, then a and bc are strongly relatively prime (multiply the two equations).

EXERCISES

1. Let R and S be commutative rings, φ a ring homomorphism from R to S, and $s_1,\ldots,s_n$ elements of S. Show that φ has a unique

extension to a homomorphism from $R[X_1,\ldots,X_n]$ to S mapping X_i to s_i.

2. An **apartness domain** is an integral domain whose inequality is an apartness. Show that the field of quotients of an apartness domain is a Heyting field. Let f and g be polynomials over an apartness domain R. Show that if $deg\ f \geq i$ and $deg\ g \geq j$, then $deg\ fg \geq i + j$. Use this result to show that if R is an apartness domain, then so is $R[X]$.

3. Let R be a ring. The **formal power series** ring $R[[X]]$ is defined to be the set of sequences $\{a_n\}$ in R, written

$$a_0 + a_1X + a_2X^2 + \cdots,$$

with addition and multiplication as suggested by the notation. Show that $R[[X]]$ is an apartness domain if R is. Hint: Suppose R is an apartness domain and $fg = h$ in $R[[X]]$ with

$$f = f_0 + f_1X + \cdots,$$
$$g = g_0 + g_1X + \cdots$$
$$h = h_0 + h_1X + \cdots.$$

Assume that $f_i \neq 0$ and $g_j \neq 0$ for some i,j; show that $h_k \neq 0$ for some $k \leq i + j$.

4. **Lagrange Interpolation.** Show that the following polynomial satisfies Theorem 5.5.

$$f(X) = \sum_{i=0}^{n} v_i \,\Pi_{j\neq i} \frac{(X - a_j)}{(a_i - a_j)}$$

5. Let k be a Heyting field and let $f \in k[X]$ be nonzero and have degree at most m. Show that if $a_0,a_1,\ldots,a_m$ are distinct elements of k, then there is i such that $f(a_i) \neq 0$.

6. Let k be a Heyting field and let $f \in k[X_1,\ldots,X_n]$ be nozero and have degree at most m in each of its variables separately. Show that if k contains $m + 1$ distinct elements, then $f(a_1,\ldots,a_n) \neq 0$ for some $a_i \in R$.

7. Let k be a discrete field. Show that any nonzero proper prime ideal in $k[X]$ is maximal.

8. Let f be a nonzero polynomial over a discrete field k, and $a \in k$. Show that there is a unique nonnegative integer n, and a unique

polynomial $u \in k[X]$, such that $f(X) = (X - a)^n u(X)$, and $u(a) \neq 0$. If $n = 1$, then a is said to be a **simple root** of f; if $n > 1$, then a is said to be a **root of multiplicity** n.

9. Find a polynomial of degree 2 over the ring of integers modulo 6 that has 3 distinct roots. Do the same for the rational quaternions.

10. Give a Brouwerian counterexample to the statement that either $deg\ f \leq deg\ g$, or $deg\ g \leq deg\ f$, for all polynomials f and g over a commutative ring.

6. MATRICES AND VECTOR SPACES

Let α be an R-module homomorphism from a free right R-module N to a free right R-module M. If $e_1,\ldots,e_n$ is a basis for N, and $f_1,\ldots,f_m$ is a basis for M, then α determines, and is determined by, the $m \times n$ matrix $A = \{a_{ij}\}$ such that

$$\alpha(e_j) = \sum_{i=1}^{m} f_i a_{ij}.$$

If β is a homomorphism from a free R-module L, with basis $d_1,\ldots,d_\ell$, to N, then we get an $n \times \ell$ matrix $B = \{b_{jk}\}$ such that

$$\beta(d_k) = \sum_{j=1}^{n} e_j b_{jk}.$$

Thus

$$\alpha\beta(d_k) = \alpha\sum_{j=1}^{n} e_j b_{jk} = \sum_{j=1}^{n} \alpha(e_j) b_{jk} = \sum_{j=1}^{n}\sum_{i=1}^{m} f_i a_{ij} b_{jk},$$

So the matrix corresponding to $\alpha\beta$ is the **matrix product** AB, which is an $m \times \ell$ matrix whose ik^{th} entry is $\Sigma_{j=1}^{n} a_{ij} b_{jk}$. If we consider only maps from N to N, then we get an isomorphism between the ring of endomorphisms of the free right R-module N and the ring $Mat_n(R)$ of $n \times n$ matrices over the ring R. The matrix corresponding to the identity endomorphism is called an **identity matrix** and is denoted by I.

If $e'_1,\ldots,e'_n$ is a new basis for N, and $f'_1,\ldots,f'_m$ is a new basis for M, then let σ and τ be the automorphisms of N and M defined by $\sigma(e_j) = e'_j$, and $\tau(f_i) = f'_i$, and let S and T be the matrices of σ and τ with respect to the old bases. Then the matrix of α with respect to the new bases is computed by

$$\alpha(e'_j) = \alpha(\sigma e_j) = \alpha(\sum_i e_i s_{ij}) = \sum_{ik} f_k a_{ki} s_{ij} = \sum_{ik} \tau^{-1}(f'_k) a_{ki} s_{ij}$$

So $\tau\alpha(e'_j) = \Sigma_{ik} f'_k a_{ki} s_{ij}$, whence the new matrix of $\tau\alpha$ is AS, so the new matrix of α is $T^{-1}AS$, where T^{-1} is the matrix of τ^{-1}.

The ith **row** $(a_{i1},\ldots,a_{in})$ of A may be considered as an element of the left R-module R^n. The **row space** of A is the submodule of R^n generated by the rows of A. An **elementary row operation** on A consists of either

(i) Interchanging two rows,
(ii) Multiplying a row by a unit of R,
(iii) Adding a multiple of a row to a different row.

The matrix resulting from applying an elementary row operation to A is the matrix of α with respect to some other basis of M. The matrix obtained by interchanging rows s and t of A is the matrix for α if we interchange the basis elements f_s and f_t. The matrix obtained by multiplying row s of A by the unit u is the matrix for α if we replace the basis element f_s by $f_s u^{-1}$. The matrix obtained by adding r times row s to row t is the matrix for α if we replace the basis element f_s by $f_s - f_t r$. The row space of A is unchanged by elementary row operations.

The jth column of $(a_{1j},\ldots,a_{mj})$ of A may be considered as an element of the right R-module R^m. An **elementary column operation** on A consists of either

(i) Interchanging two columns,
(ii) Multiplying a column by a unit of R,
(iii) Adding a multiple of a column to a different column.

The matrix resulting from applying an elementary column operation to A is the matrix of α with respect to some other basis of N. The matrix obtained by interchanging columns s and t of A is the matrix for α if we interchange the basis elements e_s and e_t. The matrix obtained by multiplying column s of A by the unit u (on the right) is the matrix for α if we replace the basis element e_s by $e_s u$. The matrix obtained by adding r times column s to column t is the matrix for α if we replace the basis element e_t by $e_t + e_s r$.

By an **elementary matrix** we mean the result of applying an elementary row operation to the identity matrix. If E is the elementary matrix obtained by applying the elementary row operation ρ to the identity matrix, then E may be obtained by applying an elementary column operation ρ' to the identity matrix. Furthermore, if A and B are matrices of the appropriate shapes, then EA is gotten by applying ρ to A, and BE is gotten by applying ρ' to B. A matrix of 0's and 1's, with exactly one 1 in each column and row is called a **permutation matrix**, and is a product of elementary matrices. It is readily seen that elementary matrices have two sided inverses, which are also elementary.

If k is a division ring, then a k-module is called a **vector space** over k. Classically every vector space over a division ring is free. This is not the case constructively, even for finitely generated vector spaces over discrete fields, as the following Brouwerian example shows.

6.1 EXAMPLE. Let a be a binary sequence, let $i^2 = -1$ and consider the sequence of subfields

$$k_n = \{s + ta_n i : s,t \in \mathbb{Q}\}.$$

of the field of Gaussian numbers $\mathbb{Q}(i)$. If $k = \cup k_n$, then k is a discrete field, and $\mathbb{Q}(i)$ is a discrete k-module generated by the two elements 1 and i. But we cannot construct a basis for $\mathbb{Q}(i)$ over k. □

If the vector space V is a free k-module of rank n, then n is referred to as the **dimension** of V and written $\mathbf{dim}_k$ V, or simply *dim* V; the space V is then called a **finite-dimensional vector space** over k. The following theorem shows, among other things, that *dim* V is well defined if k is discrete.

6.2 THEOREM. *Let* V *and* W *be vector spaces over a discrete division ring* k, *of dimensions* n *and* m *respectively. If* T *is a linear transformation from* V *to* W, *then there exist bases* $e_1,\ldots,e_n$ *of* V, *and* $f_1,\ldots,f_m$ *of* W, *and an index* $k \leq n$, *such that* $T(e_i) = f_i$, *for* $i \leq k$, *and* $T(e_i) = 0$ *for* $i > k$.

PROOF. Let $A = \{a_{ij}\}$ be the matrix of T with respect to the given bases for V and W. By elementary row and column operations we can arrange that $a_{ij} = 0$ for $i \neq j$, that $a_{ii} \in \{0,1\}$, and that $a_{ii} \geq a_{jj}$ if $i \leq j$. But this amounts to constructing the desired new bases for V and W. □

Taking f to be the identity map in (6.2) shows that the dimension of a finite-dimensional vector space is well defined. It follows immediately from (6.2) that *ker* T and *im* T are finite-dimensional summands, and *dim ker* T + *dim im* T = *dim* V.

The trouble in Example 6.1 is that the k-subspace generated by 1 is not detachable: we cannot tell whether i is in it or not. A *summand* A of a discrete vector space is detachable because $x \in A$ if and only if the projection of x onto A is equal to x. Thus the following corollary implies that finitely generated subspaces of finite-dimensional vector spaces are detachable.

6.3 COROLLARY. *Let V be a finite-dimensional vector space over a discrete division ring k. Let W be a finitely generated subspace of V. Then W is a finite dimensional summand of V.*

PROOF. Since W is a finitely generated, there exists a finite dimensional vector space F over k and a linear transformation T from F onto W. It follows from (6.2) that W is a finite-dimensional summand. □

6.4 COROLLARY. *Let V be a finite-dimensional vector space over a discrete division ring k. Then the intersection of any two finitely generated subspaces of V is a finite-dimensional summand of V.*

PROOF. Let A and B be finitely generated subspaces of V. By (6.3) we can find a complementary summand C of B in V. The projection on C, restricted to A, is a linear transformation from A to C whose kernel is $A \cap B$. Hence $A \cap B$ is a finite dimensional summand of V. □

Let V be a vector space over a division ring k. We say that $v_1,\ldots,v_n$ in V are **dependent** if there exist a_i in k such that $\Sigma_i a_i v_i = 0$, and $a_i \neq 0$ for some i. If k and V are discrete, then we say that $v_1,\ldots,v_n$ are **independent** if they are not dependent; in this case it is easily seen that $v_1,\ldots,v_n$ form a basis for V if and only if they generate V and independent.

6.5 COROLLARY. *Let V be a finite-dimensional vector space over a discrete division ring k. If $v_1,\ldots,v_n$ are in V, then either $v_1,\ldots,v_n$ are dependent, or they are independent.*

PROOF. Let T map k^n into V by taking the natural basis to $v_1,\ldots,v_n$. The kernel of T is finite-dimensional, by (6.2), and $v_1,\ldots,v_n$ are dependent if and only if the kernel of T is nonzero. □

6.6 THEOREM. *Let $k \subseteq K$ be discrete division rings such that K is a finite-dimensional vector space over k, and let V be a vector space over K. Then V is finite dimensional over K if and only if V is finite dimensional over k, in which case $\dim_k V = \dim_k K \dim_K E$.*

PROOF. If V is finite dimensional over K, then it follows from (4.3) that V is finite dimensional over k, and the product formula for the dimensions holds. Conversely suppose that V is finite dimensional over k, and that we have constructed a K-independent set $x_1,\ldots,x_m$ in V. Then $Kx_1 + \cdots + Kx_m$ is a summand of V as a vector space over k by (6.3). If $Kx_1 + \cdots + Kx_m = V$, then we are done; otherwise any element in V that is not in $Kx_1 + \cdots + Kx_m$ will extend the K-independent set $x_1,\ldots,x_m$, and we are done by induction on the dimension (over k) of a complementary summand of $Kx_1 + \cdots + Kx_m$. □

It is also true, in (6.6), that if V is finite dimensional over both k and K, and nonzero, then K is finite dimensional over k. We will not need this result, which follows immediately from the Azumaya theorem in the next chapter.

EXERCISES

1. Construct a Brouwerian example of a vector space V over $\mathbb{Q}$ that contains two finite-dimensional subspaces whose intersection is not finite dimensional. (Hint: Let $V = \mathbb{Q}^2/S$ for an appropriate subspace S of $\mathbb{Q}^2$) You can arrange for your example to be discrete.
2. Generalize Corollaries 6.3 and 6.4 to the case where V is a free module on a discrete set over a discrete division ring k.
3. Show that any type (i) row operation can be achieved by row operations of types (ii) and (iii).
4. A ring R is **von Neumann regular** if for each $a \in R$ there is $x \in R$ such that $axa = a$. Show that the ring of $n \times n$ matrices over a discrete division ring is von Neumann regular. Show that a ring is von Neumann regular if and only if every principal left ideal is generated by an idempotent.

7. DETERMINANTS

Let $Mat_n(R)$ be the ring of $n \times n$ matrices over a commutative ring R, and let $A = \{a_{ij}\}$ be an element of $Mat_n(R)$. Then the **determinant** of A is defined to be

$$\det A = \sum_\sigma sgn(\sigma) a_{1\sigma(1)} a_{2\sigma(2)} \cdots a_{n\sigma(n)}$$

where σ ranges over S_n, the set of permutations of $\{1,2,\ldots,n\}$.

7.1 THEOREM. *Let A and B be $n \times n$ matrices over a commutative ring.*

(i) $\det A = \det A^t$

(ii) *$\det A$ is a linear function of each row of A.*

(iii) *If two rows of A are equal, then $\det A = 0$.*

(iv) $\det AB = \det A \det B$.

PROOF. To verify claim (i) note that if $\tau = \sigma^{-1}$, then $\det A = \Sigma_\sigma sgn(\sigma) a_{\tau(1)1} \cdots a_{\tau(n)n} = \det A^t$ as $sgn(\sigma) = sgn(\tau)$. Claim (ii) is clear from the definition of $\det A$, as each term in the defining sum contains exactly one element from each row. As for (iii), if rows i and j of A are equal, and $i \neq j$, then the permutations may be partitioned into pairs $\{\sigma, \sigma\cdot(i,j)\}$. The terms corresponding to the elements of each pair in the defining sum are equal but of opposite sign, so the sum is zero which establishes (iii). Finally consider $\det A \det B =$

$$\left[\sum_\sigma sgn(\sigma)\, a_{1\sigma(1)} \cdots a_{n\sigma(n)}\right] \cdot \left[\sum_\tau sgn(\tau)\, b_{1\tau(1)} \cdots b_{n\tau(n)}\right] =$$

$$\sum_{\sigma,\tau} sgn(\sigma\tau)\, a_{1\sigma(1)} \cdots a_{n\sigma(n)} b_{1\tau(1)} \cdots b_{n\tau(n)} =$$

$$\sum_{\sigma,\tau} sgn(\sigma\tau)\, a_{1\sigma(1)} b_{\sigma(1)\tau\sigma(1)} \cdots a_{n\sigma(n)} b_{\sigma(n)\tau\sigma(n)} =$$

$$\sum_\sigma \sum_\pi sgn(\pi)\, a_{1\sigma(1)} b_{\sigma(1)\pi(1)} \cdots a_{n\sigma(n)} b_{\sigma(n)\pi(n)}. \tag{7.2}$$

If σ is a function from $\{1,2,\ldots,n\}$ to $\{1,2,\ldots,n\}$, rather than a permutation, and $\sigma(i) = \sigma(j)$ for $i \neq j$, then

$$\sum_\pi sgn(\pi)\, a_{1\sigma(1)} b_{\sigma(1)\pi(1)} \cdots a_{n\sigma(n)} b_{\sigma(n)\pi(n)} = 0$$

because for each permutation π the terms in the sum indexed by π and by $\pi\cdot(i,j)$ add to 0. Therefore we may let σ range over all functions from $\{1,2,\ldots,n\}$ to $\{1,2,\ldots,n\}$ in (7.2) so

$$\det A \det B = \sum_\pi sgn(\pi)\, \Pi_i \sum_j a_{ij} b_{j\pi(i)} = \det AB. \quad \square$$

The **cofactor** A_{ij} of the element a_{ij} in the matrix A is $(-1)^{i+j}$ times the determinant of the $n-1 \times n-1$ matrix obtained by deleting row i and column j from A. It follows easily from the definition of $\det A$ that, for each i,

$$a_{i1}A_{i1} + a_{i2}A_{i2} + \cdots + a_{in}A_{in} = \det A$$

whence the name *cofactor*. From (7.1.iii) we also have

$$a_{i1}A_{j1} + a_{i2}A_{j2} + \cdots + a_{in}A_{jn} = 0$$

if $i \neq j$. If we define the **adjoint** of A to be the matrix B whose ij^{th} entry is A_{ji}, then

(7.3) $$AB = (\det A)I = (\det A^t)I = (A^tB^t)^t = BA.$$

Thus we can construct an inverse for A if we can construct an inverse for $\det A$.

7.4 THEOREM. *Let R be a commutative ring and $A \in Mat_n(R)$. Then A is a unit in $Mat_n(R)$ if and only if $\det A$ is a unit in R.*

PROOF. If $AB = BA = I$, then $(\det A)(\det B) = (\det B)(\det A) = \det I = 1$, so $\det A$ is a unit in R. Conversely, if $\det A$ is a unit in R, then (7.3) shows that A is a unit in $Mat_n(R)$. □

We can now show that the rank of a finite-rank free module over a nontrivial commutative ring is an invariant; in fact, we show a little more.

7.5 THEOREM. *Let R be a commutative ring. Let $m < n$ be positive integers and let $\varphi : R^m \to R^n$ be an epimorphism of R-modules. Then $R = 0$.*

PROOF. There is an R-module map $\psi : R^n \to R^m$ such that $\varphi\psi = 1$. Extend φ to $R^n = R^m \oplus R^{n-m}$ by setting $\varphi(R^{n-m}) = 0$, and view ψ as a map into R^n. As $(\det \varphi)(\det \psi) = 1$, the map φ has a (left) inverse, which must equal ψ. As $\varphi(R^{n-m}) = 0$, we have $R^{n-m} = \psi(0) = 0$ so $R = 0$. □

7.6 LEMMA. *Let R be a commutative ring, M an R-module, and A an $n \times n$ matrix with entries in R. If U is an $n \times 1$ matrix with entries in M, and $AU = 0$, then $(\det A)U = 0$.*

PROOF. Let B be the adjoint of A. Then $BAU = 0$, so $(\det A)U = 0$. □

If A is an $n \times n$ matrix with entries in a commutative ring R, then $XI - A$ is a matrix with entries in $R[X]$. The determinant of $XI - A$ is called the **characteristic polynomial** of A. The **characteristic polynomial** of an endomorphism α of a free R-module F of rank n is the characteristic polynomial of the matrix of α with respect to a basis for F. The characteristic polynomial of α is monic of degree n. If B is the matrix of α with respect to another basis of F, then $B = S^{-1}AS$ for some invertible S in $Mat_n(R)$. Thus the characteristic polynomial of B is the determinant of $XI - S^{-1}AS = S^{-1}(XI - A)S$, which is equal to the determinant of $XI - A$, so the characteristic polynomial of α does not depend on the choice of basis of F. The Cayley-Hamilton theorem says that α satisfies its characteristic polynomial.

7.7 THEOREM (Cayley-Hamilton). *Let R be a commutative ring, and $f(X)$ be the characteristic polynomial of an endomorphism α of a finite-rank free R-module F. Then $f(\alpha) = 0$.*

PROOF. If S is the (commutative) subring of the endomorphism ring of F generated by α and R, then F is an S-module via the multiplication in F. Let $A = \{a_{ij}\}$ be the matrix of α with respect to a basis $u_1,\ldots,u_n$ of F, so

$$\alpha u_j = \sum_i a_{ij}u_i .$$

Let $U = (u_1,\ldots,u_n)^t$ and let C be the $n \times n$ matrix $\alpha I - A$ with entries in S. Then $C^tU = 0$ whence $(det\ C^t)U = 0$ by (7.6). Therefore $det\ C^t = 0$. But $det\ C^t = det\ C = f(\alpha)$. □

EXERCISES

1. Let $f : Mat_n(R) \to R$ satisfy
   (i) $f(A)$ is linear in each row of A.
   (ii) $f(A) = 0$ if two rows of A are equal.
   Show that there exists $r \in R$ such that $f(A) = r\ det\ A$. Use this to show that $det\ AB = det\ A\ det\ B$.

2. Let M be a free module of rank n over a commutative ring R, and let α be an endomorphism of M. If A and B in $Mat_n(R)$ are matrices of α with respect to bases of M, show that $det\ A = det\ B$.

3. Another proof of the Cayley-Hamilton theorem.
   (i) Show that $Mat_n(R[X])$ is isomorphic to $Mat_n(R)[X]$ for any ring R.
   (ii) Let S be a ring, $a \in S$, and $f,g \in S[X]$. Show that if $f(X) = g(X)(X - a)$, then $f(a) = 0$. (Warning: it is not true that if $f(X) = (X - a)g(X)$, then $f(a) = 0$.)
   (iii) Prove (7.7) by letting $S = Mat_n(R)$ and using (7.3) to

factor the characteristic polynomial $f(X) = g(X)(X - \alpha)$, viewed as an element of $S[X]$, and apply (ii).

4. Consider the determinant $det\ M$ of the **Vandermonde matrix** $M =$

$$\begin{matrix} 1 & X_1 & X_1^2 & \cdots & X_1^{m-1} \\ 1 & X_2 & X_2^2 & \cdots & X_2^{m-1} \\ \vdots & \vdots & \vdots & & \vdots \\ 1 & X_m & X_m^2 & \cdots & X_m^{m-1} \end{matrix}$$

with entries in the commutative ring $k[X_1,\ldots,X_m]$. Show that $det\ M = \Pi_{i<j}(X_j - X_i)$. Use this formula to prove that if $f = f_0 + f_1X + \cdots + f_{m-1}X^{m-1} \neq 0$ is a polynomial over a Heyting field k, and $a_1,\ldots,a_m$ are distinct elements of k, then $f(a_i) \neq 0$ for some i.

8. SYMMETRIC POLYNOMIALS

Let R be a commutative ring and $f \in R[X_1,\ldots,X_n]$ a polynomial with coefficients in R. We say that f is **invariant** under a permutation π of the set of indeterminates $\{X_1,\ldots,X_n\}$ if

$$f(X_1,\ldots,X_n) = f(\pi(X_1),\ldots,\pi(X_n)).$$

If f is invariant under each permutation of its indeterminates, then we say that f is a **symmetric** polynomial. If we consider

$$f = (Y + X_1)(Y + X_2) \ldots (Y + X_n)$$

as a polynomial in the indeterminates $Y,X_1,\ldots,X_n$ with coefficients in R, then it is readily seen that f is invariant under any permutation of the indeterminates $X_1,\ldots,X_n$. So if we write f as a polynomial

$$Y^n + \sigma_1Y^{n-1} + \cdots + \sigma_n$$

in Y with coefficients in $R[X_1,\ldots,X_n]$, then the coefficients σ_i are symmetric polynomials in $R[X_1,\ldots,X_n]$. Expanding f to a sum of monomials we find

$$\sigma_1 = \Sigma_i X_i, \quad \sigma_2 = \Sigma_{i<j} X_iX_j, \quad \sigma_3 = \Sigma_{i<j<k} X_iX_jX_k, \quad \ldots,$$
$$\sigma_n = X_1X_2\cdots X_n.$$

The polynomials $\sigma_1,\ldots,\sigma_n$ are called the **elementary symmetric polynomials** in n indeterminates.

Clearly each polynomial in the subring $R[\sigma_1,\ldots,\sigma_n]$ of $R[X_1,\ldots,X_n]$ is symmetric; the fact that every symmetric polynomial has a unique representation as a member of $R[\sigma_1,\ldots,\sigma_n]$ is the fundamental theorem on symmetric polynomials.

8.1 THEOREM. *Let f be a symmetric polynomial in $R[X_1,\ldots,X_n]$. Then there is a unique polynomial h in $R[Y_1,\ldots,Y_n]$ such that $f = h(\sigma_1,\ldots,\sigma_n)$.*

PROOF. We construct h by induction on n. Replacing R by $\mathbb{Z}[r_1,\ldots,r_m]$, where the r_i are indeterminates corresponding to the coefficients of the monomials in the expression for f, we may assume, for the purpose of constructing h, that R is discrete; this is a technical convenience so that we can talk about degrees. If $n = 1$, then $\sigma_1 = X_1$ and every polynomial is symmetric, so we can choose $h = f(Y_1)$. If $n > 1$, let $\tau_1,\ldots,\tau_{n-1}$ be the elementary symmetric polynomials in the indeterminates $X_1,\ldots,X_{n-1}$. Note that $\tau_i = \sigma_i(X_1,\ldots,X_{n-1},0)$. By induction we can construct $g \in R[Y_1,\ldots,Y_{n-1}]$ such that

$$f(X_1,\ldots,X_{n-1},0) = g(\tau_1,\ldots,\tau_{n-1}).$$

We may assume that g contains no monomial $Y_1^{e(1)}\cdots Y_{n-1}^{e(n-1)}$ such that $\Sigma_{i=1}^{n-1} ie(i)$ exceeds the (total) degree of $f(X_1,\ldots,X_{n-1},0)$. Then

$$f_1 = f - g(\sigma_1,\ldots,\sigma_{n-1})$$

is symmetric and $f_1(X_1,\ldots,X_{n-1},0) = 0$. Thus f_1 is divisible by X_n, so by symmetry is divisible by X_i for each i. This implies that f_1 is divisible by $\sigma_n = X_1X_2\cdots X_n$, so we can write $f_1 = \sigma_n f_2$ where f_2 is symmetric and of lower degree than f_1 or f. By induction on the degree of f we can construct $p \in R[Y_1,\ldots,Y_n]$ such that $f_2 = p(\sigma_1,\ldots,\sigma_n)$, and we set $h = Y_n p + g$.

To show that h is unique, it suffices to show that if $g(\sigma_1,\ldots,\sigma_n) = 0$ for g in $R[Y_1,\ldots,Y_n]$, then $g = 0$. We proceed by induction on n. When $n = 1$ this is trivial, so we may assume $n > 1$ and write $g = g_mY_n^m + g_{m-1}Y_n^{m-1} + \cdots + g_0$ as a polynomial in Y_n with coefficients g_i in $R[Y_1,\ldots,Y_{n-1}]$. Substituting σ_i for Y_i and then setting $X_n = 0$ we get $g_0(\tau_1,\ldots,\tau_{n-1}) = 0$, where the τ_i are the elementary symmetric polynomials in $X_1,\ldots,X_{n-1}$. By induction on n we conclude that $g_0 = 0$. Then $g = Y_n p$ for some p in $R[Y_1,\ldots,Y_n]$. As $0 = g(\sigma_1,\ldots,\sigma_n) = \sigma_n p(\sigma_1,\ldots,\sigma_n)$, we conclude that $p(\sigma_1,\ldots,\sigma_n) = 0$, and by induction on m that $p = 0$, so $g = 0$. □

8.2 COROLLARY. *The elementary symmetric polynomials* $\sigma_1,\ldots,\sigma_n$ *are algebraically independent over* R*; that is, if* $g(\sigma_1,\ldots,\sigma_n) = 0$, *then* $g = 0$. □

The ring of polynomials $R[X_1,\ldots,X_n]$ is a module over the subring $R[\sigma_1,\ldots,\sigma_n]$ of symmetric polynomials. We shall show that it is a finite-rank free module.

8.3 LEMMA. *The subring* $R[\sigma_1,\ldots,\sigma_n,X_j,\ldots,X_n]$ *of* $R[X_1,\ldots,X_n]$ *consists of those polynomials that are invariant under each permutation of* $X_1,\ldots,X_{j-1}$.

PROOF. Let $S = R[X_j,\ldots,X_n]$, and let $\tau_1,\ldots,\tau_{j-1}$ be the elementary symmetric polynomials in $X_1,\ldots,X_{j-1}$. The polynomials in $R[X_1,\ldots,X_n] = S[X_1,\ldots,X_{j-1}]$ that are invariant under each permutation of $X_1,\ldots,X_{j-1}$ constitute the subring $S[\tau_1,\ldots,\tau_{j-1}]$ by Theorem 8.1. Obviously σ_i is invariant under permutations of $X_1,\ldots,X_{j-1}$, so $S[\sigma_1,\ldots,\sigma_n] \subset S[\tau_1,\ldots,\tau_{j-1}]$. Let $f(Y) = (Y + X_1)(Y + X_2)\cdots(Y + X_n)$ and $g(Y) = (Y + X_j)(Y + X_{j+1})\cdots(Y + X_n)$. Then f and g are monic polynomials of $S[\sigma_1,\ldots,\sigma_n][Y]$, so there are polynomials q and r in $S[\sigma_1,\ldots,\sigma_n][Y]$ such that $\deg r \leq n - j$, and $f = qg + r$. Since g is cancellable, the polynomials q and r are unique in $R[X_1,\ldots,X_n]$, so $r = 0$ and $q = Y^{j-1} + \tau_1 Y^{j-2} + \cdots + \tau_{j-1}$ is in $S[\sigma_1,\ldots,\sigma_n][Y]$. Thus $S[\sigma_1,\ldots,\sigma_n] = S[\tau_1,\ldots,\tau_{j-1}]$. □

8.4 THEOREM. *The monomials* $X_1^{i(1)}X_2^{i(2)}\cdots X_n^{i(n)}$ *with* $i(k) \leq k - 1$ *form a set of* $n!$ *free generators of* $R[X_1,\ldots,X_n]$ *as a module over* $R[\sigma_1,\ldots,\sigma_n]$.

PROOF. Let $R_j = R[\sigma_1,\ldots,\sigma_n,X_j,\ldots,X_n]$. We shall show that the elements $1,X_j,X_j^2\ldots,X_j^{j-1}$ form a family of free generators of R_j over R_{j+1}. By Lemma 8.3 the polynomial $f_j(Y) = (Y - X_1)(Y - X_2)\cdots(Y - X_j)$ has its coefficients in R_{j+1}. The polynomial f_j is monic of degree j, and $f_j(X_j) = 0$. Thus $1,X_j,X_j^2\ldots,X_j^{j-1}$ generate $R_j = R_{j+1}[X_j]$ over R_{j+1}. It remains to show that $1,X_j,X_j^2,\ldots,X_j^{j-1}$ are independent over R_{j+1}. Suppose $g(X_j) = 0$ for some $g \in R_{j+1}[Y]$ such that $\deg g < j-1$. As g is invariant under permutations of $X_1,\ldots X_j$, we have $g(X_i) = 0$ for $1 \leq i \leq j$, so g has j distinct roots; thus $g = 0$ as $X_i - X_j$ is cancellable by the uniqueness part of the remainder theorem. □

EXERCISES

1. Let K be a discrete field, and let $\sigma_1,\ldots,\sigma_n$ be the elementary symmetric polynomials in $K[X_1,\ldots,X_n]$. Show that the set of $n!$ monomials $X_1^{i(1)}X_2^{i(2)}\cdots X_n^{i(n)}$ with $i(k) \leq k-1$ form a basis of the field of quotients of $K[X_1,\ldots,X_n]$ as a vector space over the field of quotients of $K[\sigma_1,\ldots,\sigma_n]$.

2. Show that the following algorithm exhibits a given symmetric polynomial f as a polynomial in the elementary symmetric polynomials. Order the monomials $X_1^{i(1)}X_2^{i(2)}\cdots X_n^{i(n)}$ lexicographically by setting
$$X_1^{i(1)}X_2^{i(2)}\cdots X_n^{i(n)} \leq X_1^{j(1)}X_2^{j(2)}\cdots X_n^{j(n)}$$
if for each k either $i(k) \leq j(k)$ or there is $m < k$ such that $i(m) < j(m)$. Let $aX_1^{i(1)}X_2^{i(2)}\cdots X_n^{i(n)}$ be the leading term of f with respect to this lexicographic ordering. Show that $i(k) \geq i(k+1)$ for $k = 1,\ldots,n-1$. Subtract
$$a\sigma_1^{i(1)-i(2)}\sigma_2^{i(2)-i(3)}\cdots\sigma_n^{i(n)}$$
from f. Show that the difference has a lower degree with respect to the lexicographic ordering. Replace f by the difference and repeat.

3. Express $\Sigma_{i=1}^n X_i^3$ in terms of the elementary symmetric polynomials $\sigma_1,\ldots,\sigma_n$.

4. Let R be a commutative ring and $E = \Pi_{i<j}(X_i - X_j) \in R[X_1,\ldots,X_n]$.
   (a) Show that E is **alternating**, that is, $E(\pi X_1,\ldots,\pi X_n) = \operatorname{sgn}(\pi)E(X_1,\ldots,X_n)$ for each permutation π, and that E^2 is symmetric. Show that if 2 is a unit in R, then any alternating polynomial is of the form fE where f is a symmetric polynomial.
   (b) Show that there is a polynomial $d \in \mathbb{Z}[Y_1,\ldots,Y_n]$ such that if $\Pi_{i=1}^n(X - r_i) = X^n + a_1X^{n-1} + \cdots + a_n$, where the r_i and a_j are in a field K, then the r_i are distinct if and only if $d(a_1,\ldots,a_n) \neq 0$ in K.
   (c) Show that E is invariant under even permutations of the X_i. Show that if 2 is a unit in R, then $R[E,\sigma_1,\ldots,\sigma_n]$ is the ring of all polynomials invariant under even permutations of the X_i.

NOTES

We leave open what a reasonable inequality on a division ring should be. In order that a division ring be a division ring in the classical sense, the inequality must be standard; but this requirement is too weak to provide any useful consequences. Many important nondiscrete fields, such as the real and complex numbers, are Heyting fields. Although discrete fields are also Heyting fields, the temptation to identify the notion of a field with that of a Heyting field is lessened by the existence of a naturally arising denial field (the residue class field of a nonarchimedean valuation) with an inequality that is neither cotransitive nor tight.

If an ideal P in a commutative ring is not detachable, then it is not clear just what it should mean for P to be prime. In the ring of real numbers, the ideal 0 is not prime according to our definition, because it is possible to construct two real numbers whose product is 0 yet we cannot tell which of the two is 0. Indeed LLPO is equivalent to the statement about real numbers that if $ab = 0$, then $a = 0$ or $b = 0$ (Exercise 3.5). In rings with a positive notion of inequality, like the real numbers, it is natural to define an ideal P to be prime if whenever $a,b \notin P$, then $ab \notin P$. Our definition of a prime ideal has the virtue of not referring to an inequality.

As P is a prime ideal if and only if P is the kernel of a homomorphism into a denial field, it is natural to define a maximal ideal to be the kernel of a homomorphism *onto* a denial field. There may be a better definition, but we won't know until we find some interesting theory involving nondetachable maximal ideals.

Free modules on nondiscrete sets are not just curiosities. They are used, for example, to construct singular homology groups and to construct tensor products of arbitrary modules.

Chapter III. Rings and Modules

1. QUASI-REGULAR ELEMENTS AND THE JACOBSON RADICAL

An element r of a ring R is called **left (right) quasi-regular** if $1-r$ has a left (right) inverse. If r is both left and right quasi-regular, then we say that r is **quasi-regular**. An element r in R is **nilpotent** if $r^n = 0$ for some positive integer n; we say that R **has no nilpotent elements** if $r = 0$ whenever r is nilpotent. Observe that if $r^n = 0$, and $s = 1 + r + r^2 + \cdots + r^{n-1}$, then $(1-r)s = s(1-r) = 1$, so nilpotent elements are quasi-regular. A left ideal L of R is called **quasi-regular** if each element of L is left quasi-regular.

1.1 THEOREM. *Let L be a quasi-regular left ideal. Then each element of L is quasi-regular.*

PROOF. If $r \in L$, then $s(1-r) = 1$ for some $s \in R$. As $-sr \in L$, the element $s = 1 + sr$ has a left inverse t. Thus $1-r = ts(1-r) = t$, so $(1-r)s = ts = 1$. □

1.2 LEMMA. *If ab is left (right) quasi-regular, then so is ba.*

PROOF. If $c(1-ab) = 1$, then $(1+bca)(1-ba) = 1 - ba + bc(1-ab)a = 1$. If $(1-ab)c = 1$, then $(1-ba)(1+bca) = 1$. □

1.3 THEOREM. *The set of all elements r such that Rr is a quasi-regular left ideal is a two-sided ideal.*

PROOF. In view of (1.2), it suffices to show that if Ra and Rb are quasi-regular, then so is $a + b$. Choose s such that $s(1 - a) = 1$, and t such that $t(1 - sb) = 1$. Then $ts(1 - (a+b)) = 1$. □

The ideal of Theorem 1.3 is called the **Jacobson radical** of R.

1.4 LEMMA (Nakayama). *Let M be a finitely generated left R-module and L a quasi-regular left ideal of R. If $LM = M$, then $M = 0$.*

PROOF. We may assume that L is the Jacobson radical of R, hence a two-sided ideal. Let $x_1,\ldots,x_n$ generate M. Then $x_1 \in LM$, so we can write $x_1 = a_1x_1 + \cdots + a_nx_n$ where each a_i is in L, as L is a two-sided ideal. Thus $x_1 = (1 - a_1)^{-1}(a_2x_2 + \cdots + a_nx_n)$, so M is generated by $x_2,\ldots,x_n$, and we are done by induction on n. □

EXERCISES

1. Use the identity in the proof of (1.2), with $a = 1$, to show that if an element of a ring has a unique left inverse, then it has a right inverse. [Rudin 1985]

2. Show that the following rings have 0 as their Jacobson radical.
   (i) The ring $\mathbb{Z}$ of integers.
   (ii) Any discrete division ring.
   (iii) The polynomial ring $R[X]$ for R a discrete integral domain.
   (iv) A von Neumann regular ring (see Exercise II.6.4).

3. Find the Jacobson radical of the subring of $\mathbb{Q}$ consisting of those rational numbers that can be written with odd denominator. Construct a Brouwerian example of a countable discrete integral domain whose Jacobson radical is not detachable.

4. An element x of an R-module M is called a **nongenerator** if whenever A is a submodule of M such that $Rx + A = M$, then $A = M$. Show that an element x of R is in the Jacobson radical of R if and only if x is a nongenerator of the left R-module R.

5. A **detachable maximal left ideal** of a ring R is a proper left ideal L such that if $x \in R$, then either $x \in L$ or $Rx + L = R$. Show that each detachable maximal left ideal contains the Jacobson radical. Show that if each element of R that is not contained in any detachable maximal left ideal has a left inverse (Zorn's lemma establishes this classically), then the Jacobson radical is equal to the intersection of the detachable maximal left ideals.

6. Let A be a square matrix with entries in the Jacobson radical of a ring. Show that $I - A$ is invertible, where I is the appropriate identity matrix.

7. If J is the Jacobson radical of R, show that the Jacobson radical of R/J is zero.

8. Use Lemma II.7.6 to show that if M is a finitely generated module over a commutative ring R, and I is an ideal in R, then $IM = M$ if and only if $(1 - r)M = 0$ for some $r \in R$. Use this to prove the Nakayama lemma for commutative rings.

9. Use Exercise 8 to show that any finitely generated module M over a commutative ring R is **Hopfian:** each epimorphism from M onto M is an automorphism. (For $f: M \to M$, consider M as an $R[f]$-module.)

2. COHERENT AND NOETHERIAN MODULES

Let R be a ring, and M an R-module. Then M is **Noetherian** if the set of finitely generated submodules of M, ordered by inclusion, satisfies the ascending chain condition. A ring R is **left Noetherian** if R is Noetherian as a left module over itself. Thus R is left Noetherian if and only if for each sequence $I_1 \subseteq I_2 \subseteq I_3 \subseteq \cdots$ of finitely generated left ideals of R there is n such that $I_n = I_{n+1}$.

It is easy to check that the ring of integers is Noetherian, as is $k[X]$ if k is a discrete field. Finite modules are Noetherian, and if R contains a discrete field k, then R-modules that are finite dimensional k-spaces are Noetherian—more generally, if S is a subring of R, and M is an R-module that is a Noetherian S-module, then M is a Noetherian R-module. The Hilbert basis theorem (see Chapter VIII) implies that polynomial rings in several variables over the integers, or over a discrete field, are Noetherian.

2.1 THEOREM. *Let N be a submodule of a module M. Then M is Noetherian if and only if both N and M/N are Noetherian.*

PROOF. Let π denote the natural map from M to M/N. If M is Noetherian, then clearly N is also Noetherian. If $J_1 \subseteq J_2 \subseteq \cdots$ is a chain of finitely generated submodules of M/N, then we can construct a chain $I_1 \subseteq I_2 \subseteq \cdots$ of finitely generated submodules of M such that $\pi I_m = J_m$. There exists n such that $I_{n+1} = I_n$, so $J_{n+1} = J_n$, whence M/N is Noetherian.

Conversely, suppose both N and M/N are Noetherian, and let $I_1 \subseteq I_2 \subseteq \cdots$ be a chain of finitely generated submodules of M. Then

$\pi I_1 \subseteq \pi I_2 \subseteq \cdots$ is a chain of finitely generated submodules of the Noetherian module M/N, so there exists n such that $\pi I_n = \pi I_{n+1}$. Therefore $I_{n+1} = I_n + K$ for some finitely generated submodule $K \subseteq N$. Iterating this construction, we construct a sequence $n(1) < n(2) < \cdots$ of positive integers, and an ascending chain of finitely generated submodules K_i of N, such that $I_{n(i)+1} = I_{n(i)} + K_i$. As N is Noetherian there exists i such that $K_i = K_{i-1} \subseteq I_{n(i)}$, so $I_{n(i)+1} = I_{n(i)}$. □

An R-module M is **finitely presented** if there is a map from a finite rank free R-module onto M with a finitely generated kernel. The following theorem shows that if M is finitely presented, then *any* map from a finitely generated R-module onto M has a finitely generated kernel.

2.2 THEOREM. *If f is a map from a finitely generated R-module M_1 onto a finitely presented R-module M_2, then the kernel of f is finitely generated.*

PROOF. Let F_i be a finite rank free R-module, and π_i a map from F_i onto M_i such that the $ker\ \pi_2$ is finitely generated. By Schanuel's trick (II.3.3) we have

$$ker\ f\pi_1 \oplus F_2 \cong ker\ \pi_2 \oplus F_1.$$

Therefore $ker\ f\pi_1$ is finitely generated. As π_1 is onto, $ker\ f = \pi_1(ker\ f\pi_1)$ is also finitely generated. □

2.3 THEOREM. *Let N be a submodule of the R-module M. Then*

(i) *If M is finitely presented, and N is finitely generated, then M/N is finitely presented.*

(ii) *If N and M/N are finitely presented, then M is finitely presented.*

PROOF. Let π be the natural map from M to M/N. To prove (i) let φ map a finite rank free module F onto M with finitely generated kernel K, and let $N' \subset F$ be finitely generated with $\varphi(N') = N$. Then $K + N' = ker\ \pi\varphi$ is finitely generated, so M/N is finitely presented.

To prove (ii) let φ_1 and φ_2 map finite rank free modules F_1 and F_2 onto N and M/N with finitely generated kernels K_1 and K_2 respectively. Let $\varphi : F_1 \oplus F_2 \to M$ be such that φ restricted to F_1 is φ_1 and $\pi\varphi$ restricted to F_2 is φ_2. Then φ maps $F_1 \oplus F_2$ onto M. If $K = ker\ \varphi$ and π_2 is the projection of $F_1 \oplus F_2$ on F_2, then $\pi_2 K = K_2$ and $K \cap F_1 = K_1$, so K is finitely generated. □

An R-module is **coherent** if every finitely generated submodule is finitely presented. A ring R is **left coherent** if it is coherent as a left R-module. Classically every left Noetherian ring is left coherent.

2.4 THEOREM. *An R-module M is coherent if and only if the following two conditions hold.*

(i) *The intersection of any two finitely presented submodules of M is finitely generated.*

(ii) *If $x \in M$, then $\{r \in R : rx = 0\}$ is a finitely generated left ideal of R.*

PROOF. Suppose that M is coherent, N_1 and N_2 are finitely generated submodules of M, and $x \in M$. Let φ_i be a map of a finite-rank free R-module F_i onto N_i, and let K be the (finitely generated) kernel of the induced map from $F_1 \oplus F_2$ into M. If π_1 is the projection of $F_1 \oplus F_2$ onto F_1, then $N_1 \cap N_2 = \varphi_1(\pi_1 K)$ is finitely generated establishing (i). The cyclic submodule Rx of M is finitely presented, so the kernel of the map taking R onto Rx is finitely generated, which shows (ii).

Conversely, suppose (i) and (ii) hold, and A is a finitely generated submodule of M. If A is cyclic, then A is finitely presented by (ii). Otherwise $A = N_1 + N_2$ where each N_i is generated by fewer elements than A is. By induction, each N_i is finitely presented, and by (i) the intersection $N_1 \cap N_2$ is finitely generated. By (2.3.ii) the direct sum $N_1 \oplus N_2$ is finitely presented. But $N_1 \oplus N_2$ maps onto $A = N_1 + N_2$ with kernel isomorphic to $N_1 \cap N_2$, so A is finitely presented by (2.3.i). □

2.5 THEOREM. *Let N be a finitely generated submodule of a module M. Then M is coherent if and only if N and M/N are coherent.*

PROOF. Let π be the natural map from M to M/N. If M is coherent, then N is clearly coherent. If B is a finitely generated submodule of M/N, then $B = \pi(A)$ for some finitely generated submodule A of M. But $A \cap N$ is finitely generated by (2.4) so B is finitely presented by (2.3.i).

Conversely suppose N and M/N are coherent, and A is a finitely generated submodule of M. Then $A \cap N$ is finitely generated by (2.2), and therefore finitely presented because N is coherent. Also πA is finitely presented because M/N is coherent. Therefore A is finitely presented by

(2.3.ii). □

2.6 COROLLARY. *A finitely presented module over a coherent ring is coherent.*

PROOF. Repeated application of (2.5) shows that finite rank free modules are coherent. One more application of (2.5), in the other direction, shows that finitely presented modules are coherent. □

A module has **detachable submodules** if each finitely generated submodule is detachable. A ring R has **detachable left ideals** if R has detachable submodules when viewed as a left module over itself. Note that R has detachable left ideals if and only if R/I is discrete for each finitely generated left ideal I of R.

Polynomial rings over discrete fields are examples of coherent Noetherian rings with detachable ideals. In classical mathematics *every* Noetherian ring is coherent with detachable ideals (but see Exercises 3 and 4).

2.7 THEOREM. *Let N be a finitely generated submodule of a coherent module M. Then M has detachable submodules if and only if N and M/N do.*

PROOF. If M has detachable submodules then clearly N and M/N do. Conversely, suppose that N and M/N have detachable submodules and let π be the natural map from M to M/N. To show that M has detachable submodules, let P be a finitely generated submodule of M and let $x \in M$. If $\pi x \notin \pi P$, then $x \notin P$. If $\pi x = \pi p$ for $p \in P$, then $x \in P$ if and only if $x - p \in P \cap N$, which is a finitely generated submodule of N because M is coherent. □

2.8 COROLLARY. *If M is a finitely presented left module over a coherent left Noetherian ring R, then M is a coherent Noetherian module. If, in addition, R has detachable left ideals, then M has detachable submodules.*

PROOF. Let $\mathcal{C}$ be the class of coherent Noetherian R-modules (with detachable submodules). As $R \in \mathcal{C}$, induction on n and (2.1) and (2.5) (and (2.7)) gives $R^n \in \mathcal{C}$. As M is finitely presented, $M \in \mathcal{C}$ by (2.1) and (2.5) (and (2.7)). □

EXERCISES

1. Show that a commutative ring R has detachable ideals if and only if R/I is discrete for each finitely generated ideal I of R.

2. Show that each nonzero finitely generated ideal in the ring $\mathbb{Z}$ of integers has finite depth in the partially ordered set of finitely generated ideals of $\mathbb{Z}$. Show that each nonzero finitely generated ideal in the ring $k[X]$, where k is a discrete field, has bounded depth.

3. Let a be a binary sequence. Let k be the two-element field and R the subring of the finite ring $k[X,Y]/(X,Y)^2$ generated by 1, X, and $\{a_nY\}$. Show that R is a Brouwerian example of a Noetherian ring with detachable ideals that is not coherent. Do the same for the ring $\mathbb{Z}/I$ where I is generated by the elements $a_n n!$.

4. Construct a Brouwerian example of a discrete coherent Noetherian ring R, and a finitely generated ideal I of R, such that I is not detachable from R. (Hint: Let R lie between $\mathbb{Z}$ and $\mathbb{Q}$)

5. Let I be a finitely generated ideal of a commutative coherent Noetherian ring R with detachable ideals. Show that the radical of I is detachable. (Hint: Given $x \in R$, consider the ascending chain of ideals $I:x^m$)

6. Show that finitely generated coherent modules over rings with detachable ideals are discrete.

7. Show that if S is a subring of R, and M is an R-module that is a Noetherian S-module, then M is a Noetherian R-module.

8. Show that if a module is bounded in number, then it is Noetherian.

9. Show that any discrete field is coherent. Show that the ring $\mathbb{Z}$ is coherent.

10. A submodule A of an R-module B is called **pure** if whenever a finite family of equations
$$\sum r_{ij}x_j = a_i,$$
with $r_{ij} \in R$ and $a_i \in A$, has a solution in B, then it has a solution in A. Show that $A \subseteq B$ is pure if and only if every map

of a finitely presented module to B/A lifts to a map to B. Show that any nonzero pure subgroup of the infinite cyclic abelian group $\mathbb{Z}$ is equal to $\mathbb{Z}$.

11. Show that every Noetherian module M is Hopfian (Exercise 1.9). (If $f(z) = 0$, construct a sequence such that $x_0 = z$ and $f(x_i) = x_{i-1}$.)

3. LOCALIZATION

Let R be a commutative ring. We construct rings of fractions of elements of R in much the same way that we construct the rational numbers from the integers. First we decide on a subset S of R whose elements we will allow as denominators; in the construction of the rational numbers, this set consists of the positive integers. In general we want S to be a **multiplicative submonoid** of R, by which we mean that $1 \in S$ and, if s_1 and s_2 are in S, then so is $s_1 s_2$.

The elements of the **ring of fractions** $S^{-1}R$ consist of pairs r/s consisting of an element r in R and an element s of S. Addition and multiplication are defined by the usual formulas:

$$r_1/s_1 + r_2/s_2 = (r_1 s_2 + r_2 s_1)/(s_1 s_2)$$
$$(r_1/s_1)(r_2/s_2) = (r_1 r_2)/(s_1 s_2)$$

but we have to be a little careful about equality of fractions because any element s of S is invertible in $S^{-1}R$, so any element r of R such that $sr = 0$ must be set equal to zero in $S^{-1}R$. With this in mind we say that two fractions r_1/s_1 and r_2/s_2 are **equal** if there exists an element s in S such that $s(r_1 s_2 - r_2 s_1) = 0$. There is a natural map $R \to S^{-1}R$ defined by taking the element r to $r/1$. We leave the verification that $S^{-1}R$ is a ring to the reader.

More generally, if M is an R-module, then we can form the **module of fractions** $S^{-1}M$ consisting of fractions m/s with m in M and s in S. Equality and addition are defined as for $S^{-1}R$, while multiplication by elements of R, or of $S^{-1}R$, is defined in the obvious way. Thus $S^{-1}M$ is an $S^{-1}R$-module. If N is a submodule of M, then $S^{-1}N$ is a submodule of $S^{-1}M$

If P is a proper prime ideal of R, then $S = R\backslash P$ is a multiplicative submonoid of R. In this case we denote the module $S^{-1}M$ by M_P. If P is detachable, then the ring R_P is **local** in the sense that for each $x \in R$, either x or $1-x$ is invertible; indeed an element r/s of R_P is invertible if $r \notin P$, while if $r \in P$, then $s-r \notin P$, so $1 - r/s = (s-r)/s$ is invertible. When R is a discrete integral domain, then $P = 0$ is a proper detachable prime ideal and the ring R_P is the field of quotients of R.

Let r be an element of a commutative ring R, and S the multiplicative submonoid of R generated by r. It follows immediately from the definitions that r is nilpotent if and only if $S^{-1}R = 0$. We illustrate the use of this fact in the proofs of (3.1), (3.2) and (3.4) below.

Our first result generalizes (II.7.5) which showed, using determinants, that the rank of a finite-rank free module over a nontrivial commutative ring is an invariant.

3.1 THEOREM. *If R is a commutative ring, and there is a monomorphism from R^m to R^n where $m > n$, then R is trivial.*

PROOF. Let A be the $m \times n$ matrix of a one-to-one map φ from R^m into R^n. We first show that the elements in the first column of A are nilpotent. If r is an element in the first column, then let $S = \{1,r,r^2,\ldots\}$ and pass to the ring $T = S^{-1}R$. It is easily seen that A is the matrix of a one-to-one map from T^m into T^n. Apply elementary row and column operations to A, which amounts to changing the bases of T^n and T^m, so that the first column and row of A are 0 except for a 1 in the upper left corner. Let $e_1,\ldots,e_m$ be the new basis for T^m. If $n = 1$ then $\varphi(e_m) = 0$, so $0 = 1$ in T because φ is one-to-one. If $n > 1$, then $0 = 1$ in T by induction applied to the matrix that results from A by deleting the first row and column. Thus r is nilpotent in R.

Let I be the ideal in R generated by the elements of the first column of A, and let $e_1,\ldots,e_m$ be the natural basis of R^m. Then $I^k = 0$ for some k. But if $I^k e_1 = 0$ for $k > 1$, then $\varphi(I^{k-1}e_1) = 0$, so $I^{k-1}e_1 = 0$ as φ is one-to-one. Thus $e_1 = 0$ so R is trivial. □

Theorem 3.1 generalizes (II.7.5) because if R^n maps onto R^m with $m > n$, then, as R^m is projective, R^m maps one-to-one into R^n. The next theorem will be used in Chapter VIII.

3.2 LEMMA. *Let $R \subseteq T$ be commutative rings, and A a matrix over R. Suppose that*

(i) *if $(r,0,\ldots,0)$ is in the row space of A over R, then $r = 0$,*

(ii) *$(1,0,\ldots,0)$ is in the row space of A over T.*

Then $R = 0$.

PROOF. Designate a finite number of rows of A as **good**, the rest being designated **bad**, so that if a row of A is good, then it contains a 1, called a **good** 1, in a column whose other entries are 0. To start we may designate all rows as bad. We induct on the number of bad rows.

Suppose r is in a bad row. Let $S = \{1, r, r^2, \ldots\}$ and consider A as a matrix over $S^{-1}R \subseteq S^{-1}T$. We can reduce the number of bad rows of A by elementary row operations that replace r by a good 1 and leave the good rows good, so $S^{-1}R = 0$ by induction; thus r is nilpotent in R. We have shown that all elements in bad rows are nilpotent.

If ρ_i denotes row i of A, then by hypothesis (ii) we have

$$t_1\rho_1 + t_2\rho_2 + \cdots + t_m\rho_m = (1,0,\ldots,0)$$

for some elements $t_1,\ldots,t_m \in T$. If all rows are bad, then 1 is nilpotent and we are done. If ρ_j has a good 1 in a column other than the first, then $t_j = 0$. So we may assume that some ρ_i has a good 1 in the first column, and the remaining $t_j\rho_j$ consist of nilpotent elements. Then every entry of ρ_i except the first is nilpotent. Let I be the ideal of R generated by these nilpotent elements. If $I^k = 0$, and $r \in I^{k-1}$, then $r\rho_i$ has zero entries except for the first, which is r, so $r = 0$ by hypothesis (i). Therefore $I = 0$, so $\rho_i = (1,0,\ldots,0)$ whereupon $1 = 0$ by hypothesis (i), so $R = 0$. □

3.3 THEOREM. *Let $R \subseteq T$ be commutative rings, and I an ideal in $R[X]$ such that $I \cap R = 0$. If $1 \in TI$, then $R = 0$.*

PROOF. Each element in TI can be written as $t_1p_1 + \cdots + t_mp_m$, with $p_i \in I$ and $t_i \in T$. Write 1 this way, and apply (3.2) to the matrix of coefficients of the p_i. □

3.4 COROLLARY. *Let $R \subseteq T$ be commutative rings, and I an ideal in $R[X]$ such that $1 \in TI$. Then the annihilator of $R \cap I$ consists of nilpotent elements.*

PROOF. Suppose $r(R \cap I) = 0$. Let S be the multiplicative submonoid of R generated by r, and pass to $S^{-1}R$. Now (3.3) applies and tells us that $S^{-1}R = 0$, so r is nilpotent. □

EXERCISES

1. Let R be a coherent commutative ring with detachable ideals, and let P be a finitely generated proper prime ideal of R. Show that R_P is discrete.

2. Let S be a multiplicative submonoid of a commutative ring R, let M be an R-module, and let N_1 and N_2 be submodules of M. Show that
$$S^{-1}(N_1 \cap N_2) = S^{-1}N_1 \cap S^{-1}N_2$$
and that
$$S^{-1}(N_1 + N_2) = S^{-1}N_1 + S^{-1}N_2.$$

3. Let S be a multiplicative submonoid of a commutative ring R, and let $\varphi : R \to S^{-1}R$ be the natural map. Show that $S^{-1}R$ is discrete if and only if $\ker \varphi$ is detachable from R. Show that if R is a discrete ring with no nilpotent elements, and S is finitely generated, then $S^{-1}R$ is discrete.

4. Let S be a multiplicative submonoid of a commutative ring R. Show that if M is a coherent R-module, then $S^{-1}M$ is a coherent $S^{-1}R$-module.

5. Let P be a detachable proper prime ideal of a commutative ring R. Show that the Jacobson radical of R_P is detachable, and that its complement consists of the units of R_P.

6. Let k be a discrete field, and a,b,c and s indeterminates. Let $R = k[a,b,c]/(ca,cb,c^2)$ and let $S = R[s]/(sa + (1-s)b)$. Let I be the ideal of $R[X]$ generated by $1+aX$ and $1+bX$. Show that $I \cap R$ is generated by c and $b-a$, and that $1 \in SI$. Thus we cannot strengthen (3.3) to conclude that the annihilator of $I \cap R$ is zero.

7. Show that an $m \times n$ matrix A over a commutative ring R has a left inverse if and only if the determinants of the set of square matrices whose rows are rows of A generate R as an ideal.

4. TENSOR PRODUCTS

Let R be a ring and B a left R-module. If R is a subring of a ring A, then there is a natural way to construct a left A-module from B. We consider formal sums $\Sigma\, a_i b_i$, and define two such formal sums to be equal if we can get from one to the other by applying left and right distributive laws, the associative law $(ar)b = a(rb)$, and computations that lie totally within A or B. The ring structure of A is used only to get the A-module structure on the formal sums; in general, all we need is for A to be a right R-module, and the result is an abelian group. Here is a precise specification of the construction.

4.1 DEFINITION. *Let R be a ring, A a right R-module, and B a left R-module. The* **tensor product** *$A \otimes B$ is defined as the quotient of the free abelian group $F(A\times B)$ on $A\times B$ by the subgroup $K(A\times B)$ of $F(A\times B)$ generated by elements of the form*

$$\begin{aligned} &\text{(i)}\ (a_1 + a_2, b) - (a_1, b) - (a_2, b)\\ &\text{(ii)}\ (a, b_1 + b_2) - (a, b_1) - (a, b_2)\\ &\text{(iii)}\ (ar, b) - (a, rb).\end{aligned}$$

We denote the image in $A \otimes B$ of the element $(a,b) \in A\times B$ by $a \otimes b$. A map f from $A\times B$ to an abelian group G is **bilinear** if

$$\begin{aligned} &\text{(i)}\ f(a_1 + a_2, b) = f(a_1, b) + f(a_2, b)\\ &\text{(ii)}\ f(a, b_1 + b_2) = f(a, b_1) + f(a, b_2)\\ &\text{(iii)}\ f(ar, b) = f(a, rb),\end{aligned}$$

that is, if the induced map from $F(A\times B)$ to G takes $K(A\times B)$ to zero. Hence each bilinear map f from $A\times B$ to G induces a unique abelian group homomorphism from $A \otimes B$ to G that takes $a \otimes b$ to $f(a,b)$. Because of this we can define a map from $A \otimes B$ by specifying its values on the elements $a \otimes b$, and checking to see if this assignment is bilinear.

The tensor product $A \otimes B$ depends on the ring R; if we want the notation to indicate that ring, we write $A \otimes_R B$. If A is an S-R-bimodule, and B is a left R-module, then the abelian group $A \otimes_R B$ admits a natural left S-module structure by setting $s(a \otimes b) = sa \otimes b$. In particular, if R is commutative, then $A \otimes B$ is an R-module.

The tensor product of cyclic modules has a simple description.

4.2 THEOREM. *Let R be a ring, I a right ideal, J a left ideal, and B a left R-module. Then*

$$\begin{aligned} &\text{(i)}\ (R/I) \otimes B \cong B/IB\\ &\text{(ii)}\ (R/I) \otimes (R/J) \cong R/(I+J).\end{aligned}$$

If I is an ideal, then these are isomorphisms of left R-modules.

PROOF. To prove (i) we define a map φ from $(R/I) \otimes B$ to B/IB by

$\varphi(r \otimes b) = rb$, and ψ from B/IB to $(R/I) \otimes B$ by $\psi(b) = 1 \otimes b$. We note that if $r - r' \in I$, then $rb - r'b \in IB$, so φ is well-defined. Also, if $b - b' \in IB$, then $1 \otimes b - 1 \otimes b' = 1 \otimes (b-b') = 0$ in $(R/I) \otimes B$, so ψ is well-defined. As $r \otimes b = 1 \otimes rb$, the two maps φ and ψ are inverses of each other. Part (ii) follows from (i) because $I(R/J) = (I+J)/J$. If I is an ideal, then R/I and B/IB are left R-modules, and the R-module structure on $(R/I) \otimes B$ is given by $r(1 \otimes b) = r \otimes b$, so φ and ψ are R-module homomorphisms □

In particular, taking $I = 0$ in (4.2.i), if B is a left R-module, then $R \otimes B \cong B$.

4.3 THEOREM. *Let R be a ring, $\{A_i\}_{i\in I}$ a family of right R-modules, and $\{B_j\}_{j\in J}$ a family of left R-modules. Then there is a natural isomorphism*

$$(\oplus_{i\in I}A_i) \otimes (\oplus_{j\in J}B_j) \cong \oplus_{i,j\in I\times J}(A_i \otimes B_j)$$

PROOF. Exercise 1. □

4.4 COROLLARY. *The tensor product of two free modules over a commutative ring R is a free module. In particular, $R^m \otimes R^n \cong R^{mn}$.*

PROOF. Theorems 4.2 and 4.3. □

The tensor product is a **bifunctor** in the sense that given R-module maps $f : A \to A'$ and $g : B \to B'$, there is a natural abelian group map $f\otimes g : A \otimes B \to A' \otimes B'$ defined by $(f\otimes g)(a \otimes b) = f(a) \otimes g(b)$. If R is commutative, then $f\otimes g$ is an R-module homomorphism.

A pair of maps $A \to B \to C$ is **exact at B** if the image of $A \to B$ is equal to the kernel of $B \to C$. A sequence of maps $A_1 \to A_2 \to \cdots \to A_n$ is **exact** if it is exact at A_i for $i = 2,\ldots,n-1$. The sequence $0 \to A \to B$ is exact if and only if $A \to B$ is one-to-one; the sequence $A \to B \to 0$ is exact if and only if $A \to B$ is onto.

4.5 THEOREM. *Let R be a ring, and let $A \to B \to C \to 0$ and $A' \to B' \to C' \to 0$ be exact sequences of right and left R-modules respectively. Then the sequence*

$$(A \otimes B') \oplus (B \otimes A') \to B \otimes B' \to C \otimes C' \to 0$$

is exact.

PROOF. Let K be the image of $(A \otimes B') \oplus (B \otimes A')$ in $B \otimes B'$. The map from $B \otimes B'$ to $C \otimes C'$ takes K to 0, so it induces $\varphi : (B \otimes B')/K \to C \otimes C'$.

We shall show that φ is an isomorphism by constructing its inverse. Define a map $\psi : C \otimes C' \to (B \otimes B')/K$ as follows. Given $c \otimes c'$ in $C \otimes C'$, choose b in B mapping onto c, and b' in B' mapping onto c'. Define $\psi(c \otimes c')$ to be the image of $b \otimes b'$ in $(B \otimes B')/K$; if b_1 and b_2 map to c, and b_1' and b_2' map to c', then

$$b_1 \otimes b_1' - b_2 \otimes b_2' \quad = \quad (b_1 - b_2) \otimes b_1' + b_2 \otimes (b_1' - b_2')$$

is in K, so ψ is well defined. It is readily seen that the bilinearity requirement is met so that we have indeed defined a map from $C \otimes C'$. Clearly φ and ψ are inverses of each other. □

4.6 COROLLARY. *Let M be a left module over a ring R, and let $A \to B \to C \to 0$ be an exact sequence of right R-modules. Then the sequence $A \otimes M \to B \otimes M \to C \otimes M \to 0$ is exact.*

PROOF. Take $A' = 0$ and $B' = C' = M$ in Theorem 4.5. □

4.7 COROLLARY. *If C and C' are finitely presented modules over a commutative ring R, then so is $C \otimes C'$.*

PROOF. We can choose A', B', A, and B to be finitely generated free R-modules in the hypothesis of Theorem 4.5. □

The equality relation on $A \otimes B$ admits the following description.

4.8 THEOREM. *Let R be a ring, A a right R-module generated by the elements $a_1,\ldots,a_m$, and B a left R-module. The element $\Sigma_{i=1}^m a_i \otimes b_i$ is 0 in $A \otimes B$ if and only if there exist elements r_{ij} in R and c_j in B such that*

$$\text{(i)}\quad b_i = \sum_j r_{ij} c_j$$

$$\text{(ii)}\quad \sum_i a_i r_{ij} = 0.$$

PROOF. Clearly the condition implies $\Sigma\, a_i \otimes b_i = 0$. Conversely, let F be a rank-m free right R-module, and map F onto A with kernel K by taking a basis $e_1,\ldots,e_m$ of F onto $a_1,\ldots,a_m$. The element $\Sigma_{i=1}^m e_i \otimes b_i \in F \otimes B$ goes to zero in $A \otimes B$, so it comes from an element of $K \otimes B$ which we can write as

$$\sum_{i=1}^m e_i \otimes b_i \quad = \quad \sum_j (\Sigma_i e_i r_{ij}) \otimes c_j$$

But the e_i are a basis of a free module, so we have $b_i = \Sigma\, r_{ij} c_j$. As $\Sigma_i e_i r_{ij} \in K$, we have $\Sigma_i a_i r_{ij} = 0$. □

EXERCISES

1. Let $\{B_i\}_{i\in I}$ be a family of left R-modules. If A is a right R-module, show that
$$A \otimes (\oplus_{i\in I} B_i) \cong \oplus_{i\in I}(A \otimes B_i).$$

2. If S is a multiplicative submonoid of a commutative ring R, and M is an R-module, show that $S^{-1}M \cong (S^{-1}R) \otimes M$ as $S^{-1}R$-modules.

3. Let a be a binary sequence, let A be the group of integers modulo 5, let C be the group of integers modulo 25, and let B be the subgroup of C generated by 5 and the set $\{a_n : n = 1,2,\ldots\}$. Show that A and B are discrete abelian groups, and that $A \otimes B$ is discrete if and only if $a_n = 1$ for some n, or $a_n = 0$ for all n.

4. Let a be a binary sequence. Let F be the ring of integers modulo 2, and let R be the subring of $F[x,y,s,t]/(sx+ty-1)$ generated by x,y and the set $\{a_n s, a_n t : n = 1,2,\ldots\}$. Let $A = R/(x)$ and $B = R/(y)$. Show that A and B are finitely presented discrete R-modules, and that $A \otimes B$ is discrete if and only if $a_n = 1$ for some n, or $a_n = 0$ for all n.

5. Show that $A \subseteq B$ is pure if and only if the map $M \otimes A \to M \otimes B$ is one-to-one for each finitely presented right module M. Show that the expression 'finitely presented' can be dropped from the preceding statement.

5. FLAT MODULES

Let R be a ring and M a left R-module. We say that M is **flat** if whenever $x_1,\ldots,x_m$ are elements of M, and $r_1,\ldots,r_m$ are elements of R such that $\Sigma r_i x_i = 0$, then there exist elements $y_1,\ldots,y_n$ in M, and elements $a_{ij} \in R$ such that $x_i = \Sigma_j a_{ij} y_j$ and $\Sigma_i r_i a_{ij} = 0$. Flatness is clearly a local property in the sense that an R-module is flat if and only if each finitely generated submodule is contained in a flat submodule.

5.1 THEOREM. *If $\{M_i\}_{i\in I}$ is a family of R-modules, then $\oplus_{i\in I} M_i$ is flat if and only if each M_i is flat.*

PROOF. Suppose each M_i is flat. Because flatness is a local property, we may assume that I is finitely enumerable, say $I = \{s(1),\ldots,s(n)\}$. Suppose $\Sigma_{\ell=1}^{m} r_\ell x_\ell = 0$, where $r_\ell \in R$ and $x_\ell \in \oplus_{i\in I} M_i$. Write $x_\ell = \Sigma_{i=1}^{n} y_{\ell i}$

with $y_{\ell i} \in M_{s(i)}$. Then $\Sigma_{i=1}^{n}(\Sigma_{\ell=1}^{m} r_\ell y_{\ell i}) = 0$, so either $\Sigma_{\ell=1}^{m} r_\ell y_{\ell i} = 0$ for $i = 1,\ldots,n$, or $s(i) = s(j)$ for some $i \neq j$. In the latter case we are done by induction on n. In the former case, as $M_{s(i)}$ is flat, there exist elements $z_{\ell ij} \in M_{s(i)}$ and elements $a_{\ell ij} \in R$ such that

$$y_{\ell i} = \Sigma_j\, a_{\ell ij} z_{\ell ij}$$
$$\Sigma_\ell\, r_\ell a_{\ell ij} = 0.$$

so $x_\ell = \Sigma_{ij}\, a_{\ell ij} z_{\ell ij}$ and $\Sigma_\ell\, r_\ell a_{\ell ij} = 0$, which shows that $\oplus_{i\in I} M_i$ is flat.

Conversely, suppose $\oplus_{i\in I} M_i$ is flat and $\Sigma_\ell r_\ell x_\ell = 0$ with $x_\ell \in M_i$. Then we can write $x_\ell = \Sigma_j a_{\ell j} y_j$, with $y_j \in \oplus_{i\in I} M_i$ and $\Sigma_\ell r_\ell a_{\ell j} = 0$. Let $s(1),\ldots,s(n)$ be an enumeration of the indices in I used in expressing the elements y_j. Then $y_j = \Sigma_{k=1}^{n} z_{jk}$ with $z_{jk} \in M_{s(k)}$, and

$$x_\ell = \Sigma_{k=1}^{n}(\Sigma_j a_{\ell j} z_{jk}).$$

So either $s(\ell) = s(k)$ for some $k \neq \ell$, in which case we are done by induction on n, or $x_\ell = \Sigma_j a_{\ell j} z_{j\ell}$ and we have shown that M_i is flat. □

The 'only if' part of (5.1) says more than just that a summand of a flat module is flat, because the M_i need not be summands.

5.2 COROLLARY. *Free modules and projective modules are flat.*

PROOF. That the left R-module R is flat follows by taking $n = 1$ and $y_1 = 1$ in the definition of flatness. Then (5.1) shows that free modules are flat. If P is projective, then the natural map of the free module $R^{(P)}$ onto P has a right inverse, so P is a summand of $R^{(P)}$, whence P is flat. □

A diagram of modules and maps, such as the square

$$\begin{array}{ccc} A & \xrightarrow{\alpha} & B \\ \beta\downarrow & & \downarrow\gamma \\ C & \xrightarrow{\delta} & D \end{array}$$

is said to be **commutative** if any two compositions of maps, beginning at the same place and ending at the same place, are equal. The square is commutative if and only if $\gamma\alpha = \delta\beta$.

5.3 THEOREM. *Let M be a left R-module. Then the following conditions are equivalent.*

(i) *M is flat.*

(ii) *For each right ideal I of R the map $I \otimes M \to M$ is one-to-*

one.

(iii) *For each finite-rank free right R-module B and submodule $A \subseteq B$ the map $A \otimes M \to B \otimes M$ is one-to-one.*

(iv) *For each right R-module B and submodule $A \subseteq B$ the map $A \otimes M \to B \otimes M$ is one-to-one.*

PROOF. Suppose (i) holds and $\Sigma\, r_i \otimes x_i \in I \otimes M$ goes to zero in M. Then $\Sigma\, r_i x_i = 0$, so there exist elements y_j in M, and a_{ij} in R such that $x_i = \Sigma_j a_{ij} y_j$ and $\Sigma_i r_i a_{ij} = 0$. Thus

$$\sum_i r_i \otimes x_i = \sum_i \Sigma_j\, r_i \otimes a_{ij} y_j = \sum_j \sum_i r_i a_{ij} \otimes y_j = 0$$

so (ii) holds.

Now suppose (ii) holds. We shall prove (iii) by induction on the rank of B. If B is rank one, then B is isomorphic to R and (ii) applies. Let $B = B_1 \oplus B_2$ where the B_i are free of rank less than the rank of B. Let $A_1 = A \cap B_1$, let A_2 be the projection of A into B_2, and consider the commutative diagram.

$$\begin{array}{ccccccc} & & 0 & & 0 & & 0 \\ & & \downarrow & & \downarrow & & \downarrow \\ & & A_1 \otimes M & \to & A \otimes M & \to & A_2 \otimes M \\ & & \downarrow & & \downarrow & & \downarrow \\ 0 & \to & B_1 \otimes M & \to & B \otimes M & \to & B_2 \otimes M \end{array}$$

The first row is exact by (4.6), and $B_1 \otimes M$ is a summand of $B \otimes M$ because B_1 is a summand of B. The first and third columns are exact by induction. An easy diagram chase shows that the second column is also exact, so (iii) holds.

Suppose (iii) holds. It suffices to verify (iv) for B a finitely generated right R-module, because if an element is zero in $B \otimes M$, then it is already zero in $B \otimes N$ for some finitely generated submodule N of M. Map a finitely generated free right R-module F onto B with kernel K, and let F' be the preimage of A in F. Consider the commutative diagram

$$\begin{array}{ccccccccc} & & & & 0 & & & & \\ & & & & \downarrow & & & & \\ 0 & \to & K \otimes M & \to & F' \otimes M & \to & A \otimes M & \to & 0 \\ & & \| & & \downarrow & & \downarrow & & \\ 0 & \to & K \otimes M & \to & F \otimes M & \to & B \otimes M & \to & 0 \end{array}$$

The rows and columns are exact because of (iii). An easy diagram chase shows that the map $A \otimes M \to B \otimes M$ is also one-to-one.

Finally suppose that (iv) holds, and $\Sigma_i r_i x_i = 0$. Let I be the right ideal of R generated by the r_i and consider the map from $I \otimes M$ to $R \otimes M$. The element $\Sigma_i r_i \otimes x_i$ goes to zero under this map, so by (iv) we have $\Sigma_i r_i \otimes x_i = 0$. That (i) holds now follows from Theorem 4.8. □

5.4 COROLLARY. *Let M be a left R-module, B a right R-module, and A a submodule of B. If B/A is flat, then the map from $A \otimes M$ to $B \otimes M$ is one-to-one.*

PROOF. Map a free module F onto M with kernel K, and consider the following commutative diagram where $C = B/A$.

$$\begin{array}{ccccccccc}
 & & & & 0 & & & & \\
 & & & & \downarrow & & & & \\
 & & A \otimes K & \to & A \otimes F & \to & A \otimes M & \to & 0 \\
 & & \downarrow & & \downarrow & & \downarrow & & \\
 & & B \otimes K & \to & B \otimes F & \to & B \otimes M & & \\
 & & \downarrow & & \downarrow & & \downarrow & & \\
0 & \to & C \otimes K & \to & C \otimes F & \to & C \otimes M & &
\end{array}$$

The middle column is exact because F is free; the last row is exact because C is flat. Let $x \in A \otimes M$ go to zero in $B \otimes M$. Then x comes from y in $A \otimes F$, which goes to z in $B \otimes F$, which comes from w in $B \otimes K$. Now w goes to zero in $C \otimes F$, hence to zero in $C \otimes K$. Therefore w comes from u in $A \otimes K$, which goes to y in $A \otimes F$ because it goes to z in $B \otimes F$. So $x = 0$. □

EXERCISES

1. Let S be a multiplicative submonoid of a commutative ring R. Show that $S^{-1}R$ is a flat R-module.

2. Let A be a flat right module over a ring R, such that AI is detachable from A for all finitely generated left ideals I of R. Let B be a finitely generated coherent left R-module. Show that $A \otimes B$ is discrete.

3. Show that the following conditions on a ring R are equivalent.
   (i) $\Pi_{i \in \mathbb{N}} M_i$ is a flat left R-module whenever each M_i is.
   (ii) $R^{\mathbb{N}}$ is a flat left R-module.
   (iii) If $\varphi : R^n \to R$ is a map of right R-modules, then each countable set of elements of $ker\ \varphi$ is contained in a finitely generated submodule of $ker\ \varphi$.

   Show that (iii) is equivalent to right coherence of R if R is countable and discrete.

4. **Finitely presented flat modules are projective.** Let F be a free left R-module of finite rank, and K a finitely generated submodule of F. For $x \in F$ let I_x denote the right ideal of R generated by the coordinates of x in F. Show that the following are equivalent.

    (i) F/K is flat.

    (ii) $x \in I_x K$ for each x in K.

    (iii) for each $x \in K$ there is $f : F \to K$ such that $x = f(x)$.

    (iv) K is a summand of F.

    Use your proof to show that if R is commutative, and finitely generated submodules of F are detachable, then we can decide whether or not F/K is projective.

5. Show that any module over a discrete division ring is flat. Show that if $k \subseteq K$ are discrete fields, and V is a vector space over k, then $v_1,\ldots,v_n$ in V are dependent over k if and only if they are dependent over K in $K \otimes_k V$.

6. Use (5.3.ii) to show that if $A \subseteq B$, and A and B/A are flat, then B is flat. Conclude that if every cyclic module is flat, then every module is flat. Show that a ring R is von Neumann regular (see Exercise II.6.4) if and only if every R-module is flat. Show that if R is von Neumann regular, then every finitely generated left ideal of R is a summand whence R is coherent.

6. LOCAL RINGS

A ring R is called **local** if for each $r \in R$, either r or $1 - r$ is a unit. An equivalent condition is that if $r_1 + r_2$ is a unit, then either r_1 or r_2 is a unit. Any Heyting field is a local ring, and many of the results in this section, and the next, are of interest because of what they say about Heyting fields. In fact a Heyting field can be characterized as a commutative local ring in which 1 cannot be equal to 0 and any element which cannot be a unit is equal to 0. A virtue of working with local rings rather than Heyting fields, in addition to the gain in generality, is that we do not concern ourselves with negative notions like the ones in the preceding sentence.

An endomorphism f of a ring R as a left R-module is given by $f(x) =$

$xf(1)$, so taking f to $f(1)$ is an isomorphism of the endomorphism ring of R as a left R-module with the opposite ring of R. As the opposite ring of a local ring is local, if R is local then the left module R has a local endomorphism ring. If e is an idempotent in a local ring, then either e is a unit, so $e = 1$, or $1 - e$ is a unit, so $e = 0$ (or both). Thus if a module M has a local endomorphism ring, then any summand of M is either 0 or M, that is, M is **indecomposable.** The next few theorems, dealing with direct decompositions involving modules with local endomorphism rings, are called **Azumaya theorems.** Keep in mind that if R is a Heyting field, or even just a local ring, then R^n is a direct sum of R-modules with local endomorphism rings.

6.1 LEMMA. *Let $B \oplus C = A_1 \oplus \cdots \oplus A_n$ be a module. If C has a local endomorphism ring, then $B \oplus C = B \oplus D$, where $D \subseteq A_i$ for some i.*

PROOF. Let π_i, π_B, and π_C be the projections on A_i, B and C respectively. Then $\pi_C(\pi_1 + \cdots + \pi_n)$ is the identity map on C, so some $\pi_C\pi_i$ is an automorphism of C. Let $D = \pi_i C$. Then π_C maps D isomorphically onto C, so $B \oplus C = B \oplus D$. □

The first Azumaya theorem shows that summands of direct sums of modules with local endomorphism rings are again direct sums of modules with local endomorphism rings.

6.2 THEOREM. *Let $A \oplus B = C_1 \oplus \cdots \oplus C_n$ be modules such that the endomorphism ring of each C_i is local. Then there exist modules D_j such that*

$$A = D_1 \oplus \cdots \oplus D_m \text{ and } B = D_{m+1} \oplus \cdots \oplus D_n,$$

and there is a permutation σ such that $C_i \cong D_{\sigma(i)}$ for each i.

PROOF. By (6.1) we can find D_1, contained in A or contained in B, such that $D_1 \oplus C_2 \oplus \cdots \oplus C_n = C_1 \oplus C_2 \oplus \cdots \oplus C_n$. We may assume that $D_1 \subseteq B$. Then $B = B' \oplus D_1$ and $A \oplus B' \cong C_2 \oplus \cdots \oplus C_n$ so we are done by induction on n. □

It follows from (6.2) that a summand of a finite-rank free module over a local ring is finite-rank free. The next Azumaya theorem shows that we can cancel modules with local endomorphism rings.

6.3 THEOREM. *If* $A \oplus C \cong B \oplus C$ *are modules, and* C *has a local endomorphism ring, then* $A \cong B$.

PROOF. We may assume that $A \oplus C' = B \oplus C$ with $C' \cong C$. By (6.1) we may assume that $C' \subseteq B$ or $C' \subseteq C$. If the latter, then C' is a summand of C, so either $C' = C$ or $C' = 0$, whence $A \cong B$. If $C' \subseteq B$, then write $B = B' \oplus C'$, so $A \oplus C' = B' \oplus C' \oplus C$. Then $A \cong B' \oplus C \cong B$. □

From (6.3) we see that if R is a local ring, and $R^m \cong R^n$, then $m = n$ or R is trivial. Recall that this is also true for R commutative (II.7.5) but is not true in general (Exercise II.4.3).

Define an **inequality** on a local ring by setting $r_1 \neq r_2$ if $r_1 - r_2$ is a unit. Using this nonstandard inequality we can develop, in a natural way, much of the theory of finite-dimensional vector spaces over Heyting fields in the more general setting of finite-rank free modules over a local ring. This inequality is symmetric and translation invariant for any ring and, for a local ring, it is also cotransitive. Exercise II.1.5 shows that addition and subtraction are strongly extensional. To show that multiplication is strongly extensional we need the following.

6.4 LEMMA. *If* a *and* b *are elements of a local ring* R, *and* ab *is a unit, then* a *and* b *are units.*

PROOF. We may assume that $ab = 1$. It suffices to show that either $-a$ or b is a unit, so we may assume that $1+a$ and $1-b$ are units. Then $a-b = (1+a)(1-b)$ is a unit, so either a or $-b$ is a unit. □

If a, b, c and d are elements of a local ring R, and $ab \neq cd$, then either $ab \neq ad$ or $ad \neq cd$ by cotransitivity; so either $b \neq d$ or $a \neq c$ by (6.4). Thus multiplication is strongly extensional. It follows that if $f(X_1,\ldots,X_n)$ is a function built up from elements of R and the variables $X_1,\ldots,X_n$ using only multiplication and addition, and if $f(0,\ldots.,0) = 0$, then from $f(r_1,\ldots,r_n) \neq 0$ it follows that $r_i \neq 0$ for some i.

A standard classical characterization of local rings is that the nonunits form an ideal. Exercise 4 outlines why we do not use this characterization, but nontrivial local rings do have the property.

6.5 THEOREM. *Let* R *be a local ring. Then* $M = \{r \in R$: *if* r *is a unit, then* R *is trivial*$\}$ *is the Jacobson radical of* R.

PROOF. Suppose Rr is a quasi-regular left ideal. If r is a unit, then

$1 \in Rr$, so 0 is a unit, whence R is trivial; thus $r \in M$. Conversely, suppose $m \in M$ and $r \in R$. Either rm or $1 - rm$ is a unit. If rm is a unit then m is a unit by (6.4), so R is trivial whence $1-rm$ is also unit. Therefore Rm is a quasi-regular left ideal. □

Let M be a module over a local ring R. The **strong inequality** on M is defined by setting $x \neq y$ if there exists a homomorphism $f:M \to R$ such that $f(x) \neq f(y)$. The strong equality on M is the smallest inequality that makes all the homomorphisms from M to R strongly extensional. On R^n the strong inequality is the natural one to impose; it can be described in terms of coordinates as follows.

6.6 THEOREM. *Let R be a local ring, and x and y elements of R^n. Then $u \neq v$ in the strong inequality on R^n if and only if $x - y$ has a coordinate that is a unit in R.*

PROOF. Let $e_1,\ldots,e_n$ be the natural basis for R^n and let $x - y = \Sigma_i a_i e_i$. If a_i is a unit, then $f(x) \neq f(y)$ where f is the projection of R^n onto its ith factor. Conversely, suppose $f(x) \neq f(y)$ for some R-module map $f : R^n \to R$. Then $\Sigma_i \varphi(a_i e_i) \neq 0$, whence $a_i\varphi(e_i) = \varphi(a_i e_i) \neq 0$ for some i. Therefore $a_i \neq 0$. □

We say that $u_1,\ldots,u_m$ in R^n are **linearly independent**, over the local ring R, if $\Sigma_{i=1}^m r_i u_i \neq 0$ whenever $r_1,\ldots,r_m$ are in R, and $r_i \neq 0$ for some i. Clearly any basis is linearly independent. We will show in (6.10) that, conversely, if $u_1,\ldots,u_m$ generate R^n, and are linearly independent, then they form a basis. We will also show that any linearly independent set in R^n can be extended to a basis.

Lemma 6.4, that $ab = 1$ implies a and b are units, extends from local rings to matrices over local rings. Moreover, the invertible matrices are all products of elementary matrices.

6.7 THEOREM. *Let A be an $n \times n$ matrix over a local ring R. If A has a left or right inverse, then A is a product of elementary matrices, so A has a two-sided inverse.*

PROOF. It suffices, by considering transposes, to assume that A has a left inverse B. Then $\Sigma_{j=1}^n b_{1j}a_{j1} = 1$, so there is j such that $b_{1j}a_{j1}$ is a unit. By (6.4) this implies that a_{j1} is a unit. Then we can find a product E of elementary matrices such that the first column of EA is all

0's except for a 1 at the top. As $(BE^{-1})(EA) = I$, and E has a two-sided inverse, we may assume that $a_{i1} = 0$ for $i \neq 1$, and $a_{11} = 1$; note that this implies that $b_{i1} = 0$ for $i \neq 1$, and $b_{11} = 1$. If M^* denotes the matrix M without its first column and row, then $B^*A^* = I^*$ so by induction on n we can find a product E of elementary matrices such that EA is the identity matrix except for $a_{12},\ldots,a_{1n}$. But these latter entries are easily made equal to 0 by elementary row operations. □

The next two lemmas concern linearly independent elements of finite-rank free modules.

6.8 LEMMA. *Let R be a local ring and M be an $m \times n$ matrix of elements of R. Let A be an invertible $m \times m$ matrix, and B be an invertible $n \times n$ matrix. Then the rows of M are linearly independent if and only if the rows of AMB are.*

PROOF. Because A and B are invertible, it suffices to prove that if the rows of M are linearly independent, then the rows of AMB are. The rows of M are linearly independent if and only if whenever X is a $1 \times m$ matrix such that $X \neq 0$, then $XM \neq 0$. If $X \neq 0$, then $XA \neq 0$ by strong extensionality of the ring operations, because $(XA)A^{-1} = X$. We then have $XAM \neq 0$ because the rows of M are linearly independent, and $XAMB \neq 0$ by strong extensionality. □

6.9 LEMMA. *Let R be a local ring and M an $m \times n$ matrix of elements of R. If the rows of M are linearly independent, then either R is trivial or there exists an invertible square matrix A, and a permutation matrix B, such that the first m columns of AMB form an $m \times m$ identity matrix.*

PROOF. The matrix A is constructed by composing elementary row operations, the matrix B by permuting columns. As the first row of M is nonzero, we can permute columns and multiply by a unit so that the first entry in the first row is 1. We can then clear the rest of the first column by elementary row operations. The rows of the resulting matrix are still linearly independent by Lemma 6.4. If $m > n = 1$, then R is trivial; otherwise, by induction on the number of rows, either R is trivial or we can perform elementary operations on rows 2 through m, and permute columns 2 through n, to get an $(m-1) \times (m-1)$ identity matrix in rows 2 through m and columns 2 through m. We can then use elementary row operations to convert the first m columns into an identity matrix. □

We can now show that any linearly independent set in a finite-rank free module over a local ring can be extended to a basis.

6.10 THEOREM. *Let R be a local ring and let $v_1,\ldots,v_m$ be linearly independent elements of a finite-rank free R-module F. Let $e_1,\ldots,e_n$ be a basis for F. If $m < n$, then there exists j such that $v_1,\ldots,v_m,e_j$ are linearly independent; if $m = n$, then $v_1,\ldots,v_m$ is a basis for F; if $m > n$, then R is trivial.*

PROOF. We may assume that F is the set of $1 \times n$ matrices of elements of R, that v_i is the ith row of an $m \times n$ matrix M, and that e_j is the $1 \times n$ matrix with a 1 in the jth column and 0 elsewhere. If $m \leq n$, then by (6.9) we can find an invertible square matrix A, and a permutation matrix B, such that the first m columns of AMB form an $m \times m$ identity matrix. If $m = n$, then M is invertible, hence its rows are a basis for F. If $m < n$, then, as B is a permutation matrix, there exists j such that $e_jB = e_{m+1}$. The rows of AMB, together with the row e_{m+1}, are linearly independent. Thus the rows of AM, together with e_j are linearly independent. As A is invertible, the rows of M together with e_j are linearly independent by (6.8).

If $m > n$, then $v_1,\ldots,v_n$ are a basis for F, so v_{n+1} can be written as a linear combination of them. But $v_1,\ldots,v_{n+1}$ are linearly independent, so $0 \neq 0$ in F, whence R is trivial. □

EXERCISES

1. Show that if $C_1 \oplus \cdots \oplus C_n \cong D_1 \oplus \cdots \oplus D_n$, and each C_i and D_i has a local endomorphism ring, then there is a permutation σ of $\{1,\ldots,n\}$ so that $C_i \cong D_{\sigma(i)}$ for each i.

2. Use the Azumaya theorems to show that if A and B are square matrices over a local ring such that $AB = I$, then A and B are invertible.

3. Show that the inequality defined on a local ring is symmetric and cotransitive.

4. Show that the following are equivalent.

   (i) Markov's principle.

   (ii) If R is a countable discrete commutative ring such that the nonunits of R form an ideal, then R is local.

Hint: To show that (i) follows from (ii) let $S = \{m \in \mathbb{Z} : m = 1$, or $m \neq 0$ and $a_n = 1$ for some $n\}$, and let $R = S^{-1}\mathbb{Z}$.

5. Give an example of elements a and b of a ring such that $ab = 1$, but neither a nor b is a unit.

6. Show that R^n is Hopfian (Exercise 1.9) if R is local.

7. Let R be a local ring and M an R-module equipped with the strong inequality. Show that addition and scalar multiplication in M are strongly extensional.

8. Generalize Theorem II.6.5 to local rings $k \subseteq K$, assuming that the K-module V comes with an inequality such that scalar multiplication is strongly extensional, and that each of the three occurrences of the term 'finite-dimensional' implies an inequality preserving isomorphism with R^n.

9. Let k be the field of integers modulo 2. Show that $K = k[X]/(X^2)$ is a local ring isomorphic to k^2, but that the inequality on K as a local ring differs from the inequality on K as a k-module. For $V = k[X,Y]/(X,Y)^2$ show that Exercise 8 fails because V is not a free K-module.

10. Show that if R is a local ring, and $u_1,\ldots,u_m$ is a basis for R^n, then either $m = n$, or $R = 0$.

11. Show that if R is a local ring, $e_1,\ldots,e_n$ is a basis for a free R-module F, and $v_1,\ldots,v_m$ generate F, then there exists i such that $v_i, e_2,\ldots,e_n$ is a basis for F.

7. COMMUTATIVE LOCAL RINGS

If R is a commutative local ring we define $R(X)$, in analogy with the rational function field over a field, by inverting all those elements in $R[X]$ which have an invertible coefficient.

7.1 LEMMA. *Let R be a commutative local ring and X an indeterminate. Let S be the set of polynomials in $R[X]$ that have an invertible coefficient. Then*

(i) *S is multiplicatively closed.*

(ii) *If $fg = 0$ for $f \in S$ and $g \in R[X]$, then $g = 0$.*

(iii) *If* $fg \in S$, *then* $f \in S$. *If* $f + g \in S$, *then* $f \in S$ *or* $g \in S$.

PROOF. Let $f = a_nX^n + a_{n-1}X^{n-1} + \ldots + a_0$ and $g = b_mX^m + b_{m-1}X^{m-1} + \ldots + b_0$. Let $fg = c_{m+n}X^{m+n} + \ldots + c_0$. Suppose a_i and b_j are invertible. Then a_ib_j is invertible, so either c_{i+j} or $c_{i+j} - a_ib_j$ is invertible. In the former case we have established (i); in the latter we have $\Sigma\, a_pb_q$ is invertible, where the sum runs over indices p,q such that $p + q = i + j$ and $p \neq i$. As R is local, there exists $p < i$ or $q < j$ such that a_pb_q is invertible, so either a_pb_j or a_ib_q is invertible, so by induction on $i + j$ we have proved (i).

To prove (ii) suppose a_i is invertible and $fg = 0$. We may assume that $a_i = 1$. Then the $(m+1)\times(m+1)$ matrix

$$\begin{bmatrix} 1 & a_{i-1} & a_{i-2} & \cdots & a_{i-m} \\ a_{i+1} & 1 & a_{i-1} & \cdots & a_{i-m+1} \\ \vdots & \vdots & & & \vdots \\ a_{i+m} & a_{i+m-1} & & \cdots & 1 \end{bmatrix}$$

where $a_j = 0$ if $j < 0$ or $j > n$, kills the vector $(b_0,\ldots,b_m)^t$. If d is the determinant of the matrix, then $db_j = 0$ for all j. Either d is invertible, in which case we are done, or $d - 1$ is a unit, in which case a_j is invertible for some $j < i$, and we are done by induction on i.

To prove (iii), first suppose $fg \in S$, so c_k is invertible for some k. As R is local, a_ib_j is invertible for some i,j with $i + j = k$. Thus a_i is invertible, so $f \in S$. Next suppose $f + g \in S$. Then $a_i + b_i$ is invertible for some i, so either a_i or b_i is invertible. □

7.2 THEOREM. *Let R be a commutative local ring and X an indeterminate. Let $R(X)$ be the ring $R[X]_S$ where S is the set of elements of $R[X]$ that have an invertible coefficient. Then $R(X)$ is a local ring containing $R[X]$. If R is a Heyting field, then so is $R(X)$.*

PROOF. If $a/s + a'/s' = (as' + a's)/ss'$ is invertible in $R(X)$, then $t(as' + a's) \in S$ for some $t \in S$, so either a or a' is in S by (7.1.iii); thus $R(X)$ is local. From (7.1.ii) we see the natural map from $R[X]$ to $R(X)$ is one-to-one. Suppose that the inequality on R is tight. If a/s cannot be a unit in $R(X)$, then a cannot be in S, so no coefficient of a can be invertible, whence all coefficients of a are zero. If the inequality on R is consistent, then 1 cannot equal 0 in R or in $R[X]$; but the map from $R[X]$ to $R(X)$ is one-to-one. □

7.3 THEOREM. *Let R be a commutative local ring and e an idempotent endomorphism of a finite-rank free R-module F. Then the kernel of e is a finite-rank free R-module.*

PROOF. The endomorphism ring of the R-module R is isomorphic to the ring R, so the Azumaya theorem (6.2) applies. □

If e is an idempotent in a local ring, then either e is a unit, in which case $e = 1$, or $1 - e$ is a unit, in which case $e = 0$. By an **impotent ring** we mean a commutative ring with no nilpotent elements, such that any idempotent is either 0 or 1. Any commutative local ring with no nilpotents is impotent.

7.4 THEOREM. *Let R be an impotent ring. If g is a factor of a monic polynomial in $R[X]$, then there is a unit λ of R so that $\lambda^{-1}g$ is monic.*

PROOF. Let $gh = X^m + c_{m-1}X^{m-1} + \cdots + c_0$, where

$$g = a_nX^n + a_{n-1}X^{n-1} + \ldots + a_0,$$

and

$$h = b_nX^n + b_{n-1}X^{n-1} + \ldots + b_0.$$

We shall show that $a_ib_j = 0$ if $i + j > m$. Proceeding by backwards induction on $i + j$, we note that $a_ib_j = 0$ trivially if $i + j > 2n$. Suppose $a_ib_j = 0$ whenever $i + j = k + 1$. If $k > m$, then

$$a_0b_k + a_1b_{k-1} + \cdots + a_kb_0 = 0.$$

Multiplying this equation by a_ib_{k-i} we get $(a_ib_{k-i})^2 = 0$, so $a_ib_{k-i} = 0$ as R has no nilpotents, completing the induction. Now consider

$$a_0b_m + a_1b_{m-1} + \cdots + a_mb_0 = 1.$$

Multiplying by a_ib_{m-i} we find that a_ib_{m-i} is idempotent, hence 0 or 1 as R is impotent. If they are all 0, then R is trivial and the theorem is trivially true. Otherwise there exist i and j such that $i + j = m$ and a_ib_j is a unit. If $k > i$, then $a_kb_j = 0$, so $a_k = 0$; thus we can choose $\lambda = a_i$. □

7.5 COROLLARY. *If g and h are polynomials with coefficients in an impotent ring, and $gh = X^d$, then there is a unit λ and $0 \leq i \leq d$ such that $g = \lambda X^i$.*

PROOF. Let λ be as in (7.4), and choose i so that $X^ig(1/X)$ and $X^{d-i}h(1/X)$ are polynomials. Then $1 = X^dg(1/X)h(1/X) = X^ig(1/X)X^{d-i}h(1/X)$, so $deg\ X^i g(1/X) \leq 0$, by (7.4), whence $g(X) = \lambda X^i$. □

7.6 LEMMA. *Let α be an endomorphism of a finite-rank free module over an impotent ring. If α is nilpotent, then the characteristic polynomial of α is a power of X.*

PROOF. Suppose $\alpha^m = 0$. Let A be an $n \times n$ matrix representing α, and I the $n \times n$ identity matrix. Then

$$X^m I = (XI - A)(X^{m-1}I + X^{m-2}A + \ldots + XA^{m-2} + A^{m-1}).$$

Taking determinants of both sides yields

$$X^{mn} = f(X)g(X)$$

where f is the characteristic polynomial of α. Then f is a power of X by (7.5). □

7.7 THEOREM. *Let α be an endomorphism of a finite-rank free module V over a commutative local ring with no nilpotents. Let $f(X)$ be its characteristic polynomial, and suppose $f(X) = (X-\lambda)^n g(X)$ where $g(\lambda)$ is a unit. Then $V = H \oplus K$ where H is a rank-n free module, $\alpha-\lambda$ is an automorphism on K, and $(\alpha-\lambda)^n H = 0$.*

PROOF. The remainder theorem shows that $X - \lambda$ and $g(X)$ are strongly relatively prime, so $(X - \lambda)^n$ and $g(X)$ are strongly relatively prime, whence there exist polynomials $s(X)$ and $t(X)$ such that

$$s(X)(X-\lambda)^n + t(X)g(X) = 1.$$

Let $e = s(\alpha)(\alpha-\lambda)^n$. Then $e^2 = e(1-t(\alpha)g(\alpha)) = e$, because $f(\alpha) = 0$. Let H be the kernel of e, and K the kernel of $1-e$; as $\alpha e = e\alpha$ it follows that $\alpha H \subseteq H$ and $\alpha K \subseteq K$. As $e = s(\alpha)(\alpha-\lambda)^n = (\alpha-\lambda)s(\alpha)(\alpha-\lambda)^{n-1}$ is the identity on K, and $\alpha K \subseteq K$, we have $\alpha-\lambda$ is an automorphism of K.

If the finite-rank free module K is zero, then $e = 0$ and $H = V$, so $g(\alpha)$ is one-to-one because $t(\alpha)g(\alpha) = 1$. In that case $(\alpha-\lambda)^n = 0$ because $0 = f(\alpha) = g(\alpha)(\alpha-\lambda)^n$. By (7.6) we have $f(X) =$ is a power of $X-\lambda$, so $g = 1$ and the theorem holds.

If the rank of K is greater than 0, then we can write $f = f_H f_K$, where f_H and f_K are the characteristic polynomials of α restricted to H and K respectively. As $f_K(\lambda)$ is the determinant of $\alpha-\lambda$ acting on K, we have $f_K(\lambda)$ is a unit. Thus $f_H = (X-\lambda)^n g^*(X)$ where $g^*(\lambda)$ is a unit. By induction on the dimension of V we have H is a rank-n free module and $(\alpha-\lambda)^n H = 0$. □

EXERCISES

1. Show that if a,b and c are polynomials over a Heyting field, and $\deg a \leq \deg b$, then $\deg ac \leq \deg bc$.

2. **Units in $R[X]$.** Let R be a commutative ring. Show that if the constant term of g is a unit, and all the other coefficients of g are nilpotent, then g is a unit in $R[X]$. Show that, conversely, if g is a unit in $R[X]$, then the constant term of g is a unit in R, and all the other coefficients of g are nilpotent (Let $S = \{1, a, a^2, \ldots\}$ where a is the highest coefficient of g not known to be nilpotent, and show that $S^{-1}R$ is trivial).

3. Show that Theorem 7.4 characterizes impotent rings (you need consider only polynomials of the form $aX + b$).

4. Give an example of a commutative local ring where (7.6) fails.

5. Give a Brouwerian example of an endomorphism α of $\mathbb{R}^2$, where $\mathbb{R}$ is the real numbers, such that the characteristic polynomial f of α has λ as a root, but cannot be written in the form given in (7.7).

6. **Jordan canonical form I.** Let α be an endomorphism of a finite-rank free module V over a commutative local ring R with no nilpotents. Suppose the characteristic polynomial of α is a product of polynomials of the form $(X - \lambda)^m$ for distinct λ (in the inequality on R). Show that V is a direct sum of submodules V_λ such that $\alpha V_\lambda \subseteq V_\lambda$ and the characteristic polynomial of α restricted to V_λ is $(X - \lambda)^m$.

7. **Jordan canonical form II.** Let R be a commutative local ring, and α an endomorphism of R^n such that $\alpha^m = 0$ and $im\ \alpha^i$ is finite-dimensional for $i = 1, \ldots, m$.
   (i) Show that $H_i = (ker\ \alpha) \cap (im\ \alpha^i)$ is finite-dimensional, hence a summand of R^n.
   (ii) Show that $ker\ \alpha = V_0 \oplus \cdots \oplus V_{m-1}$ where $H_i = V_i \oplus H_{i+1}$
   (iii) Choose a basis e_{ij} for each V_i, hence for $ker\ \alpha$. Let $e_{ij} = \alpha^i x_{ij}$ and show that $\{\alpha^k x_{ij} : k \leq i\}$ is a basis for R^n.

8. Give a Brouwerian example of a nilpotent endomorphism α of $\mathbb{R}^2$ such that $im\ \alpha$ is not finite-dimensional.

NOTES

Our theory of Noetherian rings and modules makes extensive use of the axiom of dependent choices. To attempt a theory that does not avail itself of this axiom seems too ambitious for the purposes of this book, and it appears likely that the classical theory would be significantly distorted at best.

Classically, the ascending chain condition on submodules is equivalent to the ascending chain condition on finitely generated submodules, and the latter condition admits interesting constructive examples (the former does not). The descending chain condition, on the other hand, does not seem to lend itself to a constructive treatment. A test case would be to formulate a descending chain condition that was satisfied by the abelian group $\mathbb{Z}(p^\infty)$, the p-primary component of $\mathbb{Q}/\mathbb{Z}$.

The definition of *coherent* comes from [Bourbaki 1961, §2, Exercise 11] where it is called *pseudo-coherent*, the term *coherent* being reserved for finitely generated, pseudo-coherent modules.

The classical theorem that a ring is right coherent if and only if products of flat left modules are flat involves the full axiom of choice. Exercise 5.3 is the countable version.

The proof that finitely presented flats are projective in Exercise 5.4 is from [Bourbaki 1961, §2, Exercise 23(a)]. The application to deciding whether a finitely presented module is projective is from [Baumslag et al 1981, Lemma 5.1] where the commutativity hypothesis is not stated but seems to be used; the problem is that $I_x K$ need not be a submodule. The Baumslag paper is in the context of recursive function theory and uses Markov's principle. For the relationship between constructive algebra and recursive algebra see [Bridges-Richman 1987].

In [Julian, Mines and Richman, 1978] a field is defined to be a commutative local ring, with no nilpotent elements, in which 0 cannot equal 1.

Chapter IV. Divisibility in Discrete Domains

1. DIVISIBILITY IN CANCELLATION MONOIDS

A commutative monoid is called a **cancellation monoid** if $ab = ac$ implies $b = c$. If R is a discrete integral domain, then the set of nonzero elements of R forms a discrete cancellation monoid. This example motivates our study of general cancellation monoids. Terminology introduced for cancellation monoids transfers to discrete domains by applying it to the monoid of nonzero elements.

If a and b are elements of a cancellation monoid M, then we say that a **divides** b, and write $a|b$, if there exists c in M such that $b = ca$. The divisors of 1 are the units of M and form a submonoid U of M which is a group. If U is detachable, we say that M has **recognizable units.** Two elements a and b of M are **associates,** written $a \sim b$, if each divides the other. As M has cancellation, $a \sim b$ if and only if there is a unit u such that $b = ua$. The monoid M/U is constructed from M by declaring two elements equal if they are associates. We operate in M/U when we care what the elements are only "up to a unit": in particular, the relation $a|b$ may be viewed in M/U where it constitutes a partial order.

We say that d is a **greatest common divisor**, or **GCD**, of a and b, if $d|a$ and $d|b$, and for each c such that $c|a$ and $c|b$, we have $c|d$. Observe that GCD's are unique in M/U, and that the GCD of a and b is the infimum of a and b in the partially ordered set M/U. If d is a GCD of a and b, then we let $GCD(a,b)$ denote d as an element of M/U.

Two elements a and b are **relatively prime** if $GCD(a,b) = 1$. A **GCD-monoid** is a cancellation monoid in which each pair of elements has a greatest common divisor. A **GCD-domain** is a discrete domain whose nonzero elements form a GCD-monoid.

1.1 THEOREM. *Let a, b and c be elements of a GCD-monoid M. Then*

(i) $GCD(GCD(a,b),c) = GCD(a,GCD(b,c))$.

(ii) $c \cdot GCD(a,b) = GCD(ca,cb)$.

(iii) *If* $x = GCD(a,b)$, *then* $GCD(a,bc) = GCD(a,xc)$.

(iv) *If* $a|bc$, *and* $GCD(a,b) = 1$, *then* $a|c$.

PROOF. Claim (i) is easily verified. For (ii), let $d = GCD(a,b)$ and $e = GCD(ca,cb)$. Then $cd|e$, so $e = cdx$. It remains to show that x is a unit. Now $ca = ea' = cdxa'$, so $a = dxa'$. In the same way, $b = dxb'$. Thus $dx|d$, so x is a unit.

For (iii) we have $GCD(a,bc) = GCD(GCD(a,ac),bc) = GCD(a,GCD(ac,bc)) = GCD(a,c \cdot GCD(a,b)) = GCD(a,xc)$. Claim (iv) follows immediately from (iii) upon taking x = 1 □

A nonunit p of a cancellation monoid is said to be **irreducible** if whenever $p = ab$, then either a or b is a unit. We say that a nonunit p is **prime** if whenever $p|ab$, then $p|a$ or $p|b$. Clearly every prime is irreducible.

1.2 LEMMA. *Each irreducible element in a* GCD*-monoid is prime.*

PROOF. Let p be irreducible and suppose $p|ab$. We shall show that $p|a$ or $p|b$. Let $d = GCD(p,a)$ and let $p = cd$. As p is irreducible either c is a unit or d is a unit. If c is a unit, then $p|d$, so $p|a$. If d is a unit then $GCD(p,a) = 1$ so $p|b$ by (1.1.iv). □

1.3 LEMMA. *A GCD-monoid M has recognizable units if and only if divisibility is decidable in M.*

PROOF. If divisibility is decidable in M, then we can decide whether $u|1$, so units are recognizable. Conversely suppose that M has recognizable units. Let a and b be elements of M, and let $d = GCD(a,b)$. There is s such that $a = sd$. Then $a|b$ if and only if $a|d$ if and only if s is a unit. □

1.4 DEFINITION. Let M be a cancellation monoid. An element $a \in M$ is said to be **bounded by** n if whenever $a = a_0 \cdots a_n$ with $a_i \in M$, then a_i is a unit for some i. An element of M is **bounded** if it is bounded by n for some $n \in \mathbb{N}$; the monoid M is **bounded** if each of its elements is bounded. A discrete domain is **bounded** if its nonzero elements form a bounded monoid.

The units of M are exactly those elements of M that are bounded by 0; an element of M is irreducible if and only if it is bounded by 1 but not by 0.

A **principal ideal** of a commutative monoid M is a subset I of M such that $I = Ma = \{ma : m \in M\}$ for some a in M. We say that M satisfies the **divisor chain condition** if for each ascending chain $I_1 \subseteq I_2 \subseteq I_3 \subseteq \cdots$ of principal ideals, there is n such that $I_n = I_{n+1}$. A discrete domain is said to satisfy the divisor chain condition if its monoid of nonzero elements does.

A cancellation monoid M is bounded if and only if for each a in M there exists n such that for any chain $I_0 \subseteq I_1 \subseteq \cdots \subseteq I_n$ of principal ideals, with $I_0 = Ma$, there exists $j < n$ with $I_j = I_{j+1}$; thus any bounded monoid satisfies the divisor chain condition. The ring of integers is bounded as each nonzero integer n is bounded by $|n|$. The polynomial ring $F[X]$ over a discrete field F is bounded as each nonzero polynomial f is bounded by $deg\ f$. A GCD-domain satisfying the divisor chain condition is called a **quasi-UFD.**

1.5 EXAMPLE. Let a be a binary sequence. Let $R = \cup_n \mathbb{Z}[a_n/2]$. Then R is a Brouwerian example of a quasi-UFD without recognizable units, as we cannot tell whether 2 is invertible. Clearly we cannot write the elements 6 and 10 of R as products irreducible elements. However we shall show that, given a finite set S of nonzero elements of a quasi-UFD, we can find a set P of pairwise relatively prime elements such that every element of S is an associate of a product of elements of P (Corollary 1.9).

1.6 LEMMA. *Let M be a GCD-monoid satisfying the divisor chain condition, and let $p_1,\ldots,p_m$ be pairwise relatively prime elements of M. If $a \in M$, then we can construct elements $a_0,a_1,\ldots,a_m$ in M such that*

(i) $a = a_0a_1\cdots a_m$

(ii) *for $j = 1,\ldots,m$ there exists e such that $a_j | p_j^e$*

(iii) *a_0 and p_j are relatively prime for $j = 1,\ldots,m$.*

PROOF. For each $j = 1,\ldots,m$ consider the sequence $x_n = a/GCD(a,p_j^n)$. By the divisor chain condition, there exists n such that $x_n | x_{n+1}$; set $e = n$ and $a_j = a/x_n = GCD(a,p_j^n) = GCD(a,p_j^{n+1})$. As the p_j are pairwise relatively prime, so are the a_j. Repeated application of (1.1.iv) shows there exists a_0 such that $a = a_0a_1\cdots a_m$. It remains to show that $GCD(a_0,p_j) = 1$ for each j. But

$$a_j = GCD(a,p_j^{e+1}) = GCD(a_0a_j,p_jp_j^e)$$

is divisible by $GCD(a_0,p_j)GCD(a_j,p_j^e) = GCD(a_0,p_j)a_j$, so $GCD(a_0,p_j) = 1$. □

1.7 LEMMA. *Let M be a GCD-monoid satisfying the divisor chain condition, and let $a,b \in M$. Then there are $a^+,a^-,c,b^+,b^- \in M$ such that*

(i) $a = a^+cb^-$ *and* $b = a^-cb^+$.

(ii) $a^-|a^+$ *and* $b^-|b^+$.

(iii) a^+, c, *and* b^+ *are pairwise relatively prime.*

PROOF. Let $x = a/d$ and $y = b/d$ where $d = GCD(a,b)$. Then x and y are relatively prime, so by Lemma 1.6 we can write $d = a^-cb^-$ where $a^-|x^n$, and $b^-|y^m$, and c is relatively prime to both x and y. Let $a^+ = xa^-$ and $b^+ = yb^-$, so $a = dx = a^+cb^-$ and $b = dy = b^+ca^-$. As x^{n+1}, y^{m+1} and c are pairwise relatively prime, so are a^+, b^+ and c. □

As an example of the decomposition of Lemma 1.7 consider the monoid $\mathbb{N}^+$ of positive integers. If a and b are in $\mathbb{N}^+$, then a^+ and a^- are the largest factors of a and b respectively, that consist of primes that occur more often in a than in b. Similarly b^+ and b^- are the largest factors of b and a respectively, that consist of primes that occur more often in b than in a, while c is the largest factor of a (or b) that consists of primes that occur equally often in a and in b. If $a = 560 = 2^3 \cdot 3 \cdot 5 \cdot 7$ and $b = 300 = 2^2 \cdot 3 \cdot 5^2$, then $a^+ = 56$, $a^- = 4$, $c = 3$, $b^+ = 25$, and $b^- = 5$. Note that $GCD(a,b) = a^-cb^-$.

1.8 THEOREM (Quasi-factorization). *Let $x_1,\ldots,x_k$ be elements of a GCD-monoid M satisfying the divisor chain condition. Then there is a family P of pairwise relatively prime elements of M such that each x_i is an associate of a product of elements of P.*

PROOF. Consider first the case $k = 2$. We construct sequences $r_n = a_1(n)\cdots a_{m(n)}(n)$ and $s_n = b_1(n)\cdots b_{m(n)}(n)$ as follows. Let $m(0) = 1$ and $r_0 = a_1(0) = x_1$ and $s_0 = b_1(0) = x_2$. We shall suppress the dependence of m, a_i and b_i on n for cleaner notation. Suppose we have constructed $r_n = a_1\cdots a_m$ and $s_n = b_1\cdots b_m$ with $GCD(a_i,a_j) = GCD(b_i,b_j) = GCD(a_i,b_j) = 1$ if $i \neq j$. Then we construct r_{n+1} and s_{n+1} as follows. For each i write $a_i = a_i^+c_ib_i^-$ and $b_i = a_i^-c_ib_i^+$ as in Lemma 1.7. Then set

$$r_{n+1} = (a_1^+/a_1^-)\cdots(a_m^+/a_m^-)b_1^-\cdots b_m^-$$

$$s_{n+1} = a_1^-\cdots a_m^-(b_1^+/b_1^-)\cdots(b_m^+/b_m^-).$$

We easily see that, except for the pairs $(a_i^+/a_i^-, a_i^-)$ and $(b_i^+/b_i^-, b_i^-)$, the

$2m$ factors of r_{n+1} and of s_{n+1} are pairwise relatively prime. The principal ideals Mr_0s_0, Mr_1s_1, $Mr_2s_2,\ldots$ form an ascending chain. By the divisor chain condition there is n such that $r_ns_n \sim r_{n+1}s_{n+1} = a_1^+\cdots a_m^+b_1^+\cdots b_m^+$, so the elements a_i^-, c_i and b_i^- are all units. Thus the elements $a_1,\ldots,a_m,b_1,\ldots,b_m$ are conjugates of $a_1^+,\ldots,a_m^+,b_1^+,\ldots,b_m^+$ so are pairwise relatively prime. It now suffices to show that if we can write the elements in the family

$$E = \{a_i^-,\ (a_i^+/a_i^-),\ b_i^-,\ (b_i^+/b_i^-)\ :\ i = 1,..,m\}$$

as products of pairwise relatively prime elements, then we can do the same for $a_1,\ldots,a_m,b_1,\ldots,b_m$. Suppose that Q is a finite family of pairwise relatively prime elements, and that each element of E is an associate of a product of elements of Q. We may assume that each element of Q divides some element of E. Then by Lemma 1.7 each of the elements $a_1,\ldots a_m,b_1,\ldots b_m$ is an associate of a product of elements in $Q \cup \{c_1,\ldots,c_m\}$, and the elements in the latter family are pairwise relatively prime.

If $k > 2$ we proceed by induction on k. Let $P = \{p_1,\ldots,p_m\}$ be a family of pairwise relatively prime elements such that each of $x_1,\ldots,x_{k-1}$ is an associate of a product of elements of P. By (1.6) we can write $x_n = a_0a_1\cdots a_m$ where a_i divides a power of p_i, and $GCD(a_0,p_i) = 1$ for $i = 1,\ldots,m$. From the case $k = 2$ we can construct a finite family S_i of pairwise relatively prime elements such that a_i and p_i are associates of products elements of S_i, and each element of S_i divides a power of p_i. Then $\{a_0\} \cup S_1 \cup \cdots \cup S_m$ forms a family of pairwise relatively prime elements for $x_1,\ldots,x_m$. □

EXERCISES

1. Show that the set of positive even integers, together with 1, form a (multiplicative) discrete cancellation monoid M. Find elements a and b in M that do not have a GCD.

2. Construct elements a, b, and c in a discrete cancellation monoid such that $a|bc$, and $GCD(a,b) = 1$, but a does not divide c.

3. Show that the set of positive integers that are congruent to 1 modulo 3 form a (multiplicative) discrete cancellation monoid M. Is M a GCD-monoid?

4. The **least common multiple** $LCM(a,b)$ of two elements a and b in a cancellation monoid M is an element $m \in M$ such that $a|m$ and $b|m$ and, if $a|c$ and $b|c$, then $m|c$. Show that, if $LCM(a,b)$ exists, then $GCD(a,b)$ exists and is equal to $ab/LCM(a,b)$. Show that if M is a GCD-monoid, then $LCM(a,b)$ always exists. Construct elements a and b of a discrete cancellation monoid M such that $GCD(a,b)$ exists, but $LCM(a,b)$ does not.

5. Let M be a cancellation monoid. Define what it means to be a greatest common divisor $GCD(a_1,\ldots,a_n)$ or a least common multiple $LCM(a_1,\ldots,a_n)$ of the finite family $a_1,\ldots,a_n$ of elements of M. Show that if M is a GCD-monoid, then these always exist.

6. Let R be a GCD-monoid. Show that if $GCD(a,b) = 1$, then $GCD(a,b^n c) = GCD(a,c)$ for all n.

7. Let $M_1 = \{2^n : n \in \mathbb{N}\}$, and let $M_2 = M_1 \backslash \{2\}$. Use these monoids to construct a Brouwerian example of a discrete cancellation monoid with recognizable units in which divisibility is not decidable (you can't tell if $4|8$).

8. Let M a submonoid of a multiplicative abelian group G. For x and y in G, we say that x **divides** y (relative to M), if $yx^{-1} \in M$. Define GCD in G using this notion of divides. Show that if M generates G as a group, and M is a GCD-monoid, then $GCD(a,b)$ exists for all a,b in G, and (1.1) holds. Show that every cancellation monoid can be embedded as a submonoid in an essentially unique abelian group that it generates as a group.

9. Let P be the set of principal ideals of a GCD-domain R, partially ordered by inclusion. Show that P is a distributive lattice.

10. Let a, b and c be elements of a cancellation monoid. Show that if $GCD(ca,cb)$ exists, then $GCD(a,b)$ exists, and $GCD(ca,cb) = c \cdot GCD(a,b)$.

11. Let $R = \mathbb{Z}[Y,X_1,X_2,\ldots]/I$, where I is the ideal generated by $\{X_{i+1}Y - X_i : i \geq 1\}$. Show that R is a GCD-domain that does not satisfy the divisor chain condition. Show that there is no finite family Q of pairwise relatively prime elements such that Y and X_1 are associates of products of elements of Q.

12. Show that the elements a^+, a^-, c, b^+, and b^- of Lemma 1.7 are unique up to units.

2. UFD'S AND BEZOUT DOMAINS

Questions involving factoring are touchier in constructive algebra than they are in classical algebra because we may be unable to tell whether a given element has a nontrivial factorization. Our definition of a unique factorization domain is straightforward.

2.1 DEFINITION. A discrete domain R is called a **unique factorization domain**, or **UFD**, if each nonzero element r in R is either a unit or has an essentially unique factorization into irreducible elements, that is, if $r = p_1 \cdots p_m$ and $r = q_1 \cdots q_n$ are two factorizations of r into irreducible elements, then $m = n$ and we can reindex so that $p_i \sim q_i$ for each i. We say that R is **factorial** if $R[X]$ is a UFD.

Note that a UFD has recognizable units, that is, the relation $u|1$ is decidable. Discrete fields are trivial examples of UFD's; the ring $\mathbb{Z}$ is well known to be a UFD. The somewhat peculiar looking definition of *factorial* agrees with our usage of the term as applied to discrete fields, and allows us to show that $R[X]$ is factorial if R is. The following is a Brouwerian counterexample to the classical theorem that if R is a unique factorization domain, then so is $R[X]$.

2.2 EXAMPLE. Let a be a binary sequence, $\mathbb{Q}$ the field of rational numbers, $i^2 = -1$, and $k = \bigcup_n \mathbb{Q}(ia_n)$. Then k is a discrete field, hence a unique factorization domain. However we cannot factor $X^2 + 1$ into irreducibles over $k[X]$.

The notions quasi-UFD, bounded GCD-domain, UFD, and factorial domain are classically equivalent, but the ring $k[X]$ of Example 2.2 is a Brouwerian example of a bounded GCD-domain that is not a UFD, while the field k of Example 2.2 is a Brouwerian example of a UFD that is not a factorial domain. In (3.5) we shall give a Brouwerian example of a quasi-UFD that is not a bounded GCD-domain. It is easily verified that the other implications hold.

2.3 THEOREM. *Let R be a discrete domain. Then*

(i) *If R is factorial, then R is a UFD.*

(ii) *If R is a UFD, then R is a bounded GCD-domain.*

(iii) *If R is a bounded GCD-domain, then R is a quasi-*UFD. □

A multiplicative submonoid S of a commutative ring R is called **saturated** if $xy \in S$ implies $x \in S$.

2.4 THEOREM. *Let R be a discrete domain and S a multiplicative submonoid of R not containing* 0. *Then*

(i) *If R is a GCD-domain, then so is* $S^{-1}R$.

(ii) *If S is saturated and detachable, then* $S^{-1}R$ *has recognizable units.*

PROOF. To show (i) we observe that $GCD(a/s,b/t) = GCD(a,b)/1$. To show (ii) we shall show that a/s is invertible if and only if $a \in S$. If a/s is invertible, then $ab/st = 1/1$ for some $b/t \in S^{-1}R$, so $ab = st \in S$, whence $a \in S$. Conversely, if $a \in S$, then $a(1/a) = 1/1$. □

2.5 THEOREM. *Let R be a UFD and let S be a saturated detachable multiplicative submonoid of R not containing* 0. *Then* $S^{-1}R$ *is also a UFD.*

PROOF. If $r/s \in S^{-1}R$, then r and s can be written as products of irreducibles in R. By (2.4) we can decide for each irreducible factor of r whether it is invertible in $S^{-1}R$ or not. Those irreducible factors of r that are not invertible constitute the unique irreducible factorization of r/s in $S^{-1}R$. □

2.6 COROLLARY. *If R is factorial, and S is a detachable saturated multiplicative submonoid not containing* 0, *then* $S^{-1}R$ *is also factorial.* □

The assumption of Theorem 2.5 that S be saturated is essential. Consider the following Brouwerian example. Let a be a binary sequence containing at most one 1. Let $R = S^{-1}\mathbb{Z}$ with

$$S = \{q : q = 1 \text{ or } q = 2^{mn} \text{ for some } m,n \text{ such that } a_n = 1\}.$$

Then S is a detachable multiplicative submonoid of $\mathbb{Z}$ not containing 0. But we can't tell whether or not 2 is a unit in R.

2.7 DEFINITION. A **Bézout domain** is a discrete domain such that for each pair of elements a,b there is a pair s,t such that $sa + tb$ divides a and b. A **principal ideal domain** is a Bézout domain which satisfies the divisor chain condition.

Observe that if $sa + tb$ divides a and b, then $sa + tb = GCD(a,b)$. A principal ideal domain is **Noetherian**, that is, given a sequence $I_1 \subseteq I_2 \subseteq \cdots$ of finitely generated ideals, there is n such that $I_n = I_{n+1}$.

2.8 THEOREM. *If R is a discrete domain, then the following are equivalent.*

(i) *R is a Bezout domain.*

(ii) *Every finitely generated ideal in R is principal.*

(iii) *Every finitely generated ideal in R is principal, and R is a GCD-domain.* □

2.9 COROLLARY. *If K is a discrete field, then $K[X]$ is a bounded principal ideal domain.*

PROOF. Combine Theorem 2.8 and Theorem II.4.7. □

An example of a Bézout domain that is not a principal ideal domain is constructed as follows. Let k be a discrete field, and M the monoid of nonnegative rational numbers under addition. Let R be the monoid ring $k^{(M)}$; the elements of R may be thought of as polynomials in X with coefficients in k and exponents in M. If $m_1, m_2, \ldots$ is a strictly decreasing sequence of positive rationals, then $(X^{m_1}), (X^{m_2}), \ldots$ is a strictly increasing sequence of finitely generated ideals of R. On the other hand, given a finite number of elements of R, there exists $m \in M$ such that they are all contained in $k[X^m]$, so R is a Bézout domain.

If k is a discrete field, then $k[X]$ is a principal ideal domain, but $k[X]$ but not necessarily a UFD — see Example 2.2. A related Brouwerian example of a principal ideal domain which is not a UFD is constructed as follows. Let a be a binary sequence, and let $R = \cup_n \mathbb{Z}[ia_n]$. Then R is a principal ideal domain, but we cannot factor the element 2 into irreducible factors. Note that R is a bounded GCD-domain. In general a principal ideal domain is a quasi-UFD.

In Section 4 we show that $\mathbb{Q}[X]$ is a UFD, that is, $\mathbb{Q}$ is factorial. In Chapter 6 we shall show that a discrete field k is factorial if and only if it has a root test, that is, each polynomial in $k[X]$ has a root in k or has no root in k. That will provide us with more examples of discrete fields k such that $k[X]$ is factorial.

EXERCISES

1. Let R be a discrete domain. Show that the following are equivalent.
   - (i) R is a UFD.
   - (ii) Each nonzero element of R is a unit, or is a product of primes.
   - (iii) R is a bounded GCD-domain, and each nonzero element is either a unit, or is irreducible, or has a proper factor.
   - (iv) R is a quasi-UFD, and each nonzero element is either a unit, or is irreducible, or has a proper factor.

   Remark: A proof of (iv) implies (i) requires the use of the axiom of dependent choice.

2. **Eisenstein criterion.** Let R be a discrete domain, and let $f = a_0 + \cdots + a_n X^n \in R[X]$ such that any common factor of the a_i is a unit. Let $p \in R$ be a prime element such that p does not divide a_n, and p^2 does not divide a_0, but $p | a_i$ for $i < n$. Prove that f is irreducible in $R[X]$.

3. Let R be a principal ideal domain, and let S be a multiplicative submonoid of R not containing 0. Show that $S^{-1}R$ is a principal ideal domain.

4. Show that any Bézout domain is coherent. Show that if a Bézout domain has recognizable units, then it has detachable ideals.

5. Show that if R is a Bézout domain, then each finitely generated submodule of R^n is free of rank at most n.

3. DEDEKIND-HASSE RINGS AND EUCLIDEAN DOMAINS

Discrete domains often admit maps to the nonnegative integers that can be used to study questions of divisibility. Examples are the absolute value function for the integers, the degree function for polynomials over a discrete field, and norms on rings of algebraic integers. If the map fits in with a division algorithm, or satisfies the Dedekind-Hasse condition, then the domain under consideration is a principal ideal domain.

3.1 DEFINITION. Let ν be a function from the nonzero elements of a discrete domain R to the nonnegative integers. Then ν is a

(i) **pseudonorm** if whenever a and b are nonzero elements of R such that $b|a$, then either $a \sim b$ or there exists $b' \sim b$ such that $\nu(b') < \nu(a)$.

(ii) **Dedekind-Hasse map** if for any nonzero elements a and b of R, either $a|b$, or there exists nonzero r in (a,b) such that $\nu(r) < \nu(a)$.

(iii) **Euclidean map** if for any nonzero elements a and b of R either $a|b$, or there exists nonzero r in R such that $a|(b - r)$ and $\nu(r) < \nu(a)$.

We say that ν is **multiplicative** if $\nu(ab) = \nu(a)\nu(b) > 0$ for all nonzero a and b in R. A multiplicative pseudonorm is called a **multiplicative norm.**

The notion of a pseudonorm is a technical convenience. The notion of a norm should probably lie somewhere between a pseudonorm and a multiplicative norm, possibly at one end or the other.

3.2 THEOREM. *Any Euclidean map is a Dedekind-Hasse map. Any Dedekind-Hasse map is a pseudonorm. If ν is a pseudonorm, then each nonzero element a is bounded by $\nu(a)$. If ν is a multiplicative norm, then a is a unit if and only if $\nu(a) = 1$.*

PROOF. The first claim is obviously true. To prove the second, suppose that ν is a Dedekind-Hasse map and b divides a. We proceed by induction on $\nu(a)$. Either $a|b$ or there exists nonzero r in $(a,b) = (b)$ such that $\nu(r) < \nu(a)$. In the latter case, by induction, either $r \sim b$, so we can pick $b' = r$, or there exists $b' \sim b$ such that $\nu(b') < \nu(r) < \nu(a)$.

To prove the third claim suppose that ν is a pseudonorm and a is a nonzero element. We shall show that a is bounded by $n = \nu(a)$ by induction on n. Suppose $a = a_0 b$ where $b = a_1 \cdots a_n$ if $n > 0$, and $b = 1$ if $n = 0$. Then either $a|b$, so a_0 is a unit, or there is $b' \sim b$ such that $\nu(b') < \nu(a)$. By induction b', and hence b, is bounded by $n - 1$, so a_i is a unit for some i.

Finally suppose that ν is a multiplicative norm. As $\nu(1) = \nu(1)\nu(1) > 0$, we have $\nu(1) = 1$. If a is a unit, then $ac = 1$ for some c, so $\nu(a)\nu(c) = 1$ whence $\nu(a) = 1$. Conversely, suppose $\nu(a) = 1$. As $1|a$, and ν is a pseudonorm, either $a \sim 1$, so a is a unit, or there exists b' such that $\upsilon(b') < \upsilon(a) = 1$, which is impossible. □

The Dedekind-Hasse condition provides a criterion for a ring to be a bounded principal ideal domain.

3.3 THEOREM. *A discrete domain with a Dedekind-Hasse map is a bounded principal ideal domain.*

PROOF. Suppose υ is a Dedekind-Hasse map for R. As R is bounded, by (3.2), it suffices to show that R is a Bézout domain. Given nonzero $a,b \in R$, and a nonzero element c in the ideal (a,b), we shall show by induction on $\upsilon(c)$ that there is a common divisor of a and b in (a,b). As υ is a Dedekind-Hasse map, either $c|a$ or there exists nonzero r in (c,a) such that $\upsilon(r) < \upsilon(c)$. Similarly either $c|b$ or there exists nonzero r in (c,b) such that $\upsilon(r) < \upsilon(c)$. Hence either c is the desired common divisor of a and b, or there exists nonzero r in (a,b) such that $\upsilon(r) < \upsilon(c)$, and we are done by induction. □

3.4 EXAMPLE. A multiplicative Dedekind-Hasse norm that is not a Euclidean map. Let $R = \mathbb{Z}[(1+\sqrt{-19})/2]$, which is easily seen to be a free $\mathbb{Z}$-module with basis 1 and $(1+\sqrt{-19})/2$. The function $N(a+b\sqrt{-19}) = a^2 + 19b^2$ is a multiplicative function on $\mathbb{Q}(\sqrt{-19})$ which restricts to a norm on R. If α and β are nonzero elements of R, we shall show how to construct θ in R such that

$$N(\beta/\alpha - \theta) < 1 \quad \text{or} \quad 0 < N(2\beta/\alpha - \theta) < 1.$$

This will show that N is a Dedekind-Hasse norm on R. Writing

$$\beta/\alpha - \theta = a + b\sqrt{-19}$$

we can easily find $\theta \in R$ such that $|b| \leq 1/4$ and $|a| \leq 1/2$. If $|b| \leq 3/16$, then $N(\beta/\alpha - \theta) \leq 235/256 < 1$, and we are done. If $|b| > 3/16$, then we can find $\theta' \in R$ such that

$$2\beta/\alpha - \theta = a' + b'\sqrt{-19}$$

with $|b'| \leq 1/8$ and $|a'| \leq 1/2$, so $N(2\beta/\alpha - \theta') < 1$. The only problem is that α might divide 2β but not β. However it is easy to show that 2 divides $\gamma \in R$ if and only if $N(\gamma)$ is even, so 2 is a prime in R. Therefore if $\alpha\delta = 2\beta$, then either $2|\delta$, in which case $\alpha|\beta$, or $2|\alpha$. But if 2 is a common factor of α and β, then we are done by induction because our condition depends only on β/α.

On the other hand, N is not a Euclidean norm because there is no θ in R such that $N(\beta/\alpha - \theta) < 1$ if $\beta = (1+\sqrt{-19})/2$ and $\alpha = 2$. In fact, R does not admit a Euclidean map (see Exercise 11). □

Classically any bounded domain R admits a pseudonorm: set $\upsilon(x)$ equal to the least n such that x is bounded by n. If R is a principal ideal domain, then υ is a multiplicative Dedekind-Hasse norm; constructively we must require more. The following theorem gives the construction of such a norm if the principal ideal domain is also a UFD.

3.5 THEOREM. *Any UFD admits a multiplicative norm. Let υ be a pseudonorm on a discrete domain R. If R is a Bezout domain, then υ is a Dedekind-Hasse map.*

PROOF. Let R be a UFD. For nonzero a in R define $\upsilon(a) = 2^n$ where n is the number of primes, including multiplicities, in a prime factorization of a. Clearly υ is a multiplicative norm.

Suppose R is a Bézout domain and $a,b \in R$. There exists d in R such that $(a,b) = (d)$. If $a \neq 0$, then either $a \sim d$, so a divides b, or there exists $d' \sim d$ such that $\upsilon(d') < \upsilon(a)$. Therefore υ is a Dedekind-Hasse map. □

Absolute value is a multiplicative Euclidean norm on the ring of integers. If F is a discrete field, then the degree function is a Euclidean map on the ring $F[X]$. Note that the degree function is not multiplicative, but the Euclidean norm $2^{\deg f}$ is.

A multiplicative Euclidean norm on the ring of algebraic integers $\mathbb{Z}[\sqrt{2}]$ is provided by $\upsilon(a + b\sqrt{2}) = |a^2 - 2b^2| = |(a + b\sqrt{2})(a - b\sqrt{2})|$. We easily see that υ is multiplicative. To show that υ is Euclidean, let $a + b\sqrt{2}$ and $c + d\sqrt{2} \neq 0$ be elements of $\mathbb{Z}[\sqrt{2}]$. As $\sqrt{2}$ is irrational, we can find rational numbers p and q such that $(a + b\sqrt{2})/(c + d\sqrt{2}) = p + q\sqrt{2}$, and integers m and n with $|p - m| \leq 1/2$ and $|q - n| \leq 1/2$. Then

$$(p + q\sqrt{2})(c + d\sqrt{2}) = (a + b\sqrt{2}) = (m + n\sqrt{2})(c + d\sqrt{2}) + (s + t\sqrt{2})$$

so $\upsilon(s + t\sqrt{2}) = |(p - m)^2 - 2(q - n)^2|\upsilon(c + d\sqrt{2}) \leq \upsilon(c + d\sqrt{2})/2$. Thus υ is Euclidean. □

3.6 EXAMPLE. A principal ideal domain that is not bounded. Let α be a binary sequence with at most one 1, and let

$$R = \mathbb{Q}[X, a_1Y_1, a_2Y_2, \ldots]/(a_1(X-Y_1), a_2(X-Y_2^2), \ldots).$$

Then R is a principal ideal domain, but we can't find a bound for X.

3.7 THEOREM. *Let R be a discrete domain and v a Dedekind-Hasse map on R such that $v(a) = v(b)$ if $a \sim b$. Let S be a multiplicative subset of R not containing 0. Then v extends to a Dedekind-Hasse map on $S^{-1}R$ that is Euclidean if v is.*

PROOF. Set $v(r/s) = v(r/GCD(r,s))$. This is well defined because $v(a) = v(b)$ if $a \sim b$. It is routine to check that v has the desired properties. □

3.8 EXAMPLE. A domain with a Euclidean map that does not have recognizable units. Let R be the ring of integers, and S the multiplicative subset generated by $\{2a_n : n \in \mathbb{N}\}$ for some binary sequence a. Extend the absolute value function on R to a Euclidean map on $S^{-1}R$ by (3.7).

EXERCISES

1. Call a bounded discrete domain R a **DH-ring** if for all nonzero a and b in R, with a bounded by n, either $a|b$ or there exists c in (a,b) that is bounded by $n-1$. Show that a DH-ring is a principal ideal domain.
2. Call a bounded domain R **strictly bounded** if whenever a divides an element b that is bounded by n, then either b divides a, or a is bounded by $n-1$. Show that a bounded domain with recognizable units is strictly bounded. Show that a DH-ring (see Exercise 1) is strictly bounded. Show that a strictly bounded principal ideal domain is a DH-ring.
3. Let a be a binary sequence, and let $R = \mathbb{Q}[Y^2, a_1Y, a_1/(Y^2-1), a_2Y, a_2/(Y^2-1), \ldots]$. Show that R is a Brouwerian example of a bounded principal ideal domain that is not strictly bounded (see Exercise 2) by considering the elements $a = Y^2$ and $b = Y^2(Y^2-1)$. Show that R admits a Euclidean map, but not a multiplicative one.
4. Show that the ring of Example 3.6 is a principal ideal domain.
5. Let R be a Euclidean domain. Show that for each $x \neq 0$ there is a unit u such that $v(x) \geq v(u)$.

6. Show that the ring $\mathbb{Z}[i]$ of Gaussian integers has a multiplicative Euclidean norm.

7. Show that any pseudonorm on a principal ideal domain is a Dedekind-Hasse map.

8. **A peculiar omniscience principle.** Consider the subsets C_n of the set $2^{\mathbb{N}}$ of binary sequences defined inductively as follows.

$$C_0 = \{0\}.$$
$$C_{n+1} = \{a : \text{if } a_i = 1 \text{ then either } a_j = 1 \text{ for some } j > i, \text{ or } a_{i+1}, a_{i+2}, \ldots \text{ is in } C_n\}.$$

Show that $a \in C_1$ if $\{m : a_m = 1\}$ is either finite or infinite. Show that $a \in C_n$ if $\{m : a_m = 1\}$ is bounded by n. Show that LPO is equivalent to $2^{\mathbb{N}} = C_1$. What do you make of the omniscience principle $2^{\mathbb{N}} = \cup_n C_n$?

9. **A bounded PID that does not admit a pseudonorm.** Let k be the two element field, let a be a binary sequence, and define $\varphi_n : k(X_n) \to k(X_{n+1})$ by

$$\varphi_n(X_n) = a_n(X_{n+1}^2 + n) + X_{n+1}.$$

Let R_n be a subring of $k(X_n)$ such that $R_1 = k[X_1]$, and $R_{n+1} = \varphi_n(R_n)$, if $a_n = 0$, and $R_{n+1} = S^{-1}\varphi(R_n)$, where S is the multiplicative set generated by $X_{n+1}^2 + X_{n+1} + 1$, if $a_n = 1$. Show that the direct limit R of the rings R_n is a bounded PID. Show that the omniscience principle of Exercise 8 would hold if R admitted a pseudonorm.

10. Show that $\mathbb{Z}[(1+\sqrt{-19})/2]$ is the integral closure of $\mathbb{Z}$ in $\mathbb{Q}[\sqrt{-19}]$.

11. Show that the units of R in Example 3.4 are ± 1. Let $\alpha = 2$ and $\beta = (1+\sqrt{-19})/2$. Using N, show that α, $\alpha+1$, $\alpha-1$, β, $\beta+1$, and $\beta-1$ are pairwise relatively prime. Suppose R admits a Euclidean map υ, and γ is a nonunit of R. By dividing γ into α and β, show that there is a nonunit γ' of R such that $\upsilon(\gamma') < \upsilon(\gamma)$.

12. Construct a Brouwerian example of a principal ideal domain with a Dedekind-Hasse map that does not have recognizable units or a Euclidean map. (Look at $\mathbb{Z}[(1+\sqrt{-19})/2] \subseteq \mathbb{Q}[\sqrt{-19}]$)

4. POLYNOMIAL RINGS

In this section we consider properties of a discrete domain R that are inherited by the polynomial ring $R[X]$.

4.1 DEFINITION. Let R be a GCD-domain, and let $f \in R[X]$. The GCD of the coefficients of f is called the **content** of f and is denoted by **$cont(f)$**. If $cont(f) = 1$, then f is said to be **primitive**.

4.2 LEMMA. *Let R be a GCD-domain with field of quotients K. If $f \in K[X]$, then we can find $c \in K$ and primitive $g \in R[X]$ such that $f = cg$. If $f = c'g'$ for $c' \in K$ and primitive $g' \in R[X]$, then $c = uc'$ for some unit $u \in R$.*

PROOF. Write $f = c_0g_0$ for some $c_0 \in K$ and $g_0 \in R[X]$. Let $g = g_0/cont(g_0)$ and $c = c_0 \cdot cont(g_0)$, so $f = cg$ and g is primitive. Supppose $f = c'g'$ where $c' \in K$ and g' is a primitive polynomial in $R[X]$. Choose nonzero $d \in R$ such that dc and dc' are in R. Then dc and dc' are each the content of df, so $dc = udc'$ for some unit $u \in R$. Thus $c = uc'$. □

4.3 LEMMA (Gauss's lemma). *Let R be a GCD-domain, and let $f,g \in R[X]$. Then $cont(fg) = cont(f)cont(g)$.*

PROOF. Let $m = deg\ f$ and $n = deg\ g$. We proceed by induction on $m + n$. As $cont(ah) = a \cdot cont(h)$ for constants a and polynomials h, we may divide f and g by their contents and prove the claim for primitive f and g. Let $c = cont(fg)$ and $d = GCD(c, f_m)$, where f_m is the leading coefficient of f. Then $d \mid (f - f_mX^m)g$. If $f = f_mX^m$ the lemma is clear. Otherwise, by induction, $d \mid cont(f - f_mX^m)cont(g)$. Since g is primitive we get $d \mid (f - f_mX^m)$, so $d \mid f$. As f is primitive, we get $d = GCD(c, f_m) = 1$. In the same way we prove $GCD(c, g_n) = 1$. Thus by Lemma 1.1.iii, $c = GCD(c, f_mg_n) = 1$, so fg is primitive. □

4.4 COROLLARY. *Let R be a GCD-domain with field of quotients k. Let f and g be polynomials in $R[X]$. Then f divides g in $R[X]$ if and only if f divides g in $k[X]$ and $cont(f)$ divides $cont(g)$.*

PROOF. The "only if" is immediate from Gauss's lemma. To prove the "if" we may assume that f is primitive. By (4.2) we can write $g = ahf$ where h is a primitive polynomial in $R[X]$ and $a \in k$. By Gauss's lemma fh is a primitive polynomial, so $a = cont(g) \in R$ by (4.2). Thus f divides g

in $R[X]$. □

4.5 THEOREM. *Let R be a GCD-domain with quotient field k. Let f and g be polynomials in $k[X]$, each of which has some coefficient that is a unit of R. If fg is a primitive polynomial of $R[X]$, then f and g are in $R[X]$.*

PROOF. There are $a,b \in k$ and primitive $f_1 g_1 \in R[X]$ such that $f = af_1$ and $g = bg_1$. Since f and g have units of R among their coefficients, a^{-1} and b^{-1} are in R. By Gauss's lemma $f_1 g_1 = a^{-1}b^{-1}fg$ is a primitive polynomial. So $a^{-1}b^{-1}$ is a unit of R, so a and b are in R and thus f and g are in $R[X]$. □

4.6 THEOREM. *Let R be a GCD-domain with quotient field k. Let f and g be polynomials in $k[X]$ such that $fg \in R[X]$. Then there is $b \in k$ such that bf and g/b are in $R[X]$.*

PROOF. There are $a,b \in k$ and primitive $f_1, g_1 \in R[X]$ such that $f = af_1$ and $g = bg_1$. By Gauss's lemma $f_1 g_1$ is primitive so $ab = \text{cont}(fg) \in R$ by (4.1) whence $bf = abf_1$ and $g/b = g_1$ are in $R[X]$. □

4.7 THEOREM. *Let R be a discrete domain.*

(i) *If R is a GCD-domain, then so is $R[X]$.*

(ii) *If R is bounded, then so is $R[X]$.*

(iii) *If R has recognizable units, then so does $R[X]$.*

(iv) *If R has decidable divisibility, then so does $R[X]$.*

(v) *If R satisfies the divisor chain condition, then so does $R[X]$.*

(vi) *If R is a quasi-UFD, then so is $R[X]$.*

PROOF. Let k be the field of quotients of R. To prove (i) let f and g be in $R[X]$, and let h be primitive in $R[X]$ such that h is a GCD of f and g in $k[X]$. Let $d = GCD(cont(f), cont(g))$. We shall show that $dh = GCD(f,g)$.

By (4.4) dh divides f and g in $R[X]$. Conversely, suppose q divides f and g in $R[X]$. Then q divides h in $k[X]$, and $cont(q)$ divides $cont(f)$ and $cont(g)$, hence divides d. Therefore q divides dh by (4.4) and we have shown that dh is a GCD of f and g.

To prove (ii) let $f \in R[X]$ have degree n, and let a be the leading coefficient of f. If a is bounded by m, then f is bounded by $m + n$.

Claim (iii) is trivial, as the ring $R[X]$ has the same units as R.

To prove (iv) we show by induction on $n = deg\ g$ that we can decide whether f divides g. Let a be the leading coefficient of f, and b the leading coefficient of g. We may assume that $deg\ f \leq deg\ g$. If a does not divide b, then f does not divide g. If $a|b$, then there are polynomials q and h such that $g = qf + h$ and $deg\ h \leq n - 1$. Then $f|g$ if and only if $f|h$. By induction we can decide whether f divides h, so we can decide whether f divides g.

Suppose R satisfies the divisor chain condition. Let $(f_1) \subseteq (f_2) \subseteq \cdots$ be an ascending chain of principal ideals in $R[X]$. We shall construct m such that $(f_m) = (f_{m+1})$ by induction on $deg\ f_1$. Let a_i be the leading coefficient of f_i. Then $a_{i+1}|a_i$ for each positive i, so there is m such that $a_m|a_{m+1}$. If $deg\ f_m = deg\ f_{m+1}$, then $f_m|f_{m+1}$ and we are done. Otherwise $deg\ f_{m+1} < deg\ f_m \leq deg\ f_1$, and we consider the sequence of f's starting at f_{m+1}. By induction we find the desired pause in this sequence.

Claim (vi) follows from (v) and (i). □

4.8 THEOREM (Kronecker 1). *If R is an infinite UFD with finitely many units, then so is $R[X]$. Thus R is factorial.*

PROOF. Let $f \in R[X]$ be a polynomial of degree n. It suffices to construct a finite collection of polynomials that contains all divisors of f of degree at most $n/2$. Let $a_0,\ldots,a_m$ be distinct elements of R, where $n \leq 2m$. Since R is a UFD with finitely many units, each nonzero $f(a_i)$ has a finite set of divisors. If some $f(a_i) = 0$, then $f = (X - a_i)g$ and we are done by induction on n. So we may assume that $f(a_i) \neq 0$ for each i. Note that if $g|f$, then $g(a_i)|f(a_i)$ for each i. There are only finitely many sequences $b_0,\ldots,b_m$ such that b_i divides $f(a_i)$ for each i. By the unique interpolation theorem (II.5.5) there is, for each sequence $b_0,\ldots,b_m$, a unique polynomial $g \in k[X]$ of degree at most m such that $g(a_i) = b_i$ for all i, where k is the field of quotients of R. The collection of polynomials g forms a finite set of polynomials in $k[X]$ which contains all divisors of f in $R[X]$ of degree at most $n/2$. Then a polynomial g of degree at most $n/2$ is a factor of f if and only if $g \in R[X]$ and $f/g \in R[X]$, and this is decidable since $R[X]$ is detachable from $k[X]$. □

Kronecker 1 shows that $\mathbb{Z}[X_1,\ldots,X_n]$ is factorial for each n. Using Corollary 2.6 we see that $\mathbb{Q}[X_1,\ldots,X_n]$ is also factorial. If k is a finite field, then $k[X_1,\ldots,X_n]$ is factorial, since each $f \in k[X_1,\ldots,X_n]$ has only finitely many divisors. However, not all UFD's are factorial, as (2.2) shows.

Kronecker 1 requires that the ring R be infinite. But finite UFD's are certainly factorial. In Theorem VI.6.8 we shall eliminate the requirement that R be either finite or infinite.

The algebraic closure k of $\mathbb{Q}$ — see (VI.2.5) — as well as the polynomial ring $k[X]$ is a UFD, so k is factorial. But Kronecker 1 is not enough to show that $k[X]$ is factorial, as was the case for $k = \mathbb{Q}$. Another trick of Kronecker's shows that $R[X,Y]$ is a UFD if $R[X]$ is.

4.9 THEOREM (Kronecker 2). *If R is a factorial domain, then so is $R[X]$.*

PROOF. For $m > 0$, let $\varphi_m : R[X,Y] \to R[X]$ be the ring homomorphism that is the identity on $R[X]$ and takes Y to X^m. Let $\psi : R[X] \to R[X,Y]$ be the R-module homomorphism that takes X^n to $Y^q X^r$ where $n = qm + r$, and $0 \leq r < m$. Let $R[X,Y]_m$ be those polynomials in $R[X]$ of X-degree less than m. Then

$$\varphi_m\psi \text{ is the identity on } R[X]$$
$$\psi\varphi_m \text{ is the identity on } R[X,Y]_m$$
$$R[X,Y] \text{ is closed under taking factors.}$$

To factor a polynomial f in $R[X,Y]$, of degree less than m, look at the finite number of factorizations (up to units) $\varphi_m(f) = ab$. Check to see if $\psi(a)\psi(b) = f$. Any factorization of f must have this form. □

EXERCISES

1. Show that $X^4 + 1$ is irreducible over $\mathbb{Q}[X]$ using the techniques of Kronecker 1 and (2.6). Use Kronecker 1 to factor $X^4 + 4$.
2. Factor $X^4 + 4Y^4$ into primes over the ring $\mathbb{Z}[X,Y]$.
3. Use Kronecker 2 on the polynomial $X^2 + Y$ to see that even though the image $\varphi_m(f)$ factors in $k[X]$, the original polynomial need not factor.

NOTES

The divisor chain condition is defined in terms of ascending chains of principal ideals $I_1 \subseteq I_2 \subseteq \cdots$ instead of descending chains of elements

$a_1, a_2, \ldots$ with $a_i | a_{i-1}$. The two versions are equivalent in the presence of the axiom of dependent choices. The version with ideals allows us to derive — in this chapter and in Chapter VI — the basic properties of principal ideal domains without the use of the axiom of dependent choices.

In quasi-UFD's, principal ideal domains and Euclidean domains we need not be able to factor nonzero elements into irreducible factors. Instead we can use the quasi-factorization theorem to write nonzero elements $a_1, \ldots, a_n$ of a quasi-UFD as products of pairwise relatively prime elements $p_1, \ldots, p_m$.

Kronecker 1 and 2 are found in [Kronecker 1882]. These results provide algorithms for factoring polynomials in several indeterminates over the integers as products of primes. We are not concerned here with the efficiency of these algorithms.

Chapter V. Principal Ideal Domains

1. DIAGONALIZING MATRICES

The theory of modules over a principal ideal domain is closely related to the theory of vector spaces over a field and is almost identical to the theory of abelian groups, which are modules over the integers. The analogue of a finite-dimensional vector space is a finitely presented module over a principal ideal domain. A finitely presented module is given by matrix. In this section we prove some basic facts about matrices over a principal ideal domain.

An $m \times n$ matrix $A = (a_{ij})$ is **diagonal** if $a_{ij} = 0$ whenever $i \neq j$. Two $m \times n$ matrices A and B are **equivalent** if there is an invertible $m \times m$ matrix C, and an invertible $n \times n$ matrix D, such that $A = CBD$.

1.1 LEMMA. *Each matrix over a principal ideal domain is equivalent to a diagonal matrix.*

PROOF. The key is the construction of certain invertible 2×2 matrices. If $sa + tb = d \neq 0$ is the GCD of a and b, then for all u,v there are w,x such that

$$\begin{bmatrix} a & b \\ u & v \end{bmatrix} \cdot \begin{bmatrix} s & -b/d \\ t & a/d \end{bmatrix} = \begin{bmatrix} d & 0 \\ w & x \end{bmatrix}.$$

Moreover, the right hand factor is invertible since its determinant is 1. Similarly, if a and b are in a column, we multiply on the left as follows.

$$\begin{bmatrix} s & t \\ -b/d & a/d \end{bmatrix} \cdot \begin{bmatrix} a & u \\ b & v \end{bmatrix} = \begin{bmatrix} d & w \\ 0 & x \end{bmatrix}.$$

Thus if a and b are entries in the same row (column) of a matrix B, we can multiply B on the right (left) by an invertible matrix leaving d in the position occupied by a, and 0 in the position occupied by b, and fixing the entries not in the rows or columns of a or b.

We shall operate on a given matrix A by left and right multiplication by invertible matrices to obtain a diagonal matrix. If $A = 0$ we are done. Otherwise, by row and column interchanges, we can bring a nonzero element a_1 to the upper left hand corner. By a sequence of right multiplications by invertible matrices we can replace a_1 by a_2, the GCD of all elements in the first row, leaving zeros in the remaining places in the first row. Similarly, by left multiplications, we can replace a_2 by a_3, the GCD of all elements now in the first column, leaving zeros in the rest of the first column. Continue by replacing a_3 by a_4, the GCD of all elements now in the first row, and so on. In this manner we generate a sequence $(a_1),(a_2),\ldots$ of principal ideals such that $a_{i+1}|a_i$ for each i. Hence, for some n, we have $a_n|a_{n+1}$. But that means that a_n is a GCD of the elements in the first row (or column), while the remaining elements in the first column (or row) are zero. Hence, by elementary row or column operations, we can clear both the first row and the first column, making all elements in the first row and column zero, except for the corner. By induction on the size of the matrix, we can diagonalize A. □

A matrix $A = (a_{ij})$ is in **Smith normal form** if it is diagonal and $a_{ii}|a_{i+1,i+1}$ for each i.

1.2 THEOREM. *Each matrix over a principal ideal domain is equivalent to a matrix in Smith normal form.*

PROOF. By Lemma 1.1 we may assume that we are given a diagonal matrix. Let a,b be diagonal elements and $d = sa + tb$ be the GCD of a and b. Then

$$\begin{bmatrix} s & t \\ -b/d & a/d \end{bmatrix} \cdot \begin{bmatrix} a & 0 \\ 0 & b \end{bmatrix} \cdot \begin{bmatrix} 1 & -tb/d \\ 1 & sa/d \end{bmatrix} = \begin{bmatrix} d & 0 \\ 0 & ab/d \end{bmatrix}.$$

Repeated application of this observation allows us to convert our diagonal matrix to a diagonal matrix with the property that the corner element divides all the remaining elements. Then, by induction on the size of A, we obtain a matrix in Smith normal form. □

We want to show that each matrix is equivalent to an essentially unique matrix in Smith normal form. Given a matrix A, define $\Delta_i(A)$ to be the ideal generated by the determinants of all $i \times i$ submatrices of A.

1.3 LEMMA. *Let* A *and* B *be equivalent* $m \times n$ *matrices over a GCD-domain* R. *Then* $\Delta_i(A) = \Delta_i(B)$ *for all* i.

PROOF. It suffices to show that if C is an invertible $m \times m$ matrix, then $\Delta_i(CA) = \Delta_i(A)$. The rows of CA are linear combinations of the rows of A. Hence the determinants of $i \times i$ submatrices of CA are linear combinations of the determinants of $i \times i$ submatrices of A. So $\Delta_i(A) \supseteq \Delta_i(CA)$. Since C is invertible, we also have $\Delta_i(CA) \supseteq$

$\Delta_i(C^{-1}CA) = \Delta_i(A)$. So $\Delta_i(CA) = \Delta_i(A)$. □

1.4 THEOREM. *Two $m \times n$ matrices in Smith normal form over a GCD-domain are equivalent if and only if corresponding elements are associates.*

PROOF. Let $D = (d_{ij})$ be a matrix in Smith normal form. We easily verify that $\Delta_1(D) = (d_{11})$, and $\Delta_i(D)\cdot(d_{i+1,i+1}) = \Delta_{i+1}(D)$ for each $i \leq m-1$. So the diagonal elements of D are determined, up to a unit, by the ideals $\Delta_i(D)$. By Lemma 1.3 this implies that if two matrices in Smith normal form are equivalent, then their elements are associates. □

EXERCISES

1. Find a matrix in Smith normal form, over $\mathbb{Z}$, that is equivalent to the matrix
$$\begin{matrix} 1 & 2 & 3 \\ 4 & 5 & 6 \\ 7 & 8 & 9 \end{matrix}.$$
2. A **valuation ring** is a commutative ring R such that for any two elements a and b in R, either $(a) \subseteq (b)$ or $(b) \subseteq (a)$. Show that each matrix over a valuation ring is equivalent to a matrix in Smith normal form.
3. Show that a square matrix over $\mathbb{Z}$ is invertible if and only if it is a product of elementary matrices.
4. Show that a square matrix over $\mathbb{Z}$ has determinant 1 if and only if it is a product of elementary matrices corresponding to elementary operations of type (iii).

   Hint: $\begin{pmatrix} -1 & 0 \\ 0 & -1 \end{pmatrix} = \begin{pmatrix} 1 & -1 \\ 0 & 1 \end{pmatrix}\cdot\begin{pmatrix} 1 & 0 \\ 2 & 1 \end{pmatrix}\cdot\begin{pmatrix} 1 & -1 \\ 0 & 1 \end{pmatrix}\cdot\begin{pmatrix} 1 & 0 \\ 2 & 1 \end{pmatrix}$

2. FINITELY PRESENTED MODULES

Let M be a module over a commutative ring R. A **finite presentation** of M is a triple $(M, (x_1,\ldots,x_n), A)$ where $x_1,\ldots,x_n$ is a finite set of generators of M, and A is an $m \times n$ matrix of elements of R whose rows generate the module of relations among the x's. Thus for each element $(a_1,\ldots,a_n) \in R^n$ we have $a_1x_1 + \cdots + a_nx_n = 0$ if and only if $(a_1,\ldots,a_n) = vA$ for some $v \in R^m$. We may identify a finitely presented module with its finite presentation.

The matrix A contains all the information about the structure of the module M. Given M, the matrix A depends on the choice of generators

$x_1,\ldots,x_n$ and on the choice of generators for the module of relations. If $M \cong R/(r_1) \oplus \cdots \oplus R/(r_n)$, then we can get a diagonal matrix A for M; conversely, if we can get a diagonal matrix A for M, then we have exhibited M as a direct sum of cyclic modules. Thus it behooves us to examine how to pass from one matrix for M to another.

Let $(M, (x_1,\ldots,x_n), A)$ be a finitely presented R-module. Then $(M, (x_1,\ldots,x_n), B)$ is a finitely presented R-module if and only if the rows of B generate the same submodule of R^n as the rows of A; this happens if $B = FA$ for some invertible $m \times m$ matrix F. What happens to the matrix A of relations of a finitely presented module $(M, (x_1,\ldots,x_n), A)$ when we change the generators $x_1,\ldots,x_n$?

2.1 THEOREM. *Let $(M, (x_1,\ldots,x_n), A)$ be a finitely presented module over a commutative ring R, and let E be an $n \times n$ invertible matrix over R. Then $(M, (x_1,\ldots,x_n), A) = (M, (x_1,\ldots,x_n)E^t, AE^{-1})$.*

PROOF. Let $(y_1,\ldots,y_n)^t = E(x_1,\ldots,x_n)^t$. Clearly $y_1,\ldots,y_n$ generate M. Moreover the following are equivalent.

$$(r_1,\ldots,r_n)E^{-1}(y_1,\ldots,y_n)^t = (r_1,\ldots,r_n)(x_1,\ldots,x_n)^t = 0$$

$$(r_1,\ldots,r_n) = (s_1,\ldots,s_m)A \quad \text{for some} \quad (s_1,\ldots,s_m)$$

$$(r_1,\ldots,r_n)E^{-1} = (s_1,\ldots,s_m)(AE^{-1}).$$

Hence the module of relations among $y_1,\ldots,y_n$ is generated by the rows of AE^{-1}. □

2.2 COROLLARY. *Let A be an $m \times n$ matrix over a commutative ring R, and let $(M, (x_1,\ldots,x_n), A)$ be a finitely presented R-module. Let E be an $n \times n$ invertible matrix over R, and F an $m \times m$ invertible matrix over R. Then $(M, (x_1,\ldots,x_n), A) = (M, (x_1,\ldots,x_n)E^t, FAE^{-1})$.*

PROOF. As the rows of FAE^{-1} have the same span as those of AE^{-1}, the corollary is an immediate consequence of 2.1. □

2.3 THEOREM (Structure Theorem). *Let M be a finitely presented module over a principal ideal domain R. Then there exist principal ideals $I_1 \supseteq I_2 \supseteq \cdots \supseteq I_n$ such that M is isomorphic to the direct sum $R/I_1 \oplus R/I_2 \oplus \cdots \oplus R/I_n$.*

PROOF. Let $(M, (x_1,\ldots,x_n), A)$ be a finite presentation of M. By Theorem 1.4 there exist invertible matrices E and F such that $D = FAE^{-1}$ is a diagonal matrix in Smith normal form. By Corollary 2.2 we have

$$(M,\ (x_1,\ldots,x_n),\ A) = (M,\ (x_1,\ldots,x_n)E^t,\ D).$$

If $(y_1,\ldots,y_n) = (x_1,\ldots,x_n)E^t$, then $M = Ry_1 \oplus \cdots \oplus Ry_n$, and $Ry_i \cong R/I_i$ where $I_i = (d_{ii})$. □

The decomposition in Theorem 2.3 is essentially unique over an arbitrary commutative ring.

2.4 THEOREM. *Let R be a commutative ring, $m \leq n$ positive integers, and $I_1 \supseteq I_2 \supseteq \cdots \supseteq I_m$ and $J_1 \supseteq J_2 \supseteq \cdots \supseteq J_n$ ideals of R. Suppose M is an R-module that is isomorphic to $\Sigma_{i=1}^m R/I_i$ and to $\Sigma_{j=1}^n R/J_j$. Then*

(a) $J_1 = J_2 = \cdots = J_{n-m} = R$.

(b) $I_i = J_{n-m+i}$ *for* $i = 1,\ldots,m$.

PROOF. To prove (a) it suffices to show that if $m < n$, then $J_1 = R$. Let $S = R/J_1$. Then

$$S^n = \Sigma_{j=1}^n R/(J_j + J_1) \cong M/(J_1M) \cong \Sigma_{i=1}^m R/(I_i + J_1)$$

as S-modules. We can map S^m onto $\Sigma_{i=1}^m R/(I_i + J_1)$, so $S = 0$ by (II.7.5).

As (a) holds we may assume that $m = n$. To prove (b) it suffices, by symmetry, to show that $I_k \subseteq J_k$ for $k = 1,\ldots,n$. Let $x \in I_k$. Then

$$\Sigma_{j=1}^n R/(J_j : x) \cong xM \cong \Sigma_{i=k+1}^n R/(I_i : x)$$

where $(K:x) = \{r \in R : rx \in K\}$. Applying (a) to xM we get $(J_1:x) = (J_2:x) = \cdots = (J_k:x) = R$. Thus $x \in J_k$. □

EXERCISES

1. Show that a finitely presented abelian group is a direct sum of a finite number of infinite cyclic and finite cyclic groups.

2. Give a Brouwerian example of a cyclic group that is neither finite nor infinite.

3. Show that if an $m \times n$ matrix over a commutative ring R has a left inverse, and $m < n$, then $R = 0$.

4. Let H be a detachable subgroup of a free abelian group (free $\mathbb{Z}$-module) on a discrete set. Show that for each h in H there exists x in H such that $h \in \langle x\rangle$ and $H/\langle x\rangle$ is a detachable subgroup of a free abelian group on a discrete set.

5. Use Exercise 4 to show that a detachable subgroup of a free abelian group on a countable discrete set is a free abelian group on a countable discrete set. (Construct generators x_i for H inductively so that $H/\langle x_1,\ldots,x_{n-1}\rangle$ is a detachable subgroup of a free abelian group on a countable discrete set for each $n \geq 1$.)

3. TORSION MODULES, p-COMPONENTS, ELEMENTARY DIVISORS

Let M be a module over a discrete integral domain R. The **torsion submodule** $\tau(M)$ of M is defined to be $\{m \in M : am = 0 \text{ for some } a \neq 0\}$. We easily see that $\tau(M)$ is a submodule of M. If $\tau(M) = M$, then we say that M is a **torsion** module. If d is a nonzero element of R, and $dM = 0$, then we say that M is **bounded** by d.

3.1 THEOREM. *Let M be a finitely presented module over a principal ideal domain R. Then $\tau(M)$ is a finitely presented detachable submodule of M, and $M \cong \tau(M) \oplus R^n$ for some n. Moreover $\{d \in R : d\tau(M) = 0\}$ is a nonzero principal ideal.*

PROOF. By Theorem 2.3 there exist principal ideals $I_1 \supseteq I_2 \supseteq \cdots \supseteq I_m$ such that M is isomorphic to the direct sum $R/I_1 \oplus R/I_2 \oplus \cdots \oplus R/I_m$. If $I_k = 0$ for all k, then the conclusions are clear. Otherwise choose k such that $I_k \neq 0$ and $I_i = 0$ for $i > k$. Then $\tau(M) \cong R/I_1 \oplus \cdots \oplus R/I_k$ and $M \cong \tau(M) \oplus R^{m-k}$. Moreover $I_k = \{d \in R : d\tau(M) = 0\}$. □

Let M be a module over a commutative ring R, and let $d \in R$. Then the **d-component** of M is defined by $M_d = \{m \in M : d^k m = 0 \text{ for some } k\}$. We easily verify that M_d is a submodule of M. Observe that $M_a + M_b \subseteq M_{ab}$ for all $a,b \in R$ and $M_{(a^n)} = M_a$ for all $n > 0$.

3.2 LEMMA. *Let M be a module over a commutative ring R. Let a and b be strongly relatively prime elements of R. Then $M_{ab} = M_a \oplus M_b$, and, if M is finitely generated, the projection of M_{ab} onto M_a is realized by multiplication by an element of R. The module M_{ab} is cyclic if and only if the submodules M_a and M_b are cyclic.*

PROOF. To show that $M_{ab} = M_a \oplus M_b$, it suffices to show that $K = K_a \oplus K_b$ for each finitely generated submodule of M_{ab}, so we may assume that M is finitely generated. Then $a^k b^k M_{ab} = 0$ for some positive integer k.

There exist s and t in R such that $sa^k + tb^k = 1$. Let $\pi_a = tb^k$ and $\pi_b = sa^k$. Then

(i) $\pi_a M_{ab} \subseteq M_a$ and $\pi_b M_{ab} \subseteq M_b$,

(ii) $\pi_a M_b = 0$ and $\pi_b M_a = 0$,

(iii) $\pi_a x = x$ for $x \in M_b$ and $\pi_b x = x$ for $x \in M_a$,

so $M_{ab} = M_a \oplus M_b$ and multiplication by π_a gives the projection of M_{ab} onto M_a.

If $M_{ab} = Rx$, then M_a is generated by $\pi_a x$, and M_b is generated by $\pi_b x$. Conversely suppose $M_a = Ry$ and $M_b = Rz$, and let $x = y + z$. Then $y = \pi_a x \in Rx$ and $z = \pi_b x \in Rx$, so $M_{ab} = Rx$. □

3.3 THEOREM. *Let M be a module over a commutative ring. Let $a = p(1)^{e(1)} \cdots p(m)^{e(m)}$, where the $p(i)$ are pairwise strongly relatively prime. Then $M_a = M_{p(1)} \oplus \cdots \oplus M_{p(m)}$, and, if M is finitely generated, the projection of M_a onto $M_{p(i)}$ is realized by multiplication by an element of the ring.*

PROOF. Apply Lemma 3.2 repeatedly. □

If p is a prime in a discrete integral domain R, then an R-module M is said to be **p-primary** if $M_p = M$. If $dM = 0$ for some nonzero element of R which is a product of powers of strongly relatively prime primes, then we can decompose M into a direct sum of primary submodules by (3.3).

3.4 THEOREM. *Let R be a PID, and p a prime in R. If M is a finitely presented p-primary R-module, then M is isomorphic to a finite direct sum of R-modules, each of the form $R/(p^n)$ for some $n > 0$.*

PROOF. By the structure theorem (2.3), M is isomorphic to a finite direct sum of R-modules, each of the form R/I for some principal ideal I. As M is p-primary, each I contains a positive power of p. If $p^m \in I = (a)$, then $p^m = ab$. Because p is a prime, we can write $a = up^n$ where u is a unit; so $I = (p^n)$. □

The powers of p occurring in (3.5) are called the **elementary divisors** of M. If M can be written as a direct sum of primary submodules, then the elementary divisors of M are the elementary divisors of the various primary submodules of M.

EXERCISES

1. Find the primary components of the abelian group $\mathbb{Z}/12\mathbb{Z}$.

2. Let R be a Bézout domain and p a prime in R. Show that $R/(p^m)$ is a valuation ring. Prove Theorem 3.4 for R a Bézout domain.

4. LINEAR TRANSFORMATIONS

Let V be a finite dimensional vector space over a discrete field k, and let $T : V \to V$ be a linear transformation. We can make the vector space V into a module over $k[X]$ by defining $Xv = T(v)$ for each $v \in V$. By the Cayley-Hamilton theorem, the $k[X]$-module V is bounded (by the characteristic polynomial of T). We shall show that the $k[X]$-module V is finitely presented.

4.1 LEMMA. *Let V be a vector space over a discrete field k with basis $u_1,\ldots,u_n$, and $T : V \to V$ a linear transformation such that $T(u_i) = \Sigma\, a_{ji}u_j$. Let $e_1,\ldots,e_n$ be a basis for $k[X]^n$, and let $\varphi : k[X]^n \to V$ take $\Sigma\, f_i(X)e_i$ to $\Sigma\, f_i(T)u_i$. Define $d_i \in k[X]^n$ by*

$$d_i = Xe_i - \Sigma_{j=1}^n a_{ji}e_j.$$

Then $\ker \varphi$ is a free $F[X]$-module with basis $d_1,\ldots,d_n$.

PROOF. Obviously $d_1,\ldots,d_n \in \ker \varphi$. Suppose $g_1e_1 + \cdots + g_ne_n \in \ker \varphi$, where $g_i \in k[X]$. Using the relations $Xe_i = d_i + \Sigma_{j=1}^n a_{ji}e_j$, we can write

$$g_1e_1 + \cdots + g_ne_n = h_1d_1 + \cdots + h_nd_n + b_1e_1 + \cdots + b_ne_n,$$

where $b_i \in k$. So $b_1e_1 + \cdots + b_ne_n \in \ker \varphi$. Since $u_1,\ldots,u_n$ is a basis of V as vector space over k, this implies that each $b_i = 0$. Hence $d_1,\ldots,d_n$ generate $\ker \varphi$.

If $h_1d_1 + \cdots + h_nd_n = 0$, then $\Sigma_{i=1}^n h_iXe_i = \Sigma_{i=1}^n\Sigma_{j=1}^n h_ia_{ji}e_j$. If some $h_i \neq 0$, then we may assume that the degree of h_1 is maximal among the degrees of $h_1,\ldots,h_n$. But $h_1X = \Sigma_{i=1}^n h_ia_{1i}$, so $h_1 = 0$. Thus $d_1,\ldots,d_n$ are linearly independent. □

By (2.3) the $k[X]$-module V can be written as $V = C_1 \oplus \cdots \oplus C_s$, where the C_i are cyclic $k[X]$-modules, isomorphic to $k[X]/(f_i)$ for nonzero monic polynomials f_i, with f_i dividing f_{i+1} for $i = 1,\ldots,s-1$. The polynomial f_s generates the ideal $\{g \in k[X] : gV = 0\} = \{g \in k[X] : g(T) = 0\}$, and is called the **minimal polynomial** of T. By the Cayley-Hamilton theorem, the minimal polynomial of T divides the characteristic polynomial of T; the two polynomials are equal if and only if V is a cyclic $k[X]$-module.

If λ is a root of the characteristic polynomial of T, then we say that λ is an **eigenvalue** of T. If λ is an eigenvalue of T, then there exists a nonzero $v \in V$, called an **eigenvector** of T, such that $(T - \lambda)v = 0$, that is, $Tv = \lambda v$. So $X - \lambda$ must divide the minimal polynomial of T, whence λ is also a root of the minimal polynomial of T.

The decomposition of V into a direct sum of cyclic $k[X]$-modules provides a basis of V as vector space over F relative to which T has a canonical form. Let c_i be a generator for C_i over $k[X]$, and suppose f_i has degree m. Then $c_i, Xc_i, \ldots, X^{m-1}c_i$ is a basis of C_i as a vector space over k. If $f_i = X^m - b_1X^{m-1} - \cdots - b_m$, then the matrix of T restricted to C_i, relative to the basis $c_i, Xc_i, \ldots, X^{m-1}c_i$, is

$$B_i = \begin{bmatrix} 0 & 0 & \cdots & 0 & b_m \\ 1 & 0 & \cdots & 0 & b_{m-1} \\ 0 & 1 & \cdots & 0 & b_{m-2} \\ \vdots & \vdots & & \vdots & \vdots \\ 0 & 0 & \cdots & 0 & b_2 \\ 0 & 0 & \cdots & 1 & b_1 \end{bmatrix}.$$

The matrix B_i is called the **companion matrix** of f_i. Note that f_i is the characteristic polynomial of B_i. By choosing such a basis for each C_i, we obtain an F-basis relative to which the matrix of T has the form

$$\begin{bmatrix} B_1 & & \\ & B_2 & \\ & & B_s \end{bmatrix}$$

where B_i is the companion matrix of f_i. This matrix is called the **rational canonical form** for T. The characteristic polynomial of this matrix is readily seen to be $f_1 f_2 \cdots f_s$.

4.2 THEOREM (Jordan canonical form). *Let T be an endomorphism of a finite-dimensional vector space V over a discrete field k, such that the characteristic polynomial of T is a product of linear factors. Then we can find a basis for V relative to which the matrix for T has the form*

$$A = \begin{bmatrix} J_1 & & \\ & J_2 & \\ & & J_r \end{bmatrix}$$

where each matrix J_i is a m-by-m matrix of the form

$$J(m,\lambda) = \begin{bmatrix} \lambda & 0 & \cdots & 0 & 0 \\ 1 & \lambda & \cdots & 0 & 0 \\ 0 & 1 & \cdots & 0 & 0 \\ \vdots & \vdots & & \vdots & \vdots \\ 0 & 0 & \cdots & \lambda & 0 \\ 0 & 0 & \cdots & 1 & \lambda \end{bmatrix}.$$

for various m *and* λ. *The diagonal of* A, *that is, the matrix* A *with the off-diagonal entries set to* 0, *can be written as a polynomial in* A.

PROOF. By (3.3) we can write the $k[X]$-module V as a direct sum of primary modules $V_{X-\lambda}$, and (3.4) says that $V_{X-\lambda}$ is a direct sum of modules, each of which is isomorphic to $k[X]/((X-\lambda)^m)$ for some m. This latter module has an k-basis $1,(X-\lambda),\ldots,(X-\lambda)^{m-1}$, and relative to this basis the matrix of T restricted to $V_{X-\lambda}$ has the form $J(m,\lambda)$. Performing this construction on each $V_{X-\lambda}$, we obtain a basis relative to which the matrix of T has the form A. The diagonal of A is a polynomial in A because, by (3.3), the projections of V on the various $V_{X-\lambda}$ are given by polynomials in A. □

We say that T is **diagonalizable** if V admits a basis of eigenvectors of T. Thus T is diagonalizable if and only if there is a basis for V relative to which the matrix of T is diagonal. We can express the fact that T is diagonalizable in terms of the minimal polynomial of T.

4.3 THEOREM. *Let* T *be a linear transformation of a finite-dimensional vector space* V *over a discrete field* k. *Then* T *is diagonalizable if and only if the minimal polynomial of* T *is a product of distinct monic linear factors.*

PROOF. If $\lambda_1,\ldots,\lambda_m$ are the distinct entries on the diagonal of a diagonal matrix representing T, then clearly the minimal polynomial of T is $(X-\lambda_1)(X-\lambda_2)\cdots(X-\lambda_m)$. Conversely, if the minimal polynomial of T is $(X-\lambda_1)(X-\lambda_2)\cdots(X-\lambda_m)$ where the λ_i are distinct, then by (3.3) V admits a basis of eigenvectors of T. □

Any two diagonal matrices commute, so there is no hope of getting a basis relative to which linear transformations T_1 and T_2 both have diagonal matrices, unless T_1 and T_2 commute. This condition turns out to sufficient.

4.4 THEOREM. *Let T_1 and T_2 be commuting linear transformation of a finite-dimensional vector space V over a discrete field k. If V admits a basis of eigenvectors for each of T_1 and T_2, then V admits a basis whose elements are eigenvectors of both T_1 and T_2.*

PROOF. Let $V_\lambda^i = ker\ T_i - \lambda$ be the subspace of V whose nonzero elements are the eigenvectors of T_i with eigenvalue λ. Then, for $i = 1,2$, the space V is a direct sum of the subspaces V_λ^i as λ ranges over the eigenvalues of T_i. As T_2 commutes with $T_1 - \lambda$, the subspace V_λ^1 is invariant under T_2, hence also under the projection of V onto V_μ^2, which is a polynomial in T_2. Therefore $V_\lambda^1 = \Sigma_\mu V_\lambda^1 \cap V_\mu^2$, so $V = \Sigma_{\lambda,\mu} V_\lambda^1 \cap V_\mu^2$. □

EXERCISES

1. Show that the Jordan canonical form of a linear transformation is unique in the sense that, for each pair m and λ, the number of Jordan blocks $J(m,\lambda)$ that appear is invariant.

NOTES

Smith normal form for matrices over the integers goes back to [Smith 1861]. The diagonal elements are called **invariant factors**; the proof of (1.4) shows why.

The ascending chain condition is used in Lemma 1.1 to diagonalize a matrix. The question as to whether it is really needed, or whether matrices over Bézout domains are diagonalizable, has tantalized people for many years.

The standard proofs of the uniqueness of the decomposition of a module over a PID into a direct sum of cyclics involve factoring into primes, despite the fact that Kaplansky proved the more general Theorem 2.4 in 1949 (using two proofs by contradiction). As we need not be able to factor in a PID, we were led to rediscover the better theorem. The quasi-factorization theorem (IV.1.8) was formulated in order to constructivize the inferior theorem.

Chapter VI. Field Theory

1. INTEGRAL EXTENSIONS AND IMPOTENT RINGS

Let R be a subring of a commutative ring E. An element of E is **integral** over R if it satisfies a monic polynomial in $R[X]$. The **integral closure of** R **in** E is the set of elements in E that are integral over R. If every element of E is integral over R, then we say that E is an **integral extension** of R. If R is equal to the integral closure of R in E, then we say that R is **integrally closed in** E. If R is a field, the word *integral* in the above definitions may be replaced by the word ***algebraic***.

We shall show that the integral closure of R in E is a subring of E. First a lemma.

1.1 LEMMA. *Let $R \subseteq E$ be commutative rings. If $\alpha \in E$ satisfies a monic polynomial f of degree n over R, then the ring $R[\alpha]$ is generated by $1,\alpha,\ldots,\alpha^{n-1}$ as an R-module.*

PROOF. If $\beta \in R[\alpha]$, then $\beta = g(\alpha)$ for some $g \in R[X]$. By the division algorithm (II.4.3) there exist polynomials q and r in $R[X]$, with $\deg r \leq n-1$, so that $g = qf + r$. Then $\beta = g(\alpha) = r(\alpha)$ is an R-linear combination of $1,\alpha,\ldots,\alpha^{n-1}$. □

Recall that an R-module M is **faithful** if $r = 0$ whenever $rm = 0$ for all m in M.

1.2 THEOREM. *Let E be a commutative ring, R a subring of E, and $\alpha \in E$. The following are equivalent.*

(i) *α satisfies a monic polynomial of degree n over R.*

(ii) *$R[\alpha]$ is generated by n elements as an R-module.*

(iii) *E has a faithful R-submodule M, generated by n elements, such that $\alpha M \subseteq M$.*

PROOF. If (i) holds, then $R[\alpha]$ is generated by $1,\alpha,\ldots,\alpha^{n-1}$. If (ii) holds, then $M = R[\alpha]$ satisfies (iii). Suppose (iii) holds, and $m_1,\ldots,m_n$

generate M. Then $\alpha m_j = \sum r_{ij} m_i$ with $r_{ij} \in R$. Let f be the characteristic polynomial of the matrix $\{r_{ij}\}$. Then $f(\alpha)m_j = 0$ for each j by (II.6.6), so $f(\alpha) = 0$ as M is faithful. □

1.3 COROLLARY. *Let β satisfy a monic polynomial of degree n over R. If $\alpha \in R[\beta]$, then α satisfies a monic polynomial of degree n over R.*

PROOF. Take $M = R[\beta]$ and apply Theorem 1.2. □

1.4 COROLLARY. *If α is integral over R, and β is integral over $R[\alpha]$, then $R[\alpha,\beta]$ is integral over R. Hence the elements in E that are integral over R form a subring of E.*

PROOF. From (1.2) we have $R[\alpha]$ is a finitely generated R-module, and $R[\alpha,\beta]$ is a finitely generated $R[\alpha]$-module. From (II.3.2) we have $R[\alpha,\beta]$ is a (faithful) finitely generated R-module, so (1.2) says that $R[\alpha,\beta]$ is integral over R. □

1.5 COROLLARY. *Let $R \subseteq E \subseteq F$ be commutative rings with E integral over R. Then every element of F that is integral over E is integral over R. If E is a finitely generated ring extension of R, then E is a finitely generated R-module.*

PROOF. We prove the second statement first. Let $E = R[a_1,\ldots,a_n]$. Then $R[a_1]$ is a finitely generated R-module by (1.1), and E is a finitely generated $R[a_1]$-module by induction on n. Therefore E is a finitely generated R-module by (II.4.3). To prove the first statement we may assume that E is finitely generated. If $\alpha \in F$ is integral over E, then there is a faithful finitely generated E-submodule M of F such that $\alpha M \subseteq M$. But M is a finitely generated R-module by (II.4.3), so α is integral over R. □

Corollary 1.5 shows that the integral closure of R in E is integrally closed in E. A discrete integral domain is said to be **integrally closed** if it is integrally closed in its field of quotients.

1.6 THEOREM. *If R is a GCD-domain, then R is integrally closed.*

PROOF. Each element of the field of quotients of R can be written as u/v where u and v are relatively prime elements of R. Suppose $f(X) = X^n + a_{n-1}X^{n-1} + \cdots + a_1X + a_0$ is a monic polynomial in $R[X]$ such that $f(u/v) = 0$. Then $v^n f(u/v) = 0$ so $v | u^n$. As u and v are relatively prime,

$v|u$ so $u/v \in R$. □

Recall that a commutative ring E is impotent if

(i) $a^n = 0$ for some $n \in \mathbb{N}$ implies $a = 0$, and
(ii) $a^2 = a$ implies $a = 0$ or $a = 1$.

Condition (i), which is equivalent to $a^2 = 0$ implies $a = 0$, says that E has no nilpotent elements. Condition (ii) says that E has no idempotent elements other than 0 and 1. Any Heyting field is an impotent ring. We shall show that if E is an impotent extension of a discrete field k, then the elements of E that are algebraic over k form a discrete field. As the complex numbers are a Heyting field, this shows that the algebraic numbers are a discrete field.

1.7 LEMMA. *Let a and b be elements of an impotent ring such that $a + b = 1$ and $ab = 0$. Then either $a = 0$ and $b = 1$, or $a = 1$ and $b = 0$.*

PROOF. Multiplying the equation $a + b = 1$ by b, and making use of the fact that $ab = 0$, we obtain $b^2 = b$. As the ring is impotent, it follows that $b = 0$ or $b = 1$. □

1.8 LEMMA. *Let E be an impotent ring, k a subring of E, and $\alpha \in E$. If $f,g \in k[X]$ are strongly relatively prime, and $f(\alpha)g(\alpha) = 0$, then either $f(\alpha)$ or $g(\alpha)$ is a unit, so either $g(\alpha) = 0$ or $f(\alpha) = 0$.*

PROOF. As f and g are strongly relatively prime, there exist s and t in $k[X]$ such that $s(\alpha)f(\alpha) + t(\alpha)g(\alpha) = 1$. From (1.7) it follows that $s(\alpha)f(\alpha) = 1$ or $t(\alpha)g(\alpha) = 1$. Therefore either $f(\alpha)$ or $g(\alpha)$ is a unit, so the other is 0. □

1.9 THEOREM. *Let E be an impotent ring and k a discrete subfield of E. If $\alpha \in E$ is algebraic over k, then $k[\alpha]$ is a discrete field. Thus the set of elements in E algebraic over k is a discrete field.*

PROOF. Let $\gamma \in k[\alpha]$. We may assume that $k[\alpha] = E$, and we shall show that γ is either a unit, or that $\gamma = 0$. From Corollary 1.4 it follows that γ is algebraic over k. Thus there is a monic polynomial g in $k[X]$ so that $g(\gamma) = 0$. Write $g(X) = X^m h(X)$ *with* $h(0) \neq 0$. By Lemma 1.8, either γ^m is a unit, and so γ is a unit, or $\gamma^m = 0$. In the latter case $\gamma = 0$, as E is impotent. □

1.10 COROLLARY. *Let* $k \subseteq K = k(x_1,\ldots,x_n)$ *be discrete fields. Then the following are equivalent.*

(i) K *is algebraic over* k,

(ii) $x_1,\ldots,x_n$ *are algebraic over* k,

(iii) K *is a finitely generated vector space over* k.

In addition, each of (i), (ii), *and* (iii) *implies*

(iv) $K = k[x_1,\ldots,x_n]$.

PROOF. Clearly (i) implies (ii).

Suppose (ii) holds. To prove (iii), it suffices to show that $k(x_1,\ldots,x_i)$ is a finitely generated vector space over $k(x_1,\ldots,x_{i-1})$ for each $i \geq 1$. As x_i is algebraic over k, it is integral over $k(x_1,\ldots,x_{i-1})$; the result then follows from (1.2).

Theorem 1.2 shows that (i) follows from (iii). Finally, if (i) holds, then $k[x_1,\ldots,x_n]$ is a field by (1.9), so (iv) holds. □.

In fact, if condition (iv) holds in (1.10) then so do the others; this result is sometimes called the **weak Nullstellensatz**. We prove the following somewhat stronger version.

1.11 THEOREM. *Let* $R \subseteq S$ *be discrete commutative rings with recognizable units. If* $S = R[x_1,\ldots,x_n]$, *then* S *is integral over* R, *or* S *contains a nonzero nonunit, or* R *contains a nonzero nonunit.*

PROOF. It suffices to show, for each i, that x_i is integral over R, or S contains a nonzero nonunit, or R contains a nonzero nonunit. Let

$$R(x_i) = \{fg^{-1} : f,g \in R[x_i] \text{ and } g \text{ is a unit in } S\}.$$

Then $R(x_i)$ has recognizable units, so by induction on n we may assume that S is integral over $R(x_i)$. By taking a power of the product of the denominators appearing in the monic polynomials satisfied by the x_j over $R(x_i)$ we construct a polynomial $\beta \in R[X]$ such that $\beta(x_i)$ is a unit in S, and $\beta(x_i)x_j$ is integral over $R[x_i]$ for each j. We may assume that β is monic, or else its leading coefficient is a nonzero nonunit in R. We may assume that x_i is a unit of S and, multiplying by X if necessary, we may assume that β is nonconstant. Any element of S can be multiplied by a power of $\beta(x_i)$ to make it integral over $R[x_i]$. In particular $1 - \beta(x_i)$ equals 0, in which case x_i is integral over R, or $1 - \beta(x_i)$ is a nonzero nonunit of S, or there exists m such that $\alpha(x_i) = \beta(x_i)^m(1 - \beta(x_i))^{-1}$ is integral over $R[x_i]$, and thus a root of a monic polynomial $f(X) \in$

$R[x_i][X]$. Multiplying $f(\alpha(x_i)) = 0$ by $(1 - \beta(x_i))^d$, where d is the degree of f, we get

$$\beta(x_i)^{md} = g(x_i)(1 - \beta(x_i)).$$

Since $\beta(X)^{md}$ and $1 - \beta(X)$ are strongly relatively prime in $R[X]$, this implies that $(1 - \beta(x_i))h(x_i) = 1$ for some $h \in R[X]$, whence x_i satisfies the polynomial $(1 - \beta(X))h(X) - 1$. Either the leading coefficient of this polynomial is a nonzero nonunit of R, or x_i is integral over R. □

The more familiar form of the weak Nullstellensatz is as follows.

1.12 COROLLARY. *Let $k \subseteq K = k[x_1,\ldots,x_n]$ be discrete fields. Then K is algebraic over k.*

PROOF. The discrete fields K and k have no nonzero nonunits, so K is algebraic over k by (1.11). □

If E is a commutative ring containing a field k, then E is a vector space over k. If α in E satisfies a polynomial of degree n over k, then $k[\alpha]$, being generated by $1,\alpha,\ldots,\alpha^{n-1}$, is a finitely generated vector space over k. When is $k[\alpha]$ finite dimensional?

1.13 THEOREM. *Let E be a commutative ring, k a discrete subfield of E, and $\alpha \in E$. If α satisfies an irreducible polynomial over k, then $k[\alpha]$ is a finite-dimensional vector space over k. If E is impotent, and $k[\alpha]$ is contained in a finite-dimensional k-subspace of E, then α satisfies an irreducible polynomial over k.*

PROOF. If α satisfies an irreducible polynomial of degree n over k, then $1,\alpha,\ldots,\alpha^{n-1}$ is a basis for $k[\alpha]$ over k. Conversely, suppose E is impotent and V is an n-dimensional k-subspace of E containing α. Then (1.2) says that α satisfies a monic polynomial f of degree n over k. Thus $k[\alpha]$ is generated by $1,\alpha,\ldots,\alpha^{n-1}$ as a k-subspace, so $k[\alpha]$ is finite dimensional by (II.5.3). By induction on n, we may assume that $k[\alpha] = V$. We shall show that f is irreducible. By (1.9) the ring $k[\alpha]$ is a discrete field, so if $f = gh$, then either $g(\alpha) = 0$ or $f(\alpha) = 0$. But α cannot satisfy a monic polynomial of degree less than n because $k[\alpha] = V$ is n-dimensional. □

Let E be a commutative ring that is finite dimensional over a discrete subfield k. Each $\alpha \in E$ is integral over k by (1.2). Because we can decide whether or not the elements $1,\alpha,\ldots,\alpha^m$ are linearly independent, we can construct a monic polynomial $f \in k[X]$ of smallest degree such that $f(\alpha) = 0$. If $g(\alpha) = 0$ for some $g \in k[X]$, then f must divide g lest the Euclidean algorithm construct a polynomial satisfied by α of smaller degree than f. The polynomial f is called the **minimal polynomial** of α over k. If E is impotent, then (1.13) shows that f is irreducible.

EXERCISES

1. Let $R \subseteq E$ be commutative rings with E integral over R, and let I be an ideal in R. Show that if $\alpha \in IE$, then there exist elements $r_i \in I^i$ such that $\alpha^n + r_1\alpha^{n-1} + \cdots + r_n = 0$. Conclude that $R \cap IE \subseteq \sqrt{I}$.

2. Let K be a finite-dimensional extension field of $\mathbb{Q}$. Show that an element of K is integral over $\mathbb{Z}$ if and only if its minimal polynomial over $\mathbb{Q}$ has integer coefficients. Show that the integral closure of $\mathbb{Z}$ in $\mathbb{Q}(i)$ is $\mathbb{Z}[i]$ (the Gaussian integers are the integral closure of $\mathbb{Z}$ in the Gaussian numbers).

3. In any ring show that if $a = 0$ whenever $a^2 = 0$, then $a = 0$ whenever $a^n = 0$ for some $n > 0$.

4. Show that any Heyting field, and any denial field. is an impotent ring.

5. Let k be the integers modulo 2. Let $E = k[X]/(X^2)$ and $F = k[X]/(X^2 - X)$. Show that E and F are algebraic over k, but that neither is a field. Why doesn't (1.9) apply?

6. Let k be a discrete subfield of a commutative ring E. If $\alpha \in E$ satisfies an irreducible polynomial over k of degree n, show that $1,\alpha,\ldots,\alpha^{n-1}$ is a basis for $k[\alpha]$ over k.

7. Let $\mathbb{Q}_2$ denote the ring of integers localized at 2. Use the pair $\mathbb{Q}_2 \subseteq \mathbb{Q} = \mathbb{Q}_2[1/2]$ to show that the phrase 'or R contains a nonzero nonunit' cannot be removed from the conclusion of the weak Nullstellensatz. Use the same pair to construct a Brouwerian example where the weak Nullstellensatz fails because R does not have recognizable units.

2. ALGEBRAIC INDEPENDENCE AND TRANSCENDENCE BASES

Let $k \subseteq K$ be commutative rings. The elements $x_1,\ldots,x_n$ of K are called **algebraically independent** over k if whenever $f \in k[X_1,\ldots,X_n]$, and $f(x_1,\ldots,x_n) = 0$, then $f = 0$. If k is discrete, $x_1,\ldots,x_n$ are called **algebraically dependent** over k if there exists nonzero f in $k[X_1,\ldots,X_n]$ such that $f(x_1,\ldots,x_n) = 0$; in this case, $x_1,\ldots,x_n$ are algebraically independent if and only if they are not algebraically dependent.

If K is discrete, and S is a finite subset of K, then we say that S is **algebraically dependent,** or **algebraically independent,** if its elements in some (any) order are. We may reduce the notion of algebraic dependence to that of algebraic elements as follows.

2.1 THEOREM. *If $k \subseteq K$ are discrete fields, then $x_1,\ldots,x_n$ in K are algebraically dependent over k if and only if there exists i such that x_i is algebraic over $k(x_1,\ldots,x_{i-1})$.*

PROOF. Suppose x_i satisfies a nonzero polynomial $f(X_i)$ with coefficients in $k(x_1,\ldots,x_{i-1})$. Then there are nonzero polynomials $g \in k[X_1,\ldots,X_i]$ and $h \in k[X_1,\ldots,X_{i-1}]$ such that $f(X_i) = g(x_1,\ldots,x_{i-1},X_i)/h(x_1,\ldots,x_{i-1})$ and $g(x_1,\ldots,x_i) = 0$. So $x_1,\ldots,x_i$, and hence $x_1,\ldots,x_n$, are algebraically dependent over k.

Conversely, suppose $x_1,\ldots,x_n$ are algebraically dependent over k, so there exists nonzero $g \in k[X_1,\ldots,X_n]$ such that $g(x_1,\ldots,x_n) = 0$. Then $g(x_1,\ldots,x_{n-1},X_n)$ is a polynomial in X_n with coefficients in $k[x_1,\ldots,x_{n-1}]$, satisfied by x_n. If $g(x_1,\ldots,x_{n-1},X_n) = 0$, then any nonzero coefficient of g, where g is thought of as a polynomial in X_n with coefficients in $k[X_1,\ldots,X_{n-1}]$, gives an algebraic dependence among $x_1,\ldots,x_{n-1}$, and we are done by induction on n. If $g(x_1,\ldots,x_{n-1},X_n) \neq 0$, then x_n is algebraic over $k(x_1,\ldots,x_{n-1})$. □

As a corollary we have the **exchange property** for algebraic dependence.

2.2 COROLLARY. *Let $k \subseteq K$ be discrete fields, and $x,y \in K$. If x is algebraic over $k(y)$, then either y is algebraic over $k(x)$ or x is algebraic over k.*

PROOF. By (2.1) we have y,x are algebraically dependent over k, so x,y are algebraically dependent over k. The conclusion follows from (2.1). □

Corollary 2.2 parallels the exchange property for linear dependence of vectors: if x is a linear combination $y_1,\ldots,y_n$, then either x is a linear combination of $y_1,\ldots,y_{n-1}$, or y_n is a linear combination of $y_1,\ldots,y_{n-1}$ and x (see Exercise 2).

2.3 THEOREM. *Let $k \subseteq K$ be discrete fields. Let S and T be finite subsets of K such that K is algebraic over $k(S)$. Then either T is algebraically dependent over k, or we can partition S into finite subsets S_0 and S_1 such that $\#S_0 = \#T$,* and *K is algebraic over $k(T \cup S_1)$.*

PROOF. Proceed by induction on $m = \#(T\backslash S)$. If $m = 0$, then we can take $S_0 = T$. Otherwise let $x \in T\backslash S$. Let $S = \{s_1,\ldots,s_k\}$, with the elements of $T \cap S$ listed first. Repeated application of (2.2) to x and $k(s_1,\ldots,s_j)$, for $j = k,k-1,\ldots$, results either in showing that x is algebraic over k, hence T is algebraically dependent, or in finding an element s_i that is algebraic over $k(s_1,\ldots,s_{i-1},x)$, with x algebraic over $k(s_1,\ldots,s_i)$. If $s_i \in T$, then T is algebraically dependent. If $s_i \notin T$, replace s_i in S by x, and we are done by induction on m. □

Let $k \subseteq K$ be discrete fields, and let B be a finite subset of K that is algebraically independent over k. The field extension $k \subseteq k(B)$ is said to be **purely transcendental**, and if K is algebraic over $k(B)$, we call B a **transcendence basis** of K over k. An immediate consequence of (2.3) is the following.

2.4 COROLLARY. *Any two transcendence bases of a discrete field K over a discrete field k have the same cardinality.* □

If B is a transcendence basis for K over k, then the cardinality of B is called the **transcendence degree of K over k**, and is written as $\mathbf{tr.deg}_k\, K$. The transcendence degree of an algebraic extension is zero, the empty set being a transcendence basis.

A purely constructive consequence of (2.3) is that we can decide algebraic dependence if we have a transcendence basis.

2.5 COROLLARY. *If the discrete field K has a transcendence basis over a subfield k, then any finite subset of K is either algebraically dependent, or algebraically independent, over k.*

PROOF. Let S be a transcendence basis for K of cardinality n, and T a finite subset of K. By (2.3) either T is algebraically dependent, or we can enlarge T to a set T' of cardinality n such that K is algebraic over

$k(T')$. Were T' algebraically dependent, then we could construct a set T'' of cardinality $n-1$ with K algebraic over $k(T'')$; but this would contradict (2.3) as S is algebraically independent. □

2.6 LEMMA. *Let K be a discrete field of transcendence degree n over a subfield k. If K is algebraic over $k(S)$ for some subset $S \subseteq K$, then S contains a transcendence basis of K over k. If S is a finite subset of K that is algebraically independent over k, then S can be extended to a transcendence basis.*

PROOF. Choose a finite subset B of S such that each element of the transcendence basis, and hence K itself, is algebraic over $k(B)$. By (2.5) we can decide whether or not B is algebraically dependent. If B is algebraically independent, then B is the desired transcendence basis. If B algebraically dependent, then there exists $b \in B$ such that K is algebraic over $k(B\backslash\{b\})$, in which case we are done by induction on $\#B$.

Now suppose S is a finite algebraically independent set. By (2.3) we can enlarge S to a finite subset S' of K of cardinality n such that K is algebraic over $k(S')$. By the first part of this theorem, S' contains a transcendence basis; so S' is a transcendence basis by (2.4). □

2.7 THEOREM. *Let $k \subseteq K \subseteq L$ be discrete fields. If two of the three extensions $k \subseteq L$, $K \subseteq L$ and $k \subseteq K$ have finite transcendence bases, then so does the third, and*

$$\operatorname{tr.deg}_k L = \operatorname{tr.deg}_K L + \operatorname{tr.deg}_k K.$$

PROOF. If B_0 is a transcendence basis for K over k, and B_1 is a transcendence basis for L over K, then one easily verifies that $B_0 \cup B_1$ is a transcendence basis for L over k. If B_0 is a transcendence basis for K over k, and B_1 is a transcendence basis for L over k, then by (2.3) we may assume that $B_0 \subseteq B_1$, whereupon $B_1\backslash B_0$ is a transcendence basis for L over K.

Finally, suppose that B_0 is a transcendence basis for L over K, and B_1 is a transcendence basis for L over k. We can find a finite subset S of K so that each element of B_1, and hence L, is algebraic over $k(S \cup B_0)$. By (2.3) we may assume that S is algebraically independent over k. If $x \in K$, then by repeated application of (2.2) we find that x is algebraic over $k(S)$, because B_0 is algebraically independent over K. □

Let k be a discrete field and X an indeterminate. **Lüroth's theorem** states that every field between k and $k(X)$ is of the form $k(z)$ for some $z \in k(X)$. We can't hope to prove Lüroth's theorem even for detachable subfields of $k(X)$. However Lüroth's theorem holds for finitely generated subfields, and a trivial classical argument then yields the classical Lüroth's theorem. First we show that if $t \in k(X)\backslash k$, then $k(X)$ is finite dimensional over $k(t)$.

2.8 LEMMA. *Let k be a discrete field, X an indeterminate, and $t \in k(X)\backslash k$. If $t = \frac{u(X)}{v(X)}$, with u and v relatively prime polynomials in $k[X]$, then $p(Y) = tv(Y) - u(Y)$ is an irreducible polynomial over $k(t)$ satisfied by X.*

PROOF. Note that $p(Y) \neq 0$, because $t \notin k$. Clearly $p(X) = 0$, so $k(X)$ is algebraic over $k(t)$. Therefore t is transcendental over k.

By Gauss's lemma it suffices to show that $p(Y)$ is irreducible in $k[t][Y]$. Suppose $p = gh$ with g and h in $k[t][Y]$. As the t-degree of p is 1 it follows that either g or h has t-degree 0, say g. So $g \in k[Y]$. As g divides $p(Y) = tv(Y) - u(Y)$, and u and v are relatively prime, it follows that $g \in k$. Thus $p(Y)$ is irreducible over $k(t)$. □

If K is a subfield of $k(X)$, and $t \in K\backslash k$, then Lemma 2.8 says that $k(X)$ is finite-dimensional over $k(t)$. As, classically, this immediately implies that K is finite-dimensional over $k(t)$, the classical Lüroth's theorem is a direct consequence of Lemma 2.8 and the following.

2.9 THEOREM. *Let k be a discrete field, X an indeterminate over k, and $\alpha_1,\ldots,\alpha_n \in k(X)$. Then $k(\alpha_1,\ldots,\alpha_n) = k(z)$ for some $z \in k(X)$.*

PROOF. Let $K = k(\alpha_1,\ldots,\alpha_n)$. We may assume that $\alpha_1 \notin k$. By Lemma 2.8 the field $k(X)$ is finite dimensional over $k(\alpha_1)$, so $k(X)$ is finite dimensional over $K = k(\alpha_1)[\alpha_2,\ldots,\alpha_n]$. Thus X satisfies an irreducible polynomial f in $K[Y]$. As X is transcendental over k, some coefficient z of f is not in k. We will show that $k(z) = K$.

Write $z = \frac{u(X)}{v(X)}$ with u and v relatively prime polynomials in $k[X]$, and let $p(Y) = zv(Y) - u(Y)$. Lemma 2.8 says that $p(Y)$ is irreducible over $k(z)$. We shall show that $\deg p = \deg f$, from which it follows that $k(z) = K$, as $k(z) \subseteq K$.

For any polynomial $q \in k(X)[Y]$, let q^* denote a primitive polynomial in $k[X][Y]$ such that $q/q^* \in k(X)$; in particular let $p^* = u(X)v(Y) - v(X)u(Y)$.

Let $m = deg_X\, f^*$, the maximum of the X-degrees of the coefficients of f^*. As f is a monic polynomial, $deg_X\, p^* = deg_Y\, p^* = max(deg\, u,\, deg\, v) \leq m$. As $f(X) = 0$, and f is irreducible in $K[Y]$, we can write $p = fg$ with g in $K[Y]$. Then $p^*(Y) = df^*(Y)g^*(Y)$ for some d in k, so $deg_X\, p^* = m$ and $deg_X\, g^* = 0$. Therefore $g^*(Y) \in k[Y]$; but $g^*(Y)$ divides $p(Y)$, which is irreducible over $k(z)$, so $g^*(Y) \in k$. Thus $deg_Y\, p^* = deg_Y\, f^*$, so $deg\, p = deg\, f$. □

We close this section with the construction of an example that will provide a Brouwerian example of a field that is separably factorial but not factorial, and show the necessity of the separability assumptions in Theorems VII.1.2 and VII.2.3.

2.10 THEOREM. *Let F be a discrete field of characteristic 2, and let $K = F(b,s,t)$, where b, s and t are indeterminates. Let $a = bs^2 + t^2$ and $k = F(a,b)$. Then k is algebraically closed in K.*

PROOF. Note that K has transcendence degree 3 over F, and that $t^2 \in F(a,b,s) = k(s)$, so K is algebraic over $k(s)$. It follows that a and b are algebraically independent over F, and that s is transcendental over k. By (1.6) and (IV.4.7), the ring $F[a,b]$ is integrally closed in k, hence in $k(s)$.

Now let $\theta \in K$ be algebraic over k, and let $f \in F[a,b,X]$ be a nonzero polynomial with $f(\theta) = 0$. Let $r \in F[a,b]$ be the leading coefficient of f. The polynomial $r^{n-1}f\left[\frac{X}{r}\right]$ is monic and has coefficients in $F[a,b]$, so $r\theta$ is integral over $F[a,b]$ whence $w = (r\theta)^2 \in k(s)$ is also integral over $F[a,b]$, so $w \in F[a,b]$. Write $r\alpha = p/q + (u/v)t$ where $p,q,u,v \in F[a,b,s]$. Then $w = p^2/q^2 + (u^2/v^2)(a + bs^2)$, as our fields have characteristic 2, so

$$(*) \qquad p^2v^2 + u^2q^2(a + bs^2) = wq^2v^2$$

If $u = 0$, then $r\theta = p/q$ is integral over $F[a,b]$ and in $k(s)$, so $r\theta \in F[a,b]$, and hence $\theta \in k$ as desired. Otherwise, let n be the largest power of s dividing u^2q^2. The coefficient of s^n on the left hand side of equation $(*)$ contains only even powers of b and is nonzero as it contains an odd power of a. Hence w (which is in $F[a,b]$) and so wq^2v^2, contains only even powers of b. But the coefficient of s^{n+2} on the left hand side of $(*)$ contains an odd power of b. Thus $u \neq 0$ is impossible. □

EXERCISES

1. Let $k \subseteq E$ be discrete fields, and S a finite subset of E. Show that if a subfield K of E is algebraic over k, then S is algebraically dependent over K if and only if S is algebraically dependent over k.

2. A **span operation** on a discrete set S is a map s from finite subsets of S to subsets of S such that

   (i) if $A \subseteq B$, then $sA \subseteq sB$,

   (ii) $A \subseteq sA$,

   (iii) if B is a finite subset of sA, then $sB \subseteq sA$,

   (iv) if $x \in s(A \cup \{y\})$, then either $x \in sA$ or $y \in s(A \cup \{x\})$.

   Let $k \subseteq K$ be discrete fields, and for A a finite subset of K, define $sA = \{x \in K : x \text{ is algebraic over } k(A)\}$. Let V be a discrete vector space over the discrete field k, and for A a subset of V, define sA to be the subspace of V generated by A. Show that we get a span operation in either case. Develop the theory of this section in this more general setting.

3. Give a Brouwerian example of a discrete field K with a detachable subfield k such that $K = k(\theta)$ but K does not have a transcendence basis over k.

4. Construct a Brouwerian example of a detachable subfield of $\mathbb{Q}(X)$, that is not of the form $\mathbb{Q}(z)$.

5. Let F be a discrete field of characteristic 2, let b and t be indeterminates, and let P be a statement. Let

$$k = \{x \in F(b) : x \in F(b^2) \text{ or } P\},$$

and $$K = \{x \in k(t^2 + bt) : x \in k(t^4 + b^2t^2) \text{ or } P\}.$$

   Show that K is a detachable subfield of $k(t)$, properly containing k, and that $K = k(z)$ for some z, if and only if P or not P.

3. SPLITTING FIELDS AND ALGEBRAIC CLOSURES

Let f be a nonconstant polynomial over a discrete field k; we want to construct a discrete field K containing k such that f has a root in K. If we are content that K be a discrete *ring*, then we let K be the quotient ring $k[X]/(f)$: if α denotes the image of X in K, then $K = k[\alpha]$, and α is a root of f. However K is a *field* only when f is irreducible.

3.1 THEOREM. *Let f be a nonconstant polynomial over a discrete field k. Let $K = k[X]/(f)$, and let α denote the image of X in K. Then $K = k[\alpha]$ is a discrete commutative ring containing k, and $f(\alpha) = 0$. Moreover K is a discrete field if and only if f is irreducible.*

PROOF. The division algorithm (II.5.2) allows us to decide whether or not a polynomial in $k[X]$ is divisible by f, so K is discrete. As f is nonconstant, k maps onto an isomorphic copy of itself in K, which we may identify with k. Clearly $K = k[\alpha]$ and $f(\alpha) = 0$.

Suppose K is a field. If $f = gh$, then $gh = 0$ in K so, as K is a field, either $g = 0$ in K or $h = 0$ in K. Thus either $f|g$ or $f|h$, whence f is irreducible. Conversely, if f is irreducible, and g is an arbitrary element of K, then by (II.5.7) there exist s and t in $k[X]$ such that $sf + tg$ divides both f and g. As f is irreducible, we may assume that $sf + tg$ is either f or 1. If $sf + tg = f$, then $f|g$, so $g = 0$ in K; if $sf + tg = 1$, then t is the inverse of g in K. □

If f is not irreducible, then it is more difficult to construct an extension field in which f has a root: the classical technique is to work with an irreducible factor of f, but we may be unable to find one (see Example IV.2.2). In fact, we cannot always construct such an extension field (Exercise 1). If k is countable, however, then we can construct such a field by constructing a detachable maximal ideal M in $k[X]$, containing f, without constructing a generator for M.

3.2 LEMMA. *Let R be a countable commutative ring in which finitely generated ideals are detachable. If I is a finitely generated proper ideal of R, then I is contained in a detachable maximal ideal.*

PROOF. Let $r_1, r_2, \ldots$ be an enumeration of R. We shall construct an ascending sequence of finitely generated ideals I_j starting with $I_1 = I$. If I_j has been constructed, construct I_{j+1} as follows: if $1 \in I_j + Rr_j$, then set $I_{j+1} = I_j$, otherwise set $I_{j+1} = I_j + Rr_j$. Let M be the union of the ideals I_j. As $r_j \in M$ if and only if $r_j \in I_{j+1}$, the ideal M is detachable. If $r_j \notin M$, then $r_j \notin I_{j+1}$ so $1 \in I_j + Rr_j \subseteq M + Rr_j$, so M is maximal. □

3.3 THEOREM. *Let k be a countable discrete field and f a nonconstant polynomial in $k[X]$. Then there is a countable discrete field E containing k, and $\alpha \in E$ such that $f(\alpha) = 0$.*

PROOF. Finitely generated ideals of $k[X]$ are principal, by the Euclidean algorithm, hence detachable by the division algorithm. Thus (3.2) applies and we can construct a detachable maximal ideal M of $k[X]$ containing f. Let $E = k[X]/M$, and α the image of X in E. □

Let f be a monic polynomial over a field k. An extension field K of k is called a **splitting field** for f over k if

$$f(X) = (X - \alpha_1)\cdots(X - \alpha_n)$$

and $K = k[\alpha_1,\ldots,\alpha_n]$. Note that K is countable if k is countable.

3.4 THEOREM *Let $f(X)$ be a nonzero polynomial over a countable discrete field k. Then we can construct a discrete splitting field for f over k.*

PROOF. Repeated application of Theorem 3.3. □

Splitting field over countable fields may be used when working with arbitrary discrete fields k as follows. If f is a polynomial in $k[X]$, consider the subfield k_0 of k generated by the coefficients of f. As k_0 is countable, f has a splitting field over k_0. We use the roots of f in this splitting field to obtain information about f as a polynomial in $k[X]$.

Classically the splitting field for a polynomial over a field k is unique in the sense that any two splitting fields are isomorphic over k. Constructively we may be unable to construct the isomorphism (see Exercise 2). There are problems even if f is irreducible because after we adjoin a root α of f to k we may not be able to find irreducible factors of f in $k[\alpha][X]$ (see Exercise 3).

A field Ω is **algebraically closed** if each monic polynomial of degree at least 1 in $\Omega[X]$ has a root in Ω. This implies that each monic polynomial in $\Omega[X]$ factors into linear factors in Ω. If an algebraically closed field Ω is algebraic over a subfield k, then Ω is called an **algebraic closure** of k. Constructively we may not be able to embed a discrete field in an algebraically closed field; or we may construct two algebraic closures without being able to construct an isomorphism between them. For countable discrete fields, however, we can construct an algebraic closure.

3.5 THEOREM. *Let k be a countable discrete field. Then there exists a discrete algebraic closure of k.*

PROOF. Let $f_1, f_2, \ldots$ be an enumeration of the nonconstant polynomials in $k[X]$. Let k_1 be a discrete splitting field for f_1 over k. Having constructed $k_1 \subseteq k_2 \subseteq \cdots \subseteq k_j$ we let k_{j+1} be a discrete splitting field of f_{j+1} over k_j. Finally let $\Omega = \cup k_j$, that is, the direct limit of the fields k_j. We must show that Ω is algebraically closed. Let $f \in \Omega[X]$. By Theorem 3.3, there is a discrete field E containing Ω, and a root α of f in E. Then α is algebraic over k, by Corollary 1.5, so there is a polynomial f_j in $k[X]$ with $f_j(\alpha) = 0$. As f_j is a product of linear factors in k_{j+1} it follows that α is in $k_{j+1} \subseteq \Omega$. $\square$

EXERCISES

1. Derive the world's simplest axiom of choice from the statement that if k is a discrete field, and f is a nonconstant polynomial over k, then there is an extension field of k in which f has a root. Hint: For each two-element set S, let k_S be the ring generated by S over $\mathbb{Q}$ subject to the relations $s^2 = -1$ for s in S and $\Sigma_{s \in S} s = 0$. For T a set of two element sets with at most one element, let $k = \cup_{S \in T} k_S$ and $f(X) = X^2 + 1$.

2. Let a be a binary sequence with a most one 1, and let $k = \cup_n \mathbb{Q}(ia_n)$. Enumerate $k[X]$ in such a way that if $a_{2n} = 1$, then the polynomial $X - i$ precedes $X + i$, while if $a_{2n+1} = 1$, then $X + i$ precedes $X - i$. Use this enumeration to effect the construction of a splitting field E for $X^2 + 1$ over k via (3.2) and (3.3). Show that E and $\mathbb{Q}(i)$ constitute a Brouwerian example of splitting fields for $X^2 + 1$ over k that are not isomorphic over k.

3. Construct a Brouwerian example of a field k, that lies between $\mathbb{Q}$ and $\mathbb{Q}[i\sqrt{3}]$, such that the polynomial $X^3 - 2$ does not have a unique splitting field over k, even though it is irreducible over k.

4. **Nonisomorphic splitting fields.** Let a be a binary sequence with at most one 1, and e a sequence in $\{-1,1\}$. Let P_e be the ideal in the polynomial ring $\mathbb{Q}(i)[X]$ generated by the elements $(iX - ne_n)a_n$.

   (i) Show that P_e is a detachable prime ideal.

(ii) Show that $P = P_e \cap \mathbb{Q}[X]$ is a detachable prime ideal of $\mathbb{Q}[X]$, and does not depend on e.

(iii) Let k be the field of quotients of $\mathbb{Q}[X]/P$, and K_e the field of quotients of $\mathbb{Q}(i)[X]/P_e$. Let x_e denote the image of X in K_e. Show that K_e is a splitting field for $Y^2 + 1$ over $\mathbb{Q}(x_e) \cong k$.

(iv) Let $e_n = 1$ and $f_n = (-1)^n$ for all n. Suppose $\varphi : K_e \to K_f$ is an isomorphism. Show that if $\varphi(ix_e) \neq ix_f$, then $a_n = 0$ for each even n, and if $\varphi(ix_e) \neq -ix_f$, then $a_n = 0$ for each odd n.

4. SEPARABILITY AND DIAGONALIZABILITY

If $f = a_0 + a_1X + \cdots + a_{n-1}X^{n-1} + a_nX^n$ is a polynomial over a commutative ring k, then the **formal derivative** of f is defined to be

$$f' = a_1 + 2a_2X + \cdots + (n-1)a_{n-1}X^{n-2} + na_nX^{n-1}.$$

A polynomial $f \in k[X]$ is **separable over** k if it is strongly relatively prime to its formal derivative, that is, if the ideal in $k[X]$ generated by f and f' contains 1. If k is a discrete field, then the Euclidean algorithm allows us to decide whether or not f is separable.

4.1 THEOREM. *Let k be a commutative ring and $f,g \in k[X]$. Then*

(i) $(f + g)' = f' + g'$.

(ii) $(fg)' = f'g + fg'$.

(iii) *fg is separable if and only if f and g are separable and strongly relatively prime.*

PROOF. Part (i) is clear. To show part (ii), let $f = aX^m$ and $g = bX^n$. Then

$$(fg)' = (m+n)abX^{m+n-1} = mabX^{m+n-1} + nabX^{m+n-1} = f'g + fg',$$

and the general result follows from (i). For (iii) compute $(fg)' = f'g + fg'$. Clearly $(fg, f'g + fg')$ is contained in (f,f') and (g,g') and (f,g). Conversely, $(fg, f'g + fg') \supseteq (f, f'g + fg')(g, f'g + fg') = (f,f'g)(g,fg') \supseteq (f,f')(g,g')(f,g)^2$. □

If a polynomial f with coefficients in a discrete field is separable, then we can find polynomials s and t, with coefficients in the field generated by the coefficients of f, such that $sf + tf' = 1$; indeed the

Euclidean algorithm does just that. Thus the notion of separability of polynomials over a field is absolute in the sense that it doesn't depend on the particular field in which we are considering the coefficients of the polynomial to lie. We can characterize separability of polynomials over fields in terms of the absence of multiple roots.

4.2 THEOREM. *Let f be a polynomial over a discrete field, and let k be the field generated by the coefficients of f. If f is separable then f has no multiple roots in any discrete field containing k. If f is not separable, then f has a multiple root in some extension field of k.*

PROOF. We may assume that f has degree at least 1. Suppose r is a root of f in a discrete field containing k, and $f(X) = (X - r)g(X)$. Then $f'(X) = g(X) + (X - r)g'(X)$, so $f'(r) = g(r)$. If f is separable, then $s(X)f(X) + t(X)f'(X) = 1$ so $t(r)f'(r) = 1$ whence $f'(r)$, and hence $g(r)$, is nonzero. Conversely, let K be a splitting field for f over k, and write $f(X) = a\Pi_i(X - r_i)$ where $a \neq 0$. If f is not separable, then f and f' have a nontrivial common factor, hence a common root r_j. But $f'(r_j) = a\Pi_{i\neq j}(r_j - r_i)$, so $r_j = r_i$ for some $i \neq j$. □

Let k be a discrete subfield of a commutative ring E. An element of E is **separable over** k if it satisfies a separable polynomial in $k[X]$; the ring E is a **separable extension** of k if each element of E is separable over k, and k is **separably closed** in E if each element in E that is separable over k is in k.

By the **field of definition** of a finite set of matrices over a field, we mean the (countable) field generated their entries. If matrices are linearly dependent over a discrete field k, then they are already linearly dependent over their field of definition: indeed if we treat the matrices as if they were row vectors, then (III.6.8) and (III.6.9) show that if the rows are independent over their field of definition, then they are independent over k; and rows are either linearly dependent or linearly independent (II.6.5). In particular, the coefficients of the minimal polynomial of a matrix lie in the field of definition of that matrix, so the minimal polynomial of a matrix does not depend on the particular field from which the entries of the matrix are considered to be taken. Also, if a matrix A can be written as a polynomial in a matrix B, then the polynomial may be chosen with coefficients in the field of definition of

A and B. The fundamental relationship between separability and diagonalizabilty is the following.

4.3 LEMMA. *A matrix over a discrete field is separable if and only if it is diagonalizable over some discrete field.*

PROOF. If a matrix A is diagonalizable over a field F, then the minimal polynomial of A is separable by (V.4.3) and (4.2). Conversely, if A is separable, let F be a splitting field for the minimal polynomial of A over the field of definition of A. Then A is diagonalizable over F by (V.4.3) and (4.2). □

The next lemma allows us to translate questions about algebraic elements over discrete fields to questions about matrices.

4.4 LEMMA. *Let E be a commutative ring containing a discrete field k, let f and g be polynomials in $k[X]$, and let α and β be elements of E such that $f(\alpha) = g(\beta) = 0$. Then there are square matrices A and B of the same size over k such that $f(A) = g(B) = 0$, and a ring map of $k[A,B]$ onto $k[\alpha,\beta]$ that is the identity on k and takes A to α and B to β.*

PROOF. The ring $k[X,Y]/(f(X),g(Y)) = k[x,y]$ is finite-dimensional over k and maps naturally onto $k[\alpha,\beta]$. Let T_x and T_y be the linear transformations on $k[x,y]$ given by multiplication by x and y, and let A and B be the matrices of T_x and T_y with respect to some basis of $k[x,y]$. Then the natural isomorphisms $k[A,B] \cong k[T_x,T_y] \cong k[x,y]$, together with the map from $k[x,y]$ to $k[\alpha,\beta]$, yield the desired results. □

4.5 THEOREM. *If E is a commutative ring containing a discrete field k, then the elements of E that are separable over k form a ring.*

PROOF. It suffices to show that if α and β are separable over k, then every element of $k[\alpha,\beta]$ is separable over k. Let f and g be separable polynomials satisfied by α and β respectively, and let A and B be as in (4.4). It suffices to show that any element of $k[A,B]$ is separable. Given a matrix C in $k[A,B]$, we can construct a field F, containing the field of definition of A, B and C, over which the minimal polynomials of A and B factor into linear factors. By (V.4.3) and (V.4.4) we can simultaneously diagonalize A and B, and hence C, over F, so C is separable by (V.4.3) and (4.2). □

We can characterize when an extension E of a discrete field k is separable in terms of the lack of nilpotent elements when we (possibly) extend the field k of coefficients of the commutative ring E. This characterization can serve as a definition of separability in the not necessarily algebraic case. We will establish the characterization only for countable k, leaving the formulation of the general case as an exercise.

4.6 THEOREM. *Let k be a discrete subfield of a commutative ring E, and consider the following two conditions.*

(i) *E is separable over k,*

(ii) *For each discrete extension field K of k, the ring $K \otimes_k E$ has no nilpotent elements.*

Then (i) implies (ii); conversely, if k is countable, and E is algebraic over k, then (ii) implies (i).

PROOF. The ring $K \otimes_k E$ is generated over K by elements that are separable over k; hence each element θ of $K \otimes_k E$ is separable over K by (4.5). If $\theta^n = 0$, then θ satisfies the polynomial X^n in addition to a separable polynomial $f(X)$ over K. The GCD of $f(X)$ and X^n divides X^n and is separable, hence is X, whereupon $\theta = 0$.

Conversely, suppose $\theta \in E$ satisfies a nonzero polynomial f over k. We proceed by induction on the degree n of f. If f is separable we are done. Otherwise we can construct, by (4.2), a discrete extension field K of k such that f has a multiple root in K. Write $f(X) = (X - r)g(X)$ where $g(r) = 0$, so $f \mid g^2$. Then $g(\theta) \in K \otimes_k E$ and $g(\theta)^2 = 0$, so $g(\theta) = 0$. That means that the elements $1, \theta, \ldots, \theta^{n-1}$ of $K \otimes_k E$ are linearly dependent over K, hence over k (Exercise III.5.6), so we can find a polynomial in $k[X]$, satisfied by θ, of degree less than n. □

EXERCISES

1. Let $k \subseteq E$ be commutative rings, and f a separable polynomial in $k[X]$. Show that any square factor of f in $E[X]$ is a unit in E.
2. Let k be a commutative ring, $a \in k$ and $f \in k[X]$. Show that $X - a$ is strongly relatively prime to f if and only if $f(a)$ is a unit of k.

3. Let $a_1, \ldots, a_n$ be elements of a commutative ring k. Show that $(X - a_1)(X - a_2)\cdots(X - a_n)$ is separable if and only if $a_i - a_j$ is a unit of k whenever $i \neq j$.

4. Let a, b and c be elements of a commutative ring k. Show that $aX + b$ is separable if and only if a and b are strongly relatively prime. Show that if $b^2 - 4ac$ is a unit, then $aX^2 + bX + c$ is separable.

5. Let E be a discrete commutative ring that is algebraic over a subfield F. Show that E is separable over F if and only if for each finite subset S of k, and finite subset T of E, there is a countable subfield k' of k, containing S, so that for each countable extension field K of k' the ring $K \otimes_{k'} k'[T]$ has no nilpotent elements.

5. PRIMITIVE ELEMENTS

Let E be a commutative ring containing a discrete field k. An element θ of E is a **primitive element** of E over k if $E = k[\theta]$. We shall show that if E is finitely generated and separable, then we can construct a primitive element provided either k is big enough, or E is a discrete field.

We first look at the situation where k is big enough. The key is to show that $k[A,B] = k[C]$ where A and B are commuting separable matrices.

THEOREM 5.1. *If A and B are commuting n-by-n matrices over a discrete field k of cardinality greater than $n(n-1)/2$, and B is separable, then there exists an element c in k such that A and B can be written as polynomials in $A + cB$ with coefficients in k.*

PROOF. We first treat the case where A is also separable. By (4.5) we may assume that A and B are diagonal with diagonal elements $a_1, \ldots, a_n$ and $b_1, \ldots, b_n$. Choose c distinct from $(a_j - a_i)/(b_i - b_j)$ for each pair i and j such that $b_i \neq b_j$. Then $a_i + cb_i \neq a_j + cb_j$ whenever $a_i \neq a_j$ or $b_i \neq b_j$. It follows from (V.3.3) that A and B can be written as polynomials in $A + cB$.

For the general case, we may assume that B has the block-diagonal form

$$\begin{bmatrix} B_1 & & & \\ & B_2 & & \\ & & \ddots & \\ & & & B_n \end{bmatrix}$$

where the block B_i is λ_i times an identity matrix, and the λ_i are distinct. As $AB = BA$, the matrix A has the same block structure as B, but the blocks A_i are arbitrary. We can put each block A_i in Jordan canonical form without affecting B. Choose $c \in k$, as in the special case, such that B and the diagonal of A can be written as polynomials in the diagonal of $A + cB$. Note that $A + cB$ is in Jordan canonical form, so its diagonal can be written as a polynomial in $A + cB$. As $A + cB$ agrees with A off the diagonal, we are done. □

5.2 COROLLARY. *Let E be a commutative ring containing a discrete field k, and let α and β be elements of E such that α is algebraic and β is separable over k. If k is big enough, then there exists θ such that $k[\alpha,\beta] = k[\theta]$.*

PROOF. Let f and g be polynomials over k, with g separable, such that $f(\alpha) = g(\beta) = 0$. Let A and B be matrices as in (4.4). By (5.1), if k is big enough, then there exists c in k such that A and B can be written as polynomials in $A + cB$. Set $\theta = \alpha + c\beta$. □

If the field k is too small, then (5.2) need not hold (Exercise 2) However if E is a discrete field, then (5.2) holds for any k. Classically the proof divides into two cases depending on whether k is finite or infinite. We have to be a little more careful because we may not be able to determine which of these cases we are in. The next lemma provides a primitive element when E is a finite field.

5.3 LEMMA. *Let k be a discrete field and G a finite subgroup of the multiplicative group of nonzero elements of k. Then G is cyclic.*

PROOF. Let x and y be elements of G of orders m and n respectively. We shall construct an element of G of order $q = LCM(m,n)$. Write $q = ab$ where $(a,b) = 1$, and $a|n$, and $b|m$. We shall show that $x^a y^b$ has order q. Clearly $(x^a y^b)^q = 1$; suppose $(x^a y^b)^i = 1$. Then $x^{ai} = y^{-bi}$, so $(x^{ai})^a = 1$ whence $m|a^2 i$, so $b|i$. Similarly, $a|i$, whence $q|i$.

Thus if g is an element of G of maximal order N, then $x^N = 1$ for every x in G. As the polynomial $X^N - 1$ has at most N roots (II.4.5), there are at most N elements of G, so each must be a power of g. □

The trick for the general case is to pass to a subfield that is either finite or contains a lot of elements.

5.4 THEOREM. *Let k be a finitely generated discrete field and N a positive integer. Then either k is finite, or k contains more than N distinct elements.*

PROOF. Either $0,1,2,\ldots,N$ are distinct elements of k, or k has finite characteristic at most N, so we may assume the latter. Let k_0 be the (finite) prime field of k, and let $a_1,\ldots,a_n$ generate k, so $k = k_0(a_1,\ldots,a_n)$. Let $S \subseteq k_0[X_1,\ldots,X_n]$ be the finite set of polynomials such that the degree of each variable X_i at most N; and let $\overline{S}$ be the image of S in k upon taking X_i to a_i. Either $\#\overline{S} \leq N$ or $\#\overline{S} > N$, so we may assume $\#S \leq N$. We will show that k is finite.

Let $p \in k_0[X_1,\ldots,X_n]$. We shall construct a polynomial $q \in S$ such that $p(a_1,\ldots,a_n) = q(a_1,\ldots,a_n)$. If $a_t = 0$, we can replace each positive power of X_t in p by X_t. If $a_t \neq 0$, then, as the sequence $1,a_t,\ldots,a_t^N$ of elements of $\overline{S}$ has $N+1$ terms, two of them must be equal, so the multiplicative order of a_t is at most N. Therefore if $m > 0$ is an integer, then $a_t^m = a_t^j$ for some unique $j < N$, so we can replace each occurence of X_t^m in p by X_t^j, giving us an element q of S. As each element of k is a quotient of such polynomials q, the field k is finite. □

5.5 COROLLARY. *Let $k \subseteq E$ be discrete fields. Suppose α and $\beta \in E$ are algebraic over k, and β is separable over k. Then there is $\theta \in E$ such that $k[\theta] = k[\alpha,\beta]$.*

PROOF. We may assume that k is generated by the coefficients of the separable polynomials satisfied by α and β. By (5.4) either k is big enough so that $k[\alpha,\beta] = k[\theta]$ by (5.2), or k is finite, so $k[\alpha,\beta]$ is finite, and $k[\alpha,\beta] = k[\theta]$ by (5.3). □

It suffices in (5.5) for β to be separable over $k[\alpha]$: see (6.7).

EXERCISES

1. Show that the 3-by-3 diagonal matrices over a two-element field k form a finite separable extension of k that does not admit a primitive element.

2. Let E be a commutative ring containing a discrete field k. Let α and β be elements of E such that α is separable over k, and β satisfies a separable polynomial over $k[\alpha]$. Show that β is separable over k.

3. Construct a Brouwerian example of a field that is neither finite nor infinite.

6. SEPARABILITY AND CHARACTERISTIC p

An extreme case of a polynomial that is not separable is a nonunit f such that $f' = 0$. This is possible even when k is a discrete field, if k has finite characteristic p.

6.1 THEOREM. *Let p be a prime, R a commutative ring such that $pR = 0$, and q a power of p. Then the map taking x in R to x^q is a ring homomorphism, whose image $R^q = \{x^q : x \in R\}$ is a subring of R. If $f \in R[X]$, then $f' = 0$ if and only if $f(X) = f_0(X^p)$ for some $f_0 \in R[X]$.*

PROOF. If $0 < i < p$, then the binomial coefficient $\binom{p}{i}$ is divisible by p, so $(a + b)^p = a^p + b^p$. By induction, the map taking x to x^q is a ring homomorphism; the image of a ring homomorphism is a subring.

If n is not divisible by p, then n is strongly relatively prime to p, so n is invertible in R. Thus if $f' = 0$, then the coefficient of X^n in f, for n not divisible by p, is 0; whence $f(X) = f_0(X^p)$ for some $f_0 \in R[X]$. Conversely, the derivative of $f_0(X^p)$ is clearly 0. □

The following is a criterion for the separability of an element over a discrete field of characteristic p.

6.2 THEOREM. *Let k be a discrete field of characteristic p, let q be a positive power of p, and let α be an element of a commutative ring containing k. Then α is separable over k if and only if $\alpha \in k[\alpha^q]$.*

PROOF. If $\alpha \in k[\alpha^q]$, then writing α as a linear combination of powers of α^q gives a polynomial f in $k[X]$ such that $f' = 1$ and $f(\alpha) = 0$, so α is separable over k. Conversely, suppose α satisfies a separable polynomial f in $k[X]$. It suffices to show that the image β of X in $k[X]/(f(X))$ can be written as a polynomial in β^q. Let T_β be the linear transformation of $k[\beta]$ induced by multiplication by β, and let B be a matrix for T_β. It suffices to show that B can be written as a polynomial in B^q. As B is separable, we may assume that B is diagonal. By (6.1) the map taking x in k to x^q is one-to-one, so diagonal elements of B^q are equal if and only if the corresponding diagonal elements of B are equal. Thus by (V.3.3) B can be written as a polynomial in B^q. □

The next theorem is used in situations where we would like to factor polynomials into irreducibles but can't.

6.3 THEOREM. *Let k be a discrete commutative ring with recognizable units, and S a finite set of monic polynomials in $k[X]$. Then either k has a nonzero nonunit or we can construct a finite set T of monic polynomials in $k[X]$ such that*

(i) *Each element of T is of the form $f(X^q)$ where f is separable, and $q = 1$ or q is a power of a prime that is zero in k.*

(ii) *Distinct elements of T are strongly relatively prime.*

(iii) *Every polynomial in S is a product of polynomials in T.*

PROOF. We will transform the set S into the desired set T, or find a nonzero nonuit of k along the way, proceeding by induction on the sum of the squares of the degrees of the polynomials in S. Given distinct elements $s_1, s_2 \in S$, the Euclidean algorithm constructs either a nonzero nonunit of k, or a monic polynomial h that generates the ideal (s_1, s_2). If $h \neq 1$, then we can replace s_1 and s_2 by h, s_1/h, and s_2/h, decreasing the sum of squares of degrees. So we may assume that the polynomials in S are pairwise strongly relatively prime.

If g is monic of degree $n > 0$, and $g' = 0$, then $n = 0$ in k, so either k has a nonzero nonunit or some prime p is zero in k. In the latter case we can write $g = f(X^p)$. Repeating this argument we see that if $s \in S$, then we may assume that $s = f(X^q)$ where $q = 1$ or q is a power of a prime that is zero in k, and $f' \neq 0$. We may assume that the Euclidean algorithm constructs a monic polynomial h that generates the ideal (f, f'). If $h \neq 1$, then we can replace s by $h(X^q)$ and $s/h(X^q)$, decreasing the sum of the squares of degrees. □

6.4 COROLLARY. *Let $k \subseteq E$ be discrete fields, and α in E be integral over k. Then there exists q, either equal to 1 or a power of the finite characteristic of k, such that α^q is separable over k.*

PROOF. Let g be a nonzero polynomial in $k[X]$ such that $g(\alpha) = 0$. Then by (6.3) we can write g as a product of relatively prime separable polynomials of the form $f(X^q)^n$. If $f(\alpha^q) \neq 0$ for each such polynomial, then each $f(\alpha^q)$ is a zero-divisor of E. □

6.5 LEMMA. *Let p be a prime and R a commutative ring such that $pR = 0$. Let q be a positive power of p, and let $a \in R$. If $X^q - a = f(X)g(X)$, where f and g are strongly relatively prime monic polynomials over R, then either f or g is 1.*

PROOF. As $f'g + fg' = 0$, and f and g are strongly relatively prime, it follows that there exist polynomials f_1 and g_1 such that $f(X) = f_1(X^p)$ and $g(X) = g_1(X^p)$. If $q = p$, then clearly either f_1 or g_1 is 1. Otherwise

$$X^{q/p} - a = f_1(X)g_1(X)$$

and we are done by induction on q, for if

$$s(X)f_1(X^p) + t(X)g_1(X^p) = 1,$$

then letting $s_1(X^p)$ and $t_1(X^p)$ be the sums of the monomials in s and t of degrees divisible by p, we have

$$s_1(X^p)f_1(X^p) + t_1(X^p)g_1(X^p) = 1$$

so
$$s_1(X)f_1(X) + t_1(X)g_1(X) = 1,$$

whence f_1 and g_1 are strongly relatively prime. □

6.6 THEOREM. *Let k be a discrete field of finite characteristic p, let $a \in k$, and let q be a positive power of p. If $X^q - a$ is reducible in $k[X]$, then $a \in k^q$.*

PROOF. By Lemma 6.5 the polynomial $X^q - a$ does not admit strongly relatively prime factors in $k[X]$. Thus, by (6.3) we can write $X^q - a$ as $h(X)^m$ for some monic h in $k[X]$ and $m > 1$. Note that m is a power of p since m divides q. As $h(0)^m = -a$, we can set $b = -h(0)^{m/p}$, and $a = b^q$. □

Let E be an impotent ring and k a discrete subfield of E. The **separable closure** of k in E is the subfield of E consisting of those elements that are separable over k. The field k is **separably closed** in E if the separable closure of k in E is k. The following generalizes Corollary 5.5 and shows that the separable closure is separably closed.

6.7 THEOREM. *Let $k \subseteq E$ be discrete fields. Suppose $\alpha \in E$ is algebraic (separable) over k, and $\beta \in E$ is separable over $k[\alpha]$. Then $k[\alpha,\beta] = k[\theta]$*

for some θ *(and* β *is separable over* k*).*

PROOF. Choose q by (6.4), either 1 or a power of a prime p that is zero in k, so that β^q is separable over k. Then $k[\alpha,\beta] = k[\alpha,\beta^q]$ by (6.2), and $k[\alpha,\beta^q] = k[\theta]$ for some θ in E by (5.5). If α is separable over k, then $k[\theta^p] = k[\alpha^p,\beta^p] = k[\alpha,\beta^p] = k[\alpha,\beta] = k[\theta]$, so θ is separable over k. Thus β is separable over k by (4.5). □

Let $k \subseteq E$ be discrete fields, and let $\alpha \in E$ be algebraic over k. Then, by Corollary 6.4, there exists q, either equal to 1 or a power of the finite characteristic of k, such that α^q is separable over k. If $\alpha^q \in k$, then α is **purely inseparable** over k. An extension E of k is a **purely inseparable extension** if each element of E is purely inseparable over k. The next theorem shows that any finitely generated algebraic extension of a discrete field is the composition of a finitely generated separable extension and a purely inseparable extension.

6.8 THEOREM. *Let* E *be a discrete field that is a finitely generated vector space over a subfield* k. *Then there is a subfield* K *of* E *containing* k, *such that* K *is finitely generated and separable over* k, *and* E *is purely inseparable over* K.

PROOF. Let $\alpha_1,\ldots,\alpha_n$ generate E over k. For each α_i there exists $q(i)$ so that $\alpha_i{}^{q(i)}$ is separable over k. By (4.5) the field $K = k[\alpha_1{}^{q(1)},\ldots,\alpha_n{}^{q(n)}]$ is separable over k. As each α_i is purely inseparable over K, it follows at once that $k[\alpha_1,\ldots,\alpha_n]$ is purely inseparable over K. □

EXERCISES

1. Let k be a discrete field and $f \in k[X]$ a nonconstant polynomial. Show that f is not separable if and only if either $g^2 | f$ for some nonconstant $g \in k[X]$, or $\mathrm{char}\ k = p$ and $g(X^p) | f$ for some nonconstant $g \in k[X]$.

7. PERFECT FIELDS

If k is a discrete field of characteristic p, then $k^p = \{a^p : a \in k\}$ is a subfield of k isomorphic to k. A discrete field k is **perfect** if each nonzero polynomial in $k[X]$ is a product of separable polynomials. The following adaptation of the classical characterization of perfect fields does not require that the characteristic of the field be known.

7.1 THEOREM. *A discrete field k is perfect if and only if for each prime p either $p \neq 0$ in k, or $k^p = k$.*

PROOF. Assume that k is perfect. If p is a prime such that $p = 0$ in k, and $a \in k$, then $X^p - a$ is a product of separable polynomials, and so is reducible. Therefore $a \in k^p$ by Theorem 6.6.

To prove the converse, let $f \in k[X]$. By (6.3) we may assume that $f(X) = g(X^q)$ where g is separable, and either $q = 1$, or k has finite characteristic p and $q = p^e$. If $q = 1$, then f is separable. If $q = p^e$, then the map $x \to x^q$ is an isomorphism from $k[X]$ to $(k[X])^q = k^q[X^q] = k[X^q]$. Thus there is h in $k[X]$ so that $g(X^q) = h(X)^q$. As $n^q \equiv n \pmod p$, the isomorphism preserves formal derivatives, so $g'(X^q) = h'(X)^q$. As g is separable, so is h, so f is a product of separable polynomials. □

If k is a discrete field of characteristic 0, then k is perfect, as $p \neq 0$ in k for each prime p. If k is a finite field of characteristic p, then the map $x \to x^p$ is an isomorphism, so k is perfect; but $k(X)$ is not perfect as $X \notin k(X^p)$. If k is a prime field, we don't need to know its characteristic to establish that it is perfect.

7.2 COROLLARY. *If k is a discrete prime field, then k is perfect.*

PROOF. Let p be a prime. If $p = 0$ in k, then k is the field with p elements so $k = k^p$. □

7.3 DEFINITION. *Let k be a subfield of a discrete field K. Then K is a* **perfect closure** *of k if*

(i) *K is perfect*

(ii) *If $a \in K$, then either $a \in k$, or k has finite characteristic p, and there exists $q = p^e$ such that $a^q \in k$.*

In classical treatments the perfect closure is often constructed within an algebraic closure. Although we may be unable to construct an algebraic closure, we can always construct a perfect closure as a direct limit.

If $k_1 \to k_2 \to k_3 \to \cdots$ is a sequence of rings and ring homomorphisms, then the **direct limit** k_∞ is defined as follows. The elements of k_∞ are the elements of the disjoint union $\cup\, k_i$; elements $a \in k_i$ and $b \in k_j$ are equal if there exists m such that a and b go to the same element in k_m. It is easily checked that k_∞ is a ring. If the k_i are discrete fields, then k_∞ is a discrete field and each k_i may be identified with a subfield of k_∞.

If we are given the finite characteristic of a field, then the construction in Exercise 1 gives a perfect closure. If we do not know what the characteristic of the field is, then we must refine that construction.

7.4 THEOREM. *Every discrete field has a perfect closure.*

PROOF. Let $\{p_n : 1 \leq n\}$ be the increasing enumeration of the primes. For each n, let k_n be k, and define $\varphi_n : k_n \to k_{n+1}$ by

$$\varphi_n(x) = \begin{cases} x^p & \text{if } p \in \{p_j : j \leq n\} \text{ and } p = 0 \text{ in } k \\ x & \text{otherwise.} \end{cases}$$

Let K be the direct limit, and denote the image of $x \in k_n$ in K by x_n. We may identify the image of k_1 in K with k. To show that K is perfect, suppose $\alpha \in K$ and p is a prime equal to 0 in K. There exists $n \in \mathbb{N}$ and $x \in k$ so that $\alpha = x_n$. We may assume that n large enough so that $p_n \geq p$. Then $\varphi_n(x) = x^p$, so

$$\alpha = x_n = \varphi_n(x)_{n+1} = (x^p)_{n+1} = x_{n+1}^p.$$

Thus K is perfect. To verify condition (ii) of (7.3), let $\alpha \in K$ and choose $n \in \mathbb{N}$ and $x \in k$ so that $\alpha = x_n$. If $p_1,\ldots,p_n$ are all nonzero in k, then $\alpha = x_n = x_1 \in k$. Otherwise there exists $j \leq n$ so that $p_j = 0$ in k. Then $\varphi_{n-1}\circ\cdots\circ\varphi_1(x) = x^q$, where $q = p^{n-j}$. Thus

$$\alpha^q = x_n^q = (\varphi_{n-1}\circ\cdots\circ\varphi_1(x_1))_n = x_1 \in k. \quad \square$$

The perfect closure is essentially unique.

7.5 THEOREM. *If K and L are perfect closures of the discrete field k, then there is a unique isomorphism of K and L over k.*

PROOF. Define a function f from K to L as follows. Given $\alpha \in K$, either $\alpha \in k$, in which case we set $f(\alpha) = \alpha$, or there exists $q = p^e$, where p is the finite characteristic of k, such that $\alpha^q \in k$.

In the latter case there is β in L such that $\beta^q = \alpha^q$. As the map $x \to x^q$ is an isomorphism of L, the element β is unique and does not depend on the choice of q. Setting $f(\alpha) = \beta$ we obtain a homomorphism from K to L. Similarly, interchanging the roles of K and L, we can construct a homomorphism from L to K. It is easy to see that these are inverse maps and so are isomorphisms. □

7.6 THEOREM. *A discrete field k is perfect if and only if each algebraic element of an arbitrary discrete field extension of k is separable over k.*

PROOF. If k is perfect, then each polynomial in $k[X]$ is a product of separable polynomials, so each algebraic element of a discrete extension field of k is separable. Conversely let p be a prime such that $p = 0$ in k. Given $a \in k$, let $K \supseteq k$ be a perfect field and choose b in K with $b^p = a$. As b is separable over k, it follows, by Theorem 6.2, that $b \in k[b^p] = k$. By Theorem 7.1 we conclude that k is perfect. □

7.7 THEOREM. *Let E be a discrete field containing the perfect field k. If α in E is algebraic over k, then $k[\alpha]$ is perfect.*

PROOF. Let p be a prime such that $p = 0$ in k. Then $k[\alpha]^p = k^p[\alpha^p]$. As k is perfect, α is separable over k, and $k^p[\alpha^p] = k[\alpha^p] = k[\alpha]$ by Theorem 6.2. Thus $k[\alpha]$ is perfect. □

EXERCISES

1. Let k be a discrete field of finite characteristic p. Show that the direct limit of the sequence $k \to k \to k \to \cdots$, where each map $k \to k$ takes a to a^p, is a perfect closure of k.
2. Construct a Brouwerian example of a discrete field k of finite characteristic p such that neither $k^p = k$ nor $k^p \neq k$.

8. GALOIS THEORY

Let $k \subseteq K$ and $k \subseteq E$ be commutative rings. A ring homomorphism $\sigma: K \to E$ is a **k-homomorphism** if σ is the identity on k. We will often denote the image of an element x under the homomorphism σ by x^σ, and the image of K under σ by K^σ. We extend σ to a $k[X]$-homomorphism of $K[X]$ into $E[X]$ by setting $X^\sigma = X$. The fundamental technique for extending a k-homomorphism of discrete fields is the following.

8.1 LEMMA. *Let $k \subseteq K \subseteq E$ be discrete fields, and $\sigma: K \to E$ a k-homomorphism. Let $f \in K[X]$ be irreducible, $\alpha \in E$ a root of f, and $\beta \in E$ a* *root of* f^σ. *Then* σ *extends to a homomorphism from* $K[\alpha]$ *to* E *that takes* α *to* β.

PROOF. Define a map from $K[\alpha]$ to E by taking $g(\alpha)$ to $g^\sigma(\beta)$ for each $g \in K[X]$. To show that this map is well defined, it suffices to show that if $g(\alpha) = 0$, then $g^\sigma(\beta) = 0$. Suppose $g(\alpha) = 0$. Then f divides g because f is irreducible and $f(\alpha) = 0$; so f^σ divides g^σ, whence $g^\sigma(\beta) = 0$. □

If S is a set that is bounded in number, and $\sigma: S \to S$ is one-to-one, then σ is an isomorphism (see Exercise I.2.11). There is an analogous result in the context of fields.

8.2 LEMMA. *Let* $k \subseteq K$ *be an algebraic extension of discrete fields. Then each* k*-endomorphism of* K *is an automorphism.*

PROOF. Let σ be a k-endomorphism of K. As K is a field, σ is one-to-one. If $\alpha \in K$, then α is a root of a polynomial f in $k[X]$. Let S be the set of roots of f in K. If $\beta \in S$, then $f(\beta^\sigma) = f^\sigma(\beta^\sigma) = f(\beta)^\sigma = 0$, so $\beta^\sigma \in S$. Therefore σ induces a (one-to-one) map from S to S. But S is bounded by the degree of f, so σ maps S onto S, so there is $\beta \in S \subseteq K$ such that $\beta^\sigma = \alpha$. As $\alpha \in K$ was arbitrary, σ is onto. □

Let $k \subseteq K$ be discrete fields. If each element of K satisfies a polynomial in $k[X]$ that is a product of linear polynomials in $K[X]$, then K is said to be **normal** over k.

8.3 THEOREM. *Let* $k \subseteq K$ *be discrete fields with* K *a finitely generated vector space over* k. *Then* K *is normal over* k *if and only if* K *is a splitting field of a polynomial in* $k[X]$.

PROOF. Suppose K is normal and generated by $\alpha_1,\ldots,\alpha_n$ over k. Choose monic $f_i \in k[X]$ such that $f_i(\alpha_i) = 0$ and f_i is a product of linear polynomials in $K[X]$. Then K is a splitting field for $f_1 f_2 \cdots f_n$.

Conversely, suppose K is a splitting field for $f \in k[X]$. Then $f(X) = \Pi_{i=1}^n (X - \alpha_i)$ with $\alpha_i \in K$, and given $x \in K$, there is $p \in k[X_1,\ldots,X_n]$ so that $x = p(\alpha_1,\ldots,\alpha_n)$. Let S be the group of permutations of the indeterminates $X_1,\ldots,X_n$, and set

$$q(X,X_1,\ldots,X_n) = \Pi_{\sigma\in S}\,(X - p(X_1^\sigma,\ldots,X_n^\sigma)).$$

As q is symmetric in $X_1,\ldots,X_n$, Theorem II.8.1 says that the coefficients of $q(X,\alpha_1,\ldots,\alpha_n)$ are polynomials in the coefficients of f, so

$q(X,\alpha_1,\ldots,\alpha_n) \in k[X]$. Clearly $q(X,\alpha_1,\ldots,\alpha_n)$ is a product of linear polynomials in $K[X]$, and $q(x,\alpha_1,\ldots,\alpha_n) = 0$. □

Let $k \subseteq K$ be discrete fields. The set of k-automorphisms of K forms a group $\mathcal{G}(K/k)$ called the **Galois group** of K over k. If K is normal and separable over k, then K is said to be a **Galois extension** of k. The group $\mathcal{G}(K/k)$ is a subset of the set of all functions from K to K, so has the natural inequality given by $\sigma_1 \neq \sigma_2$ if there exists $\alpha \in K$ such that $\sigma_1(\alpha) \neq \sigma_2(\alpha)$. When we say that $\mathcal{G}(K/k)$ is finite, we mean finite with respect to that inequality.

8.4 THEOREM. *Let $k \subseteq K$ be discrete fields. If K is a finite-dimensional Galois extension of k, then $\mathcal{G}(K/k)$ is finite and $\#\mathcal{G}(K/k) = \dim_k K$.*

PROOF. As K is separable over k, there is θ such that $K = k[\theta]$. The minimal polynomial f of θ over k has degree $n = \dim_k K$ and is separable. As K is normal, we can write $f(X) = (X-\theta_1)\cdots(X-\theta_n)$ with $\theta_i \in K$ for each i. As f is separable, the θ_i are distinct. Lemmas 8.1 and 8.2 say that, for each $i = 1,\ldots,n$, there is $\sigma_i \in \mathcal{G}(K/k)$ such that $\sigma_i(\theta) = \theta_i$. As $K = k[\theta]$, this completely determines σ_i. On the other hand, if $\sigma \in \mathcal{G}(K/k)$, then $f(\theta^\sigma) = f^\sigma(\theta^\sigma) = f(\theta)^\sigma = 0$, so $\theta^\sigma = \theta_i$ for some i, whence $\sigma = \sigma_i$. Thus $\mathcal{G}(K/k) = \{\sigma_1,\ldots,\sigma_n\}$, and the σ_i are distinct since the θ_i are distinct. □

We construct k-automorphisms of a normal extension of k by extending k-homomorphisms of subfields.

8.5 LEMMA. *Let $k \subseteq K \subseteq E$ be discrete fields with E normal over k and finite dimensional over K. Then any k-homomorphism of K into E can be extended to a k-automorphism of E.*

PROOF. Let $\sigma: K \to E$ be a k-homomorphism. As E is finite dimensional over K it suffices, by induction, to show that if $\alpha \in E$, then E is finite dimensional over $K[\alpha]$, and σ can be extended to a k-homomorphism of $K[\alpha]$ into E.

As E is finite dimensional over K, Theorem 1.13 says that $K[\alpha]$ is finite dimensional over K, so E is finite dimensional over $K[\alpha]$ by (II.6.6). As E is normal over k, there is a polynomial g in $k[X]$ so that $g(\alpha) = 0$ and g is a product of linear polynomials in $E[X]$. As E is finite dimensional over K, Theorem 1.13 says that there is an irreducible polynomial f in $K[X]$ with $f(\alpha) = 0$. Clearly f divides g in $K[X]$, so f^σ divides $g^\sigma = g$. As g is a product of linear polynomials in $E[X]$, it follows that f^σ has a root β in E. By (8.1) we can extend σ to a k-homomorphism from $K[\alpha]$ to E that takes α to β. □

Normality is related to invariance under automorphisms.

8.6 THEOREM. *Let $k \subseteq K \subseteq E$ be discrete fields. If K is normal over k, then every k-endomorphism of E maps K onto K. If E is finite dimensional and normal over k, and if every k-automorphism of E maps K into K, then K is normal over k.*

PROOF. Suppose K is normal over k and σ is a k-endomorphism of E. By (8.2) it suffices to show that $K^\sigma \subseteq K$. Let $\alpha \in K$, and let $f \in k[X]$ be a product of linear polynomials in $K[X]$ such that $f(\alpha) = 0$. As $f(\alpha^\sigma) = f^\sigma(\alpha^\sigma) = f(\alpha)^\sigma = 0$, and f is a product of linear polynomials in $K[X]$, it follows that $\alpha^\sigma \in K$.

Now suppose E is finite dimensional and normal over k, and every k-automorphism of E maps K into K. If $\alpha \in K$, then α is a root of a polynomial $f \in k[X]$ that is a product of linear polynomials in $E[X]$. As E is finite dimensional over k we may assume that f is irreducible. It suffices to show that each root β of f in E is in K. By (8.1) there is a k-homomorphism $\sigma: k[\alpha] \to E$ taking α to β. By (8.5) we can extend σ to a k-automorphism of E. But each k-automorphism of E maps K into K. Thus $\beta = \alpha^\sigma \in K$. □

A key property of finite-dimensional Galois extensions $k \subseteq K$ is that each element of $K \backslash k$ is moved by some element of $\mathcal{G}(K/k)$. In fact, this is a characterizing property.

8.7 THEOREM. *Let K be a finite-dimensional extension of a discrete field k. Then the following are equivalent.*

(i) *K is a Galois extension of k,*

(ii) *If $\alpha \in K \backslash k$, then there is $\sigma \in \mathcal{G}(K/k)$ such that $\sigma(\alpha) \neq \alpha$.*

(iii) *If $\alpha \in K$, and the minimal polynomial f of α over k has degree n, then there are $\sigma_1, \ldots, \sigma_n$ in $\mathcal{G}(K/k)$ such that $f(X) = (X - \sigma_1(\alpha)) \cdots (X - \sigma_n(\alpha))$, and $\sigma_1(\alpha), \ldots, \sigma_n(\alpha)$ are distinct.*

PROOF. Suppose that (i) holds, and let f be the minimal polynomial over k of the element α in $K\backslash k$. As K is separable and normal over k, and the degree of f is greater than 1, there is a root $\beta \neq \alpha$ of f in K. By (8.1) we can then construct a k-homomorphism $\sigma: k[\alpha] \to K$ such that $\sigma(\alpha) = \beta$, and by (8.5) we can extend σ to an automorphism of K.

Now suppose that (ii) holds, and let α and f be as in (iii). We show by induction how to construct the desired elements $\sigma_1,\ldots,\sigma_n$ of $\mathcal{G}(K/k)$. Suppose that we have constructed $\sigma_1,\ldots,\sigma_s$, with $s < n$, such that $g(X) = (X - \sigma_1(\alpha))\cdots(X - \sigma_s(\alpha))$ is a factor of f, and $\sigma_1(\alpha),\ldots,\sigma_s(\alpha)$ are distinct. As f is irreducible, and g is a proper factor of f, some coefficient of g is in $K\backslash k$. By (ii) there is τ in $\mathcal{G}(K/k)$ such that $g^\tau \neq g$. As $\tau\sigma_i(\alpha)$ is a root of g^τ, for $i = 1,\ldots,s$, there is i with $\tau\sigma_i(\alpha) \notin \{\sigma_1(\alpha),\ldots,\sigma_s(\alpha)\}$. Set $\sigma_{s+1} = \tau\sigma_i$.

That (iii) implies (i) is clear. □

If K is a discrete field, and S is a set of automorphisms of K, then $\{x \in K : x^\sigma = x \text{ for each } \sigma \in S\}$ is a field called the **fixed field** of S. Theorem 8.7 implies that if K is a finite-dimensional Galois extension of k, then k is the fixed field of $\mathcal{G}(K/k)$. Exercise 8 shows that (8.7.ii) is stronger than the condition that k be the fixed field of $\mathcal{G}(K/k)$.

Let $k \subseteq E$ be a finite-dimensional Galois extension. The fundamental theorem of Galois theory concerns the correspondence between subfields K of E containing k that are finite dimensional over k, and finite subgroups of $\mathcal{G}(E/k)$. The subgroup of $\mathcal{G}(E/k)$ associated with K is $\mathcal{G}(E/K)$.

8.8 THEOREM. *Let E be a finite-dimensional Galois extension of a discrete field k. Let K be a subfield of E, containing k, that is finite dimensional over k. Then*

(i) *E is a finite-dimensional Galois extension of K.*

(ii) *$\mathcal{G}(E/K)$ is a finite subgroup of $\mathcal{G}(E/k)$.*

(iii) *The fixed field of $\mathcal{G}(E/K)$ is K.*

(iv) *If K is normal over k, then restriction to K gives a map from $\mathcal{G}(E/k)$ onto $\mathcal{G}(K/k)$ with kernel $\mathcal{G}(E/K)$. So $\mathcal{G}(E/K)$ is a normal subgroup of $\mathcal{G}(E/k)$*

(v) *If $\mathcal{G}(E/K)$ is a normal subgroup of $\mathcal{G}(E/k)$, then K is normal over k.*

PROOF. As E is Galois over k, it is Galois over K; so (i) holds.

Condition (ii) is clear, while (iii) follows from (8.7.ii).

To prove (iv) suppose that K is normal over k and $\sigma \in \mathcal{G}(E/k)$. Lemma 8.6.i says that $K^{\sigma} = K$, so the restriction of σ to K is an element of $\mathcal{G}(K/k)$. Therefore the restriction map defines a homomorphism from $\mathcal{G}(E/k)$ to $\mathcal{G}(K/k)$. The kernel of this map is clearly $\mathcal{G}(E/K)$. By (8.5), any element of $\mathcal{G}(K/k)$ can be extended to a k-automorphism of K. Thus the homomorphism is onto.

To prove (v) note that if $\sigma,\tau \in \mathcal{G}(E/k)$, then $\sigma \in \mathcal{G}(E/K)$ if and only if $\tau\sigma\tau^{-1} \in \mathcal{G}(E/K^{\tau})$; thus $\tau\mathcal{G}(E/K)\tau^{-1} = \mathcal{G}(E/K^{\tau})$. So if $\mathcal{G}(E/K)$ is normal in $\mathcal{G}(E/k)$, then $\mathcal{G}(E/K) = \mathcal{G}(E/K^{\tau})$. By (8.7.ii) this implies that $K^{\tau} = K$, so K is normal. □

It remains to show that every finite subgroup of $\mathcal{G}(E/k)$ is of the form $\mathcal{G}(E/K)$. In fact we shall show that if G is a finite group of automorphisms of a discrete field E, and K is the fixed field of G, then $\dim_K E = \#G$, and E is a Galois extension of K with Galois group G.

Let E be a discrete field and M a monoid. A monoid homomorphism from M to the multiplicative monoid E is called a **character** of M in E. Any automorphism of E is a character of the multiplicative monoid E in E. The set E^M of all functions from M to E is a vector space over E under pointwise addition and scalar multiplication. The natural inequality on E^M is defined by setting $f \neq g$ if there is m in M with $f(m) \neq g(m)$.

The functions $f_1,\ldots,f_n$ in E^M are **linearly independent** if $a_1f_1 + a_2f_2 + \ldots + a_nf_n \neq 0$ whenever $a_1,\ldots,a_n$ are elements of E that are not all zero. The following lemma gives a simple criterion for the linear independence of characters.

8.9 LEMMA. *Let M be a monoid and E a discrete field. Let $\sigma_1,\ldots,\sigma_n$ be characters of M in E such that $\sigma_i \neq \sigma_j$ if $i \neq j$. Then $\sigma_1,\ldots,\sigma_n$ are linearly independent.*

PROOF. We proceed by induction on n. Suppose that $a_1,\ldots,a_n$ are elements of E that are not all zero. If $n = 1$, then $\Sigma_{i=1}^{n} a_i\sigma_i(1) = a_1 \neq 0$, so we may assume, after renumbering, that $a_2 \neq 0$. As $\sigma_1 \neq \sigma_2$, there is x in M so that $\sigma_1(x) \neq \sigma_2(x)$. Let m be an arbitrary element of M and subtract

$$(*)\qquad \sigma_1(x)\left[a_1\sigma_1(m) + \ldots + a_n\sigma_n(m)\right]$$

from

(**) $\quad a_1\sigma_1(xm) + \ldots + a_n\sigma_n(xm) = a_1\sigma_1(x)\sigma_1(m) + \ldots + a_n\sigma_n(x)\sigma_n(m).$

The terms involving a_1 cancel, leaving the term $a_2(\sigma_2(x) - \sigma_1(x))\sigma_2(m)$ plus a linear combination of the characters $\sigma_3,\ldots,\sigma_n$. By induction there is m so that this expression is nonzero, because $a_2(\sigma_2(x) - \sigma_1(x)) \neq 0$. As this expression is the difference of (*) and (**), one of (*) or (**) must be nonzero. Thus either m or xm shows that $\Sigma\, a_i\sigma_i \neq 0$. □

From (8.9) we see that distinct automorphisms of a discrete field E are linearly independent over E. We refine this result a bit.

8.10 LEMMA. *Let K be a discrete field and $\sigma_1,\ldots,\sigma_n$ distinct automorphisms of K. Then there exist $\omega_1,\ldots,\omega_n$ in K such that the $n\times n$ matrix, whose ij^{th} element is $\sigma_i(\omega_j)$, is invertible.*

PROOF. We proceed by induction on n. The case $n = 1$ is handled by taking $\omega_1 = 1$. By induction there exist $\omega_1,\ldots,\omega_{n-1}$ in K such that the $(n-1)\times(n-1)$ matrix Σ_{n-1}, whose ij^{th} element is $\sigma_i(\omega_j)$, is invertible. So there are $a_1,\ldots,a_{n-1}$ in K so that

$$a_1\sigma_1(\omega_j) + \ldots + a_{n-1}\sigma_{n-1}(\omega_j) = \sigma_n(\omega_j)$$

for $j = 1,\ldots,n-1$. By Lemma 8.9, there is $\omega_n \neq 0$ in K so that

$$(*) \qquad a_1\sigma_1(\omega_n) + \ldots + a_{n-1}\sigma_{n-1}(\omega_n) \neq \sigma_n(\omega_n).$$

Let Σ_n be the $n\times n$ matrix whose ij^{th} element is $\sigma_i(\omega_j)$. To show Σ_n is invertible, let the ith-row of Σ_n be denoted by v_i. Replacing v_n by $v_n - (a_1v_1 + \ldots + a_{n-1}v_{n-1})$ we get a matrix with the same determinate as Σ_n, whose last row is all zeros except for the last entry, which is nonzero by (*). The determinant of the modified matrix is $det\ \Sigma_{n-1}$ times the nonzero entry in the lower right corner. Thus $det\ \Sigma_n \neq 0$. □

We can now complete the theory of the Galois correspondence between subgroups of $\mathcal{G}(E/k)$, and fields between k and E.

8.11 THEOREM. *Let E be a discrete field, G a set of n distinct automorphisms of E, and K the fixed field of G. Then there are n elements of E which are linearly independent over K. If G is a group then $dim_K E = n$, and E is a Galois extension of K with Galois group G.*

PROOF. Let $G = \{\sigma_1,\ldots,\sigma_n\}$. By (8.10) there exist $\omega_1,\ldots,\omega_n$ in E such that the matrix Σ_n, whose ij^{th} element is $\sigma_i(\omega_j)$, is invertible. Let $a_1,\ldots,a_n$ be elements of K such that $a_1\omega_1 + \ldots + a_n\omega_n = 0$. Then

$a_1\sigma_i(\omega_1) + \ldots + a_n\sigma_i(\omega_n) = 0$ for each i. Because Σ_n is invertible, this implies that each a_j is zero, so the ω_j's are linearly independent.

Now assume that G is a group. We will show that the ω_j's span E over K. Let $\omega \in E$. As Σ_n is invertible, there are $a_1,\ldots,a_n$ in K so that

$$(*) \qquad a_1\sigma_i(\omega_1) + \ldots + a_n\sigma_i(\omega_n) = \sigma_i(\omega)$$

for $i = 1,\ldots,n$. As G is a group, $\sigma G = G$ for each $\sigma \in G$, so applying σ to (*) we obtain

$$\sigma(a_1)\sigma_i(\omega_1) + \ldots + \sigma(a_n)\sigma_i(\omega_n) = \sigma_i(\omega)$$

for $i = 1,\ldots,n$. Subtracting this from (*) gives

$$(a_1 - \sigma(a_1))\sigma_i(\omega_1) + \ldots + (a_n - \sigma(a_n))\sigma_i(\omega_n) = 0.$$

But Σ_n is invertible, so $a_j = \sigma(a_j)$ for all σ and j, whence $a_1,\ldots,a_n$ are in K. Taking σ_i to be the identity in (*) we get

$$a_1\omega_1 + \ldots + a_n\omega_n = \omega.$$

Thus $\omega_1,\ldots,\omega_n$ span E over K.

Theorem 8.7 shows that E is a Galois extension of K. That G is its Galois group follows from (8.4). □

EXERCISES

1. Let K be an algebraic extension of a discrete field k. Show that K is normal if and only if each polynomial in $k[X]$ having a root in K has a factor in $k[X]$ that is a product of linear polynomials in $K[X]$.

2. Define what a splitting field of a set of polynomials is. Show that an extension K/k is normal if and only if there is a set of polynomials in $k[X]$ whose splitting field over k is K.

3. Formulate and prove (8.9) and (8.10) for K a commutative local ring.

4. Let k be a discrete field and $K = k(X)$. Define two k-automorphisms of K by $\sigma(X) = 1/X$ and $\tau(X) = 1 - X$.

   (i) Show that the group G generated by σ and τ is the symmetric group on three letters.

   (ii) Let $I = (X^2 - X + 1)^3/X^2(X - 1)^2$. Show that X is a solution to the polynomial equation

$$f(Y) = (Y^2 - Y + 1)^3 - IY^2(Y - 1)^2 = 0.$$

(iii) Show that $k(I)$ is the fixed field for G.

5. Let $x_1,\ldots,x_n$ be distinct elements of a discrete field k, and $a_1,\ldots,a_n$ elements of k that are not all zero. Use (7.9) to show that there exists $m \in \mathbb{N}$ so that $a_1x_1^m + \ldots + a_nx_n^m \neq 0$.

6. Let $k \subseteq K$ be discrete fields. Suppose K is factorial, and finite dimensional over k. Show that $\mathcal{G}(K/k)$ is finite and that $\#\mathcal{G}(K/k) \leq dim_k K$.

7. Let K be a splitting field of $X^3 - 2$ over $\mathbb{Q}$. Show that if ω is the quotient of any two distinct roots of $X^3 - 2$ in K, then $\omega^2 + \omega + 1 = 0$. By letting k be a field between $\mathbb{Q}$ and $\mathbb{Q}[\omega]$, construct a Brouwerian example of discrete fields $k \subseteq K$ such that $dim_k K = 3$, but $\mathcal{G}(K/k)$ is not finite.

8. Use Exercise 7 to construct a Brouwerian example, using Markov's principle, of discrete fields $k \subseteq K$ such that K is finite dimensional and separable over k, and k is the fixed field of $\mathcal{G}(K/k)$, but condition 8.7.ii does not hold.

NOTES

A classically equivalent form of the weak Nullstellensatz is that if I is a proper ideal in $k[X_1,\ldots,X_n]$, then there is an algebraic extension field K of k such that the polynomials in I all vanish on some point of K^n. This follows from our version by putting I in a maximal ideal M (Zorn's lemma) and setting $K = k[X_1,\ldots,X_n]/M$. With a suitable (constructive) restriction on k, we can prove this theorem directly.

For a comparison of the constructive and the recursive approaches to splitting fields and algebraic closures see [Bridges-Richman 1987].

Chapter VII. Factoring Polynomials

1. FACTORIAL AND SEPARABLY FACTORIAL FIELDS

In this section we investigate the problem of factoring a polynomial over a discrete field into irreducible factors. We have already seen a Brouwerian example of a discrete field over which this is not possible (IV.2.2). A discrete field k having the property that each nonconstant polynomial over k can be written as a product of irreducible polynomials is called **factorial**. Equivalently, k is factorial if every nonconstant polynomial over k is either irreducible or has a nontrivial factor. As $k[X]$ is a principal ideal domain with recognizable units for any discrete field k, this is equivalent to saying that k is factorial in the sense of definition IV.2.1.

We say that k is **separably factorial** if each *separable* polynomial can be written as a product of irreducible polynomials. Clearly any factorial field is separably factorial. In some respects the notion of a separably factorial field is superior to that of a factorial field: Theorem 2.3 shows that a finite-dimensional extension of a separably factorial field is separably factorial, while Exercise 1.5 gives an example of a finite-dimensional extension of a factorial field that is not factorial.

Finite fields are easily seen to be factorial: given a polynomial, divide it by each polynomial of lower degree and see if the remainder is zero. Algebraically closed discrete fields are also factorial for trivial reasons. The first nontrivial theorem about factorial fields is due to Kronecker who showed that the field $\mathbb{Q}$ of rational numbers is factorial. This follows immediately from Kronecker 1 (Theorem IV.4.8), which implies that the ring $\mathbb{Z}$ of integers is factorial, and (IV.2.6), which implies that the field of quotients of a factorial domain is factorial.

Many fields of interest are of the form $k(\alpha_1,\ldots,\alpha_n)$ where k is either a finite field or $\mathbb{Q}$. As the finite fields and $\mathbb{Q}$ are factorial, we are led to investigate simple extensions $k(\alpha)$ of a factorial field k; the most important cases occur when α is algebraic or when α is transcendental. We first show that an extension of a countable factorial field by a separable element is factorial.

1.2 THEOREM. *Let $k \subseteq E$ be discrete fields. Let $\alpha,\beta \in E$ be such that α is algebraic over k and β is separable over $k[\alpha]$. If k is factorial, then α satisfies an irreducible polynomial over $k[\beta]$.*

PROOF. There exists θ in E such that $k[\alpha,\beta] = k[\theta]$ by (6.7). If k is factorial, then $k[\beta]$ and $k[\theta]$ are both finite dimensional over k by (VI.1.13). Hence $k[\theta]$ is finite dimensional over $k[\beta]$ by (II.6.6), so by (VI.1.13) we can find an irreducible polynomial over $k[\beta]$ that is satisfied by α.

1.3 COROLLARY. *Let E be a discrete field and k a countable factorial subfield. Let β in E be separable over k. Then $k[\beta]$ is factorial.*

PROOF. Let $g(X)$ be a nonconstant polynomial with coefficients in $k[\beta]$, and let $E' \supseteq k[\beta]$ be a countable discrete field containing a root α of g (Theorem VI.3.3). Then α satisfies an irreducible polynomial over $k[\beta]$, which must be a factor of $g(X)$ in $k[\beta][X]$. □

Corollary 1.3 shows that a finitely generated separable extension of a countable factorial field is factorial. Exercise 5 shows that the separability condition is necessary. To remove the countability condition we will show that a discrete field is (separably) factorial if and only if each nonconstant (separable) polynomial either has a root in the field, or is nonzero when evaluated at any element of the field. To this end we construct, from a given polynomial f, a polynomial q so that any coefficient of a monic factor of f is a root of q.

To assure that the polynomial q is a product of separable polynomials in the case that f is separable we prove the following.

1.4 LEMMA. *Let K be a separable extension field of a discrete field k. If $q \in k[X]$ factors into linear factors in $K[X]$, then q is a product of separable polynomials.*

PROOF. Let $\alpha \in K$ be a root of q. The element α satisfies a separable polynomial f in $k[X]$. Let $h = \mathrm{GCD}(q,f)$. Then $h(\alpha) = 0$ and h, being a factor of f, is separable. The polynomial q/h is a product of separable polynomials by induction on degree, so q is is a product of separable polynomials. □

We use a splitting field to construct q, although we could dispense with this device as the coefficients of q can be written directly in terms of the coefficients of f.

1.5 THEOREM. *Let k be a discrete field, and let $f \in k[X]$ be a monic (separable) polynomial. Then there exists a polynomial q in $k[X]$ (which is a product of separable polynomials) such that, for any extension E of k, the coefficients of any monic factor of f in $E[X]$ are roots of q.*

PROOF. Let k_0 be the countable subfield of k generated by the coefficients of f, let K_f be a splitting field for f over k_0, and write

$$f(X) = (X - r_1)(X - r_2)\cdots(X - r_n)$$

in K_f. Let $S \subseteq k[Y_1,\ldots,Y_n]$ be the set of elementary symmetric polynomials in the finite subsets of the indeterminates $\{Y_1,\ldots,Y_n\}$, and set

$$q(X) = \Pi_{\sigma\in S}(X - \sigma(r_1,\ldots,r_n)).$$

As the coefficients of f are the elementary symmetric polynomials in all the r_i, it follows from (II.8.1) that $q \in k_0[X]$, and that q is independent of the splitting field K_f.

If f is separable, then q is a product of separable polynomials by (1.4). If g is a monic factor of f in $E[X]$, then construct a splitting field K_f for f over the subfield of E generated by the coefficients of g and f. Clearly K_f is a splitting field for f over k_0 that contains the coefficients of g. Each of coefficient of g is an elementary symmetric polynomial in some of the roots of f in K_f, hence is a root of q. □

1.6 COROLLARY. *If k is separably closed in K, and the monic separable polynomial f in $k[X]$ has a monic factor g in $K[X]$, then $g \in k[X]$. If k is algebraically closed in K and the monic polynomial f in $k[X]$ has a monic factor g in $K[X]$, then $g \in k[X]$.*

PROOF. The coefficients of any factor of f in $k[X]$ satisfy a monic (separable if f is separable) polynomial in $k[X]$. As k is algebraically (separably) closed in K, these coefficients must be in k. □

Theorem 1.5 guarantees that the coefficients of a monic factor of a monic separable polynomial are separable. We can drop the separability requirement if we demand that the two factors be relatively prime.

1.7 COROLLARY. *Let $k \subseteq E$ be discrete fields. If $f \in k[X]$ and $g,h \in E[X]$ are monic polynomials such that $f = gh$ and $GCD(g,h) = 1$, then the coefficients of g and h are separable over k.*

PROOF. By (VI.6.3) we can write f as a product of monic polynomials in $k[X]$ of the form $F(X^q)$, where F is separable and q is 1 or a power of the characteristic of k. Then $F(X^q) = a(X)b(X)$ where $a(X) = GCD(F(X^q),g(X))$ and $b(X) = GCD(F(X^q),h(X))$. As g is the product of the a's and h is the product of the b's, we may assume that $f = F(X^q)$. We now induct on q. If $q = 1$ we are done by (1.5). If $q > 1$, then $0 = f'(X) = g'h + gh'$ so g divides $g'h$, whence g divides g' so $g' = 0 = h'$. Therefore $g = G(X^p)$ and $h = H(X^p)$, so $F(X^{q/p}) = G(X)H(X)$. By induction the coefficients of G and H, which are the same as the coefficients of g and h, are separable over k. □

We shall show that an arbitrary finite-dimensional extension of a separably factorial field is separably factorial. First we reduce the problem of showing that a field is (separably) factorial to that of deciding whether or not a (separable) polynomial has a root.

1.8 THEOREM (the root test). *A discrete field k is (separably) factorial if and only if for each (separable) f in $k[X]$, either there exists $a \in k$ with $f(a) = 0$, or $f(a) \neq 0$ for all $a \in k$.*

PROOF. If k is (separably) factorial, then we can construct all monic linear factors of f, so the condition is clearly necessary. To prove the converse we use Theorem 1.5 to construct a polynomial q in $k[X]$ so that the coefficients of any monic factor of f in $k[X]$ are roots of q. Using induction on the degree of q, and the condition of the theorem, we can construct the finite set of elements of k that are roots of q. Thus we can obtain a finite set of polynomials containing all the monic factors of f in $k[X]$. The elements of this set can now be tested to see which are factors of f. □

The prototype of a root test is the **rational root test** that says that if s/t is a root in $\mathbb{Q}$, in lowest terms, of the polynomial $a_0 + a_1X + \cdots + a_nX^n$ in $\mathbb{Z}[X]$, then s must divide a_0, and t must divide a_n. This narrows down the possible roots to a finite number, each of which can be tested.

The rational root test, together with Theorem 1.8, provides another proof of Kronecker's theorem that the rational number field is factorial. In fact we get the following slightly improved version of Kronecker 1.

1.9 THEOREM. *If R is a unique factorization domain with finitely many units, then so is $R[X]$. Thus R is factorial.*

PROOF. Let K be the field of quotients of R. To show that each element of $R[X]$ can be factored into irreducibles in $R[X]$, it suffices, by Gauss's Lemma, to show that K is factorial. By Theorem 1.8, it is enough to show that K has a root test. The argument sketched above for the rational root test applies here because R is a unique factorization domain with a finite number of units. □

If R is a unique factorization domain with a finite number of units, then (1.9) and induction show that $R[X_1,\ldots,X_n]$ is a unique factorization domain. So if F is a finite field, then $F(X_1,\ldots,X_n)$ is factorial by (IV.2.6). Also $\mathbb{Q}(X_1,\ldots,X_n)$ is factorial because the ring of integers is a unique factorization domain with two units. This does not cover all the cases of interest. Kronecker 2 (Theorem IV.4.9) shows that if k is a factorial field, then so is $k(X)$. We present here a less elegant proof of this theorem, based on the root test, that includes information about separably factorial fields.

1.10 THEOREM. *If k is a (separably) factorial field, then so is $k(X)$.*

PROOF. Let $f(X,Y) \in k[X,Y]$ (be separable as a polynomial in Y over $k[X]$); it suffices, by the root test, to decide whether or not there exists $\alpha \in k(X)$ such that $f(X,\alpha) = 0$. Write

$$f(X,Y) = a_0(X) + a_1(X)Y + \cdots + a_n(X)Y^n.$$

If we substitute $Y = Z/a_n(X)$ and multiply by $a_n(X)^{n-1}$ we get a monic polynomial in Z, so we may assume that $a_n(X) = 1$, and we need only look for roots α in $k[X]$ that divide $a_0(X)$.

In the separable case, let $f'(X,Y)$ denote the derivative of $f(X,Y)$ as a polynomial in Y, and choose $s(X,Y)$ and $t(X,Y)$ in $k[X,Y]$ such that

$$s(X,Y)f(X,Y) + t(X,Y)f'(X,Y) = g(X)$$

where $g(X)$ is not zero. In the nonseparable case, let $g(X) = 1$. If k is finite, then $k(X)$ is factorial (1.9) so we are done. So (VI.5.4) says that either we are done, or we can find distinct elements $x_1,\ldots,x_m$ in k such that m exceeds the degree of $a_0(X)$, and $g(x_i) \neq 0$ for each i. The polynomial $f(x_i,Y)$ is separable, if necessary, so we can construct the finite set $A_i = \{y \in k : f(x_i,y) = 0\}$. If $f(X,\alpha(X)) = 0$, then $\alpha(x_i) \in A_i$ so we can construct, by unique interpolation, a finite set of candidates for $\alpha(X)$. □

EXERCISES

1. Show that a discrete field k is factorial if and only if each nonconstant polynomial over k is either irreducible or has a nontrivial factor.
2. Give a Brouwerian example of a simple extension $k(\alpha)$ with α neither algebraic nor transcendental.
3. A **prime field** is a discrete field k such that every subfield of k is k. Let $p_1,p_2,p_3,p_4,\ldots$ be the sequence of prime numbers congruent to 1 mod(4), and let a be a binary sequence with at most one 1. Define the ring R to be the ring of integers with the following equality relation: $d = e$ if there is n such that $a_n p_n | d-e$. Show that R is a discrete integral domain, and that the field of quotients k of R is a Brouwerian example of prime field that is not factorial.
4. Show that if $F_1 \subseteq F_2 \subseteq F_3 \subseteq \cdots$ are factorial fields, and F_i is algebraically closed in F_{i+1} for each i, then $F = \bigcup F_i$ is factorial. Conclude that if k is factorial, then so is the field $k(X_1,X_2,\ldots)$ in countably many indeterminates.
5. Let a be a nondecreasing binary sequence. Let $F_i = k$ of Theorem VI.2.10, if $a_i = 0$, and $F_i = K$ of the same theorem if $a_i = 1$. Use Exercise 4 to show that $F = \bigcup F_i$ is factorial. Let θ be an element of a discrete extension field of F such that $\theta^2 = a$. By considering the polynomial $X^2 - b$, show that $F[\theta]$ is a Brouwerian example of a finite-dimensional extension of a factorial field that is not factorial; and therefore a separably factorial field that is not factorial.

2. EXTENSIONS OF (SEPARABLY) FACTORIAL FIELDS

Corollary 1.3 says that if k is a countable factorial field, and α is separable over k, then $k[\alpha]$ is factorial. In this section we remove the countability assumption from that theorem, show that if k is separably factorial and α is algebraic, then $k[\alpha]$ is separably factorial, and construct unique splitting fields for separable polynomials over separably factorial fields.

2.0 THEOREM. *Let $k \subseteq E$ be discrete fields. Let $\alpha,\beta \in E$ be such that α is algebraic over k and β is separable over $k[\alpha]$. If k is separably factorial, then β satisfies an irreducible polynomial over $k[\alpha]$.*

PROOF. By Corollary VI.6.7, there is a primitive element θ for $k[\alpha,\beta]$ over k, which is separable over $k[\alpha]$ by (VI.4.5). To show that β satisfies an irreducible polynomial it suffices, by Theorem VI.1.13, to show that $k[\alpha,\beta]$ is a finite-dimensional vector space over $k[\alpha]$. Choose $q = p^e$, with $e \geq 0$, so that $\lambda = \theta^q$, and hence α^q, is separable over k. Then $k[\lambda]$ and $k[\alpha^q]$ are finite dimensional over k, as k is separably factorial. Therefore, by (II.6.6), $k[\lambda]$ is finite dimensional over $k[\alpha^q]$. Let $1,\lambda,\lambda^2,\ldots,\lambda^s$ be a basis for $k[\lambda]$ over $k[\alpha^q]$. As $\theta^q = \lambda \in k[\lambda,\alpha]$, and θ is separable over $k[\alpha]$ we have, by Theorem VI.6.2, $\theta \in k[\lambda,\alpha]$, so $k[\theta] = k[\lambda,\alpha]$. Thus $1,\lambda,\lambda^2,\ldots,\lambda^s$ generate $k[\theta]$ over $k[\alpha]$. Suppose $\Sigma_{i=0}^{s} a_i\lambda^i = 0$ with $a_i \in k[\alpha]$. Then

$$0 = (\Sigma_{i=0}^{s} a_i\lambda^i)^q = \Sigma_{i=0}^{s} (a_i)^q\lambda^{iq}$$

where $(a_i)^q \in k[\alpha^q]$. As λ is separable over k it follows, by Theorem VI.6.2, that $k[\alpha^q,\lambda] = k[\alpha^q,\lambda^q]$. Thus $1,\lambda^q,\lambda^{q2},\ldots,\lambda^{qs}$ is a basis for $k[\alpha^q,\lambda]$ over $k[\alpha^q]$. Therefore $(a_i)^q = 0$, so $a_i = 0$ for all i. Thus $1,\lambda,\ldots,\lambda^s$ is a basis for $k[\theta]$ over $k[\alpha]$, so $k[\theta]$ is finite dimensional over $k[\alpha]$. □

2.1 LEMMA. *Let k be a discrete subfield of a ring E. Suppose θ in E satisfies a monic polynomial in $k[X]$ of degree $n > 0$. If $\alpha \in k[\theta]$, then either $\theta \in k[\alpha]$, or α satisfies a polynomial in $k[X]$ of degree less than n.*

PROOF. Write $\alpha^i = \Sigma_{j=1}^{n-1} a_{ij}\theta^j$ for $i = 0,\ldots,n-1$. Using row operations we can put the matrix $\{a_{ij}\}$ into upper triangular form. If there is a zero on the diagonal, then $1,\alpha,\ldots,\alpha^{n-1}$ are linearly dependent over k, so

α satisfies a polynomial of degree less than n. If all the diagonal elements are nonzero, then the determinant of $\{a_{ij}\}$ is nonzero, so $\theta \in k[\alpha]$. □

The following theorem, which has classical content, will be used to remove the countability hypothesis from the classically trivial Corollary 1.3.

2.2 THEOREM. *Let $K \subseteq E$ be impotent rings and k a discrete subfield of K that is separably closed in K. Let α in E be algebraic over k. Then $k[\alpha]$ is separably closed in $K[\alpha]$. If, moreover, α is separable and k is algebraically closed in K, then $k[\alpha]$ is algebraically closed in $K[\alpha]$.*

PROOF. Let $\beta \in K[\alpha]$ be separable over $k[\alpha]$. By Theorem VI.6.7, there is θ in $k[\alpha,\beta]$ so that $k[\theta] = k[\alpha,\beta]$. To show that $\beta \in k[\alpha]$ it is enough to show that $\theta \in k[\alpha]$. By Corollary VI.6.4, there exists q so that θ^q, which is in $K[\alpha^q]$, is separable over k, where $q = 1$, or k has finite characteristic p and $q = p^e$. Then, by Theorem VI.4.5, $k[\alpha^q] \subseteq k[\theta^q]$ is separable over k. Let α^q satisfy the separable polynomial $f \in k[X]$, of degree n. We proceed by induction on n to prove that $\theta^q \in k[\alpha^q]$; so $\theta \in k[\alpha]$, by (VI.6.2), as θ is separable over $k[\alpha]$.

If $n = 1$, then $\alpha^q \in k$, so $\theta^q \in K$ and thus $\theta^q \in k$ since k is separably closed in K. Now suppose that $n > 1$. As $\theta^q \in K[\alpha^q]$, it follows, by Corollary VI.1.3, that θ^q satisfies a polynomial of degree n over K. But θ^q satisfies a separable polynomial over k. Taking GCD's of these two polynomials, and using Theorem VI.4.1.iii, we obtain a separable polynomial of degree at most n in $k[X]$ that θ^q satisfies. As $\alpha^q \in k[\theta^q]$, Lemma 2.1 says either $\theta^q \in k[\alpha^q]$, and we are done, or else α^q satisfies a polynomial of degree less than n and we are done by induction. Thus $\theta^q \in k[\alpha^q]$.

Now suppose that α is separable over k, and that k is algebraically closed in K. Let $\theta \in K[\alpha]$ be algebraic over $k[\alpha]$; we want to show that $\theta \in k[\alpha]$. By Theorem VI.5.5, we may assume that $k[\alpha,\theta] = k[\theta]$. Let α satisfy a monic polynomial of degree n in $k[X]$. We proceed by induction on n to show that $\theta \in k[\alpha]$. As $\theta \in K[\alpha]$ it follows, by Corollary VI.1.3, that θ satisfies a monic polynomial in $K[X]$ of degree n. The element θ is also algebraic over k. Applying Corollary 1.6 to the GCD of the polynomials that θ satisfies over k and K we obtain a polynomial of degree at most n in $k[X]$ satisfied by θ. By Lemma 2.1 either $\theta \in k[\alpha]$, and we are done, or α satisfies a monic polynomial of degree less than n. In the latter case induction shows that $\theta \in k[\alpha]$. □

We can now remove the countability restriction of Corollary 1.3 and prove the analogous result for separably factorial fields.

2.3 THEOREM. *Let k be a separably factorial subfield of a discrete field E, and let $\alpha \in E$ be algebraic over k. Then $K = k[\alpha]$ is separably factorial. If α is separable and k is factorial, then K is factorial.*

PROOF. Let $f \in K[X]$ be separable. Let k_0 be the separable closure within k of the countable field generated by the coefficients of a nonzero polynomial in $k[X]$ satisfied by α, together with the coefficients of the powers of α that occur in the coefficients of f. As k is separably factorial, k_0 is a countable field. As we can determine whether separable polynomials over k_0 have roots in k, and hence in k_0, it follows from Theorem 1.8 that k_0 is a separably factorial field. To complete the proof of the first statement it suffices, by Theorem 1.8, to find the roots of f that lie in $k_0[\alpha]$, as Theorem 2.2 says that $k_0[\alpha]$ is separably closed in $k[\alpha]$, and the coefficients of f lie in $k_0[\alpha]$.

As $k_0[\alpha]$ is countable, we can construct, by Theorem VI.3.4, a splitting field L of f over $k_0[\alpha]$. Let $r_1,\ldots,r_s$ be the roots of f in L. By Theorem 2.0, we can find irreducible polynomials g_i in $k_0[\alpha][X]$ that are satisfied by the r_i. If g_i is linear, then $r_i \in k_0[\alpha]$, and we have found a root of f in $k_0[\alpha]$. If no g_i is linear, then f has no roots in $k_0[\alpha]$.

The second statement is proved in exactly the same way as the first, except we take k_0 to be the algebraic closure instead of the separable closure. □

In Section 1 we showed how to construct a splitting field for a polynomial over a countable field. For a separable polynomial over a separably factorial field we can drop the countability assumption.

2.4 COROLLARY. *Let k be a separably factorial field, and let $f \in k[X]$ be a separable polynomial. Then there is a separably factorial splitting field K for f over k.*

PROOF. We proceed by induction on the degree n of f. If $n = 1$, take $K = k$. Suppose $n > 1$. As k is separably factorial, f has an irreducible factor p. Let $F = k[Y]/(p(Y))$. Then F is a field and the element $\alpha = Y$ in F is a root of f, so $f(X) = (X - \alpha)q(X)$ in $F[X]$. By Theorem 2.3, the field F is separably factorial. Thus, by induction, q has a separably factorial splitting field over F. This splitting field is the desired splitting field for f over k. □

Finally we show that any two splitting fields for a separable polynomial over a separably factorial field k are isomorphic over k.

2.5 THEOREM. *Let k_1 and k_2 be separably factorial fields, and $\varphi : k_1 \to k_2$ be an isomorphism. Let p_1 be a separable polynomial in $k_1[X]$, and p_2 be the image of p_1 under φ. Let K_1 and K_2 be splitting fields for p_1 over k_1 and p_2 over k_2 respectively. Then φ can be extended to an isomorphism $K \to L$.*

PROOF. It is clear that the polynomial p_2 is separable over k_2. If $\deg p_1 = 0$, there is nothing to prove, so let α_1 be a root of p_1 in K_1. As k_1 is separably factorial, α_1 is a root of an irreducible factor q_1 of p_1 in $k_1[X]$. Let q_2 be the image of q_2 under φ. Then q_2 is a factor of p_2, so $q_2(\alpha_2) = 0$ for some $\alpha_2 \in K_2$. The isomorphism φ extends to an isomorphism, also denoted by φ, from $k_1[\alpha_1]$ to $k_2[\alpha_2]$ such that $\varphi(\alpha_1) = \alpha_2$. As $k_1[\alpha_1]$ is separably factorial, we can extend this isomorphism, by induction on the degree of p_1, to an isomorphism $K \to L$. □

EXERCISES

1. Let k be a countable separably factorial field. Show that there exists a countable separable extension K of k such that each separable polynomial over K has a root in K. The field K is called a **separable closure** of k.

2. Let k be a countable separably factorial field. Show that if K and L are separable closures of k, then there is an isomorphism of K and L over k.

3. Use Theorem VI.2.10 to construct a classical example showing that the condition that α be separable is needed in Theorem 2.2.

4. Let k be a separably factorial field contained in a discrete commutative ring R. Suppose that R is a finitely generated, separable, algebraic extension of k. Show that R is finite-

dimensional over k. (Look for nontrivial idempotents $e \in R$; note that ke is a subfield of the ring Re and is isomorphic to k)

3. SEIDENBERG FIELDS

Let k be a discrete field of finite characteristic p. Seidenberg's condition P arises from considering the problem of whether a given element of k has a pth root in k, that is, whether the subfield k^p of pth powers of elements of k is detachable from k. The condition introduced by Seidenberg is the first condition in the following theorem.

3.1 THEOREM. *Let k be a discrete field of finite characteristic p. Then the following conditions are equivalent.*

(i) *Given $a_{ij} \in k$ for $1 \leq i \leq m$ and $1 \leq j \leq n$, either there exist $x_j \in k^p$, not all zero, with $\Sigma\, a_{ij}x_j = 0$ for all i; or, whenever $\Sigma\, a_{ij}x_j = 0$ for all i, with $x_j \in k^p$, then $x_j = 0$ for all j.*

(ii) *If K is a finite-dimensional extension of k, then K^p is detachable from K.*

(iii) *If K is a finite-dimensional purely inseparable extension of k, then K^p is detachable from K.*

(iv) *Every finitely generated extension field K of k with $K^p \subseteq k$ is finite dimensional.*

(v) *Every finitely generated k^p-subspace of k is finite dimensional.*

PROOF. To derive (ii) from (i) let $\omega_1,\ldots,\omega_n$ be a basis for K over k, and let $\omega_i^p = \Sigma_{j=1}^n b_{ij}\omega_j$. Then an arbitrary element $\Sigma\, a_j\omega_j$ of K is in K^p if and only if there exist x_i in k^p such that

$$\sum_{j=1}^n a_j\omega_j = \sum_{i=1}^n x_i\omega_i^p = \sum_{j=1}^n\sum_{i=1}^n x_i b_{ij}\omega_j$$

that is, $a_j = \Sigma_{i=1}^n x_i b_{ij}$ for all j or, since the ω_i^p are independent over k^p, that the system of equations

$$x_0 a_j - \sum_{i=1}^n x_i b_{ij} = 0$$

has a nontrivial solution in k^p.

Obviously (iii) follows from (ii). To show that (iv) follows from (iii), let K be a finitely generated extension of k such that $K^p \subseteq k$. Let ω be one of the generators of K over k. Since k^p is detachable from k, either $\omega^p \in k^p$ whence $\omega \in k$ and we are done by induction on the number of generators, or $X^p - \omega$ is irreducible over k (Theorem VI.6.6) so $k(\omega)$ is a

p-dimensional extension of k. Clearly in the latter case $k(\omega)$ also satisfies (iii). But K is a finitely generated extension of $k(\omega)$, so K is finite dimensional over $k(\omega)$, whence over k, by induction on the number of generators.

To derive (v) from (iv) let V be a finitely generated k^p-subspace of k. Then V generates a subfield F of k which is finitely generated over k^p. Let $K = F^{-p} \supseteq k$. Then K is finitely generated, hence finite dimensional over k; so $F = K^p$ is finite dimensional over k^p and therefore the finitely generated subspace V of F is finitely generated over k^p.

To prove (i) from (v) choose a basis for the k^p-subspace of k generated by the a_{ij}. Then the question in (i) reduces to the existence of a nontrivial solution in k^p to a set of homogeneous linear equations with coefficients in k^p, which is decidable by (II.6.2). □

We say that a discrete field k satisfies **Seidenberg's condition *P***, or that k is a **Seidenberg field**, if, whenever p is a prime that is 0 in k, then the conditions in Theorem 3.1 hold.

From (ii) of Theorem 3.1 we see that condition P is inherited by finite dimensional extension fields. We show next that it is inherited by purely transcendental extensions.

3.2 THEOREM. *If k is a Seidenberg field, then so is $k(X)$.*

PROOF. We shall show that (i) of Theorem 3.1 holds for $k(X)$. Suppose

$$\sum_j a_{ij}x_j = 0$$

is a system of equations over $k(X)$ to which we are seeking a nontrivial solution in $k(X)^p = k^p(X^p)$. We may assume that the a_{ij} lie in $k[X]$, and that we are seeking a nontrivial solution in $k^p[X^p]$. Separating the powers of X into residue classes modulo p, we get an equivalent system of equations in which the a_{ij} lie in $k[X^p]$.

The coefficients of the a_{ij} generate a k^p-subspace of k which, by (v) of Theorem 3.1, is finite-dimensional. Let $\lambda_1,\ldots,\lambda_m$ be a basis for that subspace, and write $a_{ij} = \Sigma_k a_{ijk}\lambda_k$ with $a_{ijk} \in k^p[X^p]$. Then we can find a nontrivial solution in $k^p(X^p)$ to the original system if and only if we can find a nontrivial solution in $k^p[X^p]$ to the system

$$\sum_j a_{ijk}x_j = 0,$$

but now the coefficients and the sought-for solution both live in the same field $k^p(X^p)$, so the result follows from (II.6.2). □

Exercise 1.5 shows that a finite-dimensional extension of a factorial field need not be factorial; the missing ingredient is condition P. Call a discrete field k **fully factorial** if any finite-dimensional extension k is factorial.

3.3 THEOREM. *Let k be a discrete field. Then k is fully factorial if and only if k is separably factorial and satisfies condition P.*

PROOF. Suppose k is fully factorial. As k is a finite-dimensional extension of itself, k is (separably) factorial. To show that k satisfies condition P we will verify (ii) of Theorem 3.1. Let K be a finite-dimensional extension of k. Then K is factorial, so for each $a \in K$ the polynomial $X^p - a \in K[X]$ is either irreducible, in which case $a \notin K^p$, or it is reducible, in which case $a \in K^p$ by (VI.6.6).

Conversely, suppose k is separably factorial and satisfies condition P. Let K be a finite-dimensional extension field of k. Then K is separably factorial, by (2.3), and satisfies condition P by (3.1). Let f be a monic polynomial in $K[X]$; we will show that either f has a root in K or it doesn't, so K is factorial by (1.8). By (VI.6.3) we can write f as a product of monic polynomials of the form $g(X^q)$ where g is separable and q is either 1 or a power of a prime that is zero in k, so by induction on degree we may assume that f is of this form. As K is separably factorial, we may assume that g is irreducible. Thus if $g(X^q)$ has a root in K, then g is linear, whence $g(X^q) = X^q - a$. We can decide whether $X^q - a$ has a root in K by (ii) of Theorem 3.1. □

As a corollary we have that every purely transcendental extension of a fully factorial field is fully factorial.

3.4 COROLLARY. *If k is a fully factorial field. then so is $k(X)$.*

PROOF. Theorem 3.3 shows that k is a Seidenberg field, so $k(X)$ is a Seidenberg field by (3.2). But $k(X)$ is factorial by (1.10), so $k(X)$ is fully factorial by (3.2). □

EXERCISES

1. Show directly that the field F of Exercise 1.5 does not satisfy condition P.

2. Show that any two algebraic closures of a countable fully factorial field k are isomorphic over k.

3. Show that a countable discrete field is fully factorial if and only if every finitely generated algebraic field extension is finite dimensional.

4. Construct a Brouwerian example of a discrete commutative ring extension of $\mathbb{Q}$ that is finitely generated and algebraic, but not finite-dimensional, over $\mathbb{Q}$.

5. Show that if k satisfies condition P, then so does the field $k(X_1, X_2, \ldots)$ in countably many indeterminates.

4. THE FUNDAMENTAL THEOREM OF ALGEBRA

The complex numbers that are algebraic over $\mathbb{Q}$ form a discrete field $\mathbb{C}^a$ (Theorem VI.1.9), called the field of **algebraic numbers.** In this section we will show that $\mathbb{C}^a$ is algebraically closed, using many of the results of Galois theory and, without proof, the existence of 2-Sylow subgroups of finite groups. A proof of the corollary, that every monic polynomial of positive degree over $\mathbb{C}$ has a root in $\mathbb{C}$, is outlined in the exercises.

4.1 LEMMA (Intermediate value theorem for $\mathbb{Q}[X]$). Let $f \in \mathbb{Q}[X]$. If a and b are rational numbers with $f(a) < 0 < f(b)$, then there exists c in $\mathbb{R}$ with $f(c) = 0$.

PROOF. We may assume that $a < b$. Define, by induction, two sequences $\{a_n\}$ and $\{b_n\}$, of rational numbers. Let $a_0 = a$ and $b_0 = b$. To define a_n and b_n, for $n > 0$, let $c_n = (a_{n-1} + b_{n-1})/2$, and set.

(i) $a_n = b_n = c_n$ if $f(c_n) = 0$,
(ii) $a_n = c_n$ and $b_n = b_{n-1}$ if $f(c_n) < 0$,
(iii) $a_n = a_{n-1}$ and $b_n = c_n$ if $f(c_n) > 0$.

Clearly $a_{n-1} \le a_n \le b_n \le b_{n-1}$, and $b_n - a_n \le (b - a)/2^n$ for each $n > 0$. Thus the sequence $\{c_n\}$ is Cauchy, so represents a real number c. To show that $f(c) = 0$, we must show that $f(c_n)$ converges to 0.

Applying the remainder theorem to $f(X)$ as a polynomial over $\mathbb{Q}[Y]$, we get $f(X) = (X - Y)g(X,Y) + f(Y)$ for some polynomial $g \in \mathbb{Q}[X,Y]$. It is easy to get a bound M on $\{|g(x,y)| : a \le x,y \le b\}$, so $|f(x) - f(y)| \le M|x - y|$ whenever $a \le x,y \le b$. As $c_n - a_n$ and $b_n - c_n$ converge to 0, it follows that $f(c_n) - f(a_n)$ and $f(b_n) - f(c_n)$ converge to 0. As $f(a_n) \leq 0 \leq f(b_n)$ for each n, it follows that $f(c_n)$ converges to 0. □

4.2 COROLLARY. *Let $f \in \mathbb{Q}[X]$ be of odd degree. Then f has a root in $\mathbb{R}$.*

PROOF. As f is of odd degree there are rational numbers a and b satisfying the conditions of (4.1). □

4.3 LEMMA. If $a + bi \in \mathbb{C}^a$, then there exist $c + di$ in $\mathbb{C}^a$ so that $(c + di)^2 = a + bi$.

PROOF. We first prove the real case, $b = 0$. As $\mathbb{C}^a$ is discrete, either $a = 0$ or $a > 0$ or $a < 0$. If $a > 0$, then by Lemma 4.1 the polynomial $X^2 - a$ has a root in $\mathbb{R}^a$, which we denote by $\sqrt{a}$. If $a < 0$ then $i\sqrt{-a}$ is a root of $X^2 - a$. For the general case we take c and d to be roots of

$$X^2 - \frac{a + \sqrt{a^2 + b^2}}{2} \quad \text{and} \quad X^2 - \frac{-a + \sqrt{a^2 + b^2}}{2}.$$

Then $c + di$ is the desired square root of $a + bi$. □

To prove the final theorem we will need two results about finite groups.

(i) If G is a finite group, and 2^n divides $\#G$, then G contains a subgroup of order 2^n.

(ii) If G is a finite group, then G has a subgroup T, such that $\#T$ is a power of 2 and the index of T in G is odd.

Any subgroup T of G satisfying (ii) is called a **2-Sylow subgroup.** These are standard results in the theory of finite groups, and present no constructive problems.

4.4 LEMMA. *Let $f \in \mathbb{Q}[X]$. Then f has a root in $\mathbb{C}^a$.*

PROOF. Let E be a splitting field of f over $\mathbb{Q}$. It suffices to embed E in $\mathbb{C}^a$. As E is finite dimensional over $\mathbb{Q}$, the Galois group G of E over $\mathbb{Q}$ is finite. Let T be a 2-Sylow subgroup of G, and let K be the fixed field of T. By (VI.5.5) there is an element α in K so that $K = \mathbb{Q}[\alpha]$. Then α satisfies an irreducible polynomial g of degree $\dim_{\mathbb{Q}}K$, which is odd as it is equal to the index of T in G. By (4.2) the polynomial g has a root θ in $\mathbb{C}^a$. By Lemma VI.8.1 there is an embedding $\sigma: K \to \mathbb{C}^a$ that takes α to θ.

Suppose $\#T = 2^n$. By (i) we can find subgroups $T_0 \subseteq \cdots \subseteq T_n = T$ such that $\#T_i = 2^i$. Let K_i be the fixed field of T_i. We shall inductively extend σ, which is defined on K_n, to an embedding of $K_0 = E$ into $\mathbb{C}^a$.

Suppose σ is defined on K_{i+1}. The dimension of K_i over K_{i+1} is 2, so $K_i = L[\beta]$ where β satisfies an irreducible quadratic polynomial h over K_{i+1}. By Lemma 4.3, and the quadratic formula, h^σ has a root in $\mathbb{C}^a$, so we can extend σ to K_i by (VI.8.1). □

4.5 THEOREM (The discrete fundamental theorem of algebra). *The field $\mathbb{C}^a$ is algebraically closed.*

PROOF. Let $f \in \mathbb{C}^a[X]$ be a nonconstant polynomial; we will show that f has a root in $\mathbb{C}^a$. Let K be the subfield of $\mathbb{C}^a$ generated by the coefficients of f. By Theorem VI.3.3, construct a countable discrete field containing K that contains a root θ of f, and let $g \in \mathbb{Q}[X]$ have θ as a root. Induction on the degree of f allows us to assume that f divides g. By Theorem VI.3.4, construct a countable splitting field E for g over $\mathbb{Q}$. As E is a finitely generated separable extension of $\mathbb{Q}$, we can find $\alpha \in E$ so that $E = \mathbb{Q}(\alpha)$. Let $h \in \mathbb{Q}[X]$ be the minimal polynomial of α, and let β be a root of h in $\mathbb{C}^a$. Then g is a product of linear factors in $\mathbb{Q}(\beta)$, as $\mathbb{Q}(\beta) \cong E$ is normal. As f divides g, there is a root of f in $\mathbb{C}^a$. □

A proof of the fundamental theorem of algebra, in the form that every monic polynomial of positive degree over $\mathbb{C}$ has a root in $\mathbb{C}$, is outlined in the exercises.

EXERCISES

1. Let f be a monic polynomial in $\mathbb{C}^a[X]$ of degree $n > 0$. Show that for each $x \in \mathbb{C}^a$ there exists a root r of f in $\mathbb{C}^a$ such that $|r - x|^n \leq |f(x)|$.
2. Let f be a monic polynomial in $\mathbb{C}[X]$ of degree $n > 0$.
   (i) Use Exercise 1 to show that for each $x \in \mathbb{C}$, and $\epsilon > 0$, there exists $r \in \mathbb{Q}(i)$ such that $|f(r)| < \epsilon$ and $|r - x|^n < |f(x)| + \epsilon$.
   (ii) Use (i) to construct a Cauchy sequence $r_1, r_2, \ldots$ in $\mathbb{Q}(i)$ such that $f(r_i)$ converges to 0. Conclude that f has a root in $\mathbb{C}$.

NOTES

Kronecker (1882) showed that algebraic number fields are factorial. The result that finite separable extensions of factorial fields are factorial is found in [van der Waerden 1953], but the proof is incomplete (see the discussion in [Mines-Richman 1982]). An example showing that the separability condition is necessary (Exercise 1.5) first appeared in [Seidenberg 1974].

Condition P was introduced in [Seidenberg 1970]. In that paper it was shown that finite-dimensional, and purely transcendental, extensions of (factorial) fields satisfying condition P also satisfy condition P (and are factorial). Theorem 3.1 is from [Richman 1981].

Brouwer and de Loor (1924) gave a constructive proof that every nonconstant monic polynomial over $\mathbb{C}$ has a root in $\mathbb{C}$. Bishop (1967) proved this assuming only that the polynomial had a nonzero coefficient for some positive power of X.

Chapter VIII. Commutative Noetherian Rings

1. THE HILBERT BASIS THEOREM

The Hilbert basis theorem states that $R[X]$ is Noetherian whenever R is. No one has given a constructive proof of this theorem for our present definition of *Noetherian*, but other definitions have led to proofs. Standard classical proofs of the Hilbert basis theorem are constructive, if by *Noetherian* we mean that every ideal is finitely generated, but only trivial rings are Noetherian in this sense from the constructive point of view. The first proof of a constructively interesting Hilbert basis theorem was given by Jon Tennenbaum; we will present some of his ideas in Section 4. In this section we prove the Hilbert basis theorem for coherent Noetherian rings; in the classical context these are just the Noetherian rings.

If R is a ring, then we denote the R-module $\{f \in R[X] : \deg f < n\}$ by $R[X]_n$. Clearly $R[X]_n$ is a rank-n free R-module. If I is a left ideal of $R[X]$, then $I \cap R[X]_n = \{f \in I : \deg f < n\}$. If M is an R-submodule of $R[X]_n$, then $X^m M$ is an R-submodule of $R[X]_{n+m}$.

1.1 LEMMA. *Let R be a coherent Noetherian ring and let I be the left ideal of $R[X]$ generated by $f_1,\ldots,f_s$. If $f_i \in R[X]_n$ for each i, then there is a finitely generated R-module $M \subseteq R[X]_n$ such that $XM \cap R[X]_n \subseteq M$ and $I \cap R[X]_m = \Sigma_{i=0}^{m-n} X^i M$ for each $m \geq n$.*

PROOF. Corollary III.2.8 says that $R[X]_m$ is a coherent Noetherian R-module. Construct a chain $N_1 \subseteq N_2 \subseteq \cdots$ of finitely generated submodules of $I \cap R[X]_n$ as follows. Let $N_1 = Rf_1 + \cdots + Rf_s$, and let $N_{k+1} = N_k + XN_k \cap R[X]_n$. As $R[X]_{n+1}$ is coherent, the modules N_k are finitely generated; as $R[X]_n$ is Noetherian there is k such that $N_k = N_{k+1}$. Set $M = N_k$. Clearly $XM \cap R[X]_n \subseteq M$.

As $M \subseteq I \cap R[X]_n$ we have $\Sigma_{i=0}^{m-n} X^i M \subseteq I \cap R[X]_m$ for each $m \geq n$. To show that $I \cap R[X]_m \subseteq \Sigma_{i=0}^{m-n} X^i M$, suppose $f \in I \cap R[X]_m$. Write $f = \Sigma_{i=1}^{s} g_i f_i$, where $g_i \in R[X]_d$ for each i, and proceed by induction on d. If $d = 1$, then $f \in M$ and we are done. If $d > 1$, define $h_i \in R[X]$ by $g_i = g_i(0) + Xh_i$ and set $f^* = \Sigma_{i=1}^{s} h_i f_i \in I$. Note that $h_i \in R[X]_{d-1}$. Then $f = \Sigma_{i=1}^{s} g_i(0)f_i + Xf^*$, so $Xf^* \in I \cap R[X]_m$ whence $f^* \in R[X]_{m-1}$. If $m = n$, then induction on d gives $f^* \in M$ so $Xf^* \in XM \cap R[X]_n \subseteq M$ whereupon $f \in N_1 + M = M$. If $m > n$, then induction on d gives $f^* \in \Sigma_{i=0}^{m-1-n} X^i M$, so $f \in N_1 + X\Sigma_{i=0}^{m-1-n} X^i M \subseteq \Sigma_{i=0}^{m-n} X^i M$. □

1.2 THEOREM. *Let R be a coherent Noetherian ring and m a positive integer. If I is a finitely generated left ideal of $R[X]$, and k is a positive integer, then $I \cap R[X]_k$ is a finitely generated R-module. In particular, $I \cap R$ is a finitely generated left ideal of R.*

PROOF. Let $n \geq k$ and $M = I \cap R[X]_n$ be as in Lemma 1.1. Then $I \cap R[X]_k = M \cap R[X]_k$ is finitely generated because $R[X]_n$ is coherent and M is finitely generated. □

1.3 LEMMA. *If R is a coherent Noetherian ring, then $R[X]$ is a coherent ring. If, in addition, R has detachable left ideals, then so does $R[X]$.*

PROOF. Let I be a finitely generated left ideal of the polynomial ring $R[X]$ and let $f \in R[X]$. Choose n such that $R[X]_n$ contains f and a finite family of generators of I. By Lemma 1.1 we have $I \cap R[X]_n = M$ is finitely generated. If R has detachable left ideals, then M is detachable from $R[X]_n$ by Corollary III.2.8, so we can decide whether $f \in I \cap R[X]_n$, whence I is detachable from $R[X]$.

Let $g_1,\ldots,g_k$ generate $M \cap R[X]_{n-1}$, set $g_{k+i} = Xg_i$ for $i = 1,\ldots,k$, and let $g_{2k+1},\ldots,g_\ell$ generated M. Then $g_1,\ldots,g_\ell$ generates I as an $R[X]$-ideal. Let $e_1,\ldots,e_\ell$ be the natural $R[X]$-module basis for $R[X]^\ell$, and let φ be the map from $R[X]^\ell$ onto I that takes e_i to g_i. We shall construct a finite family of generators for $ker\ \varphi$, thus showing that $R[X]$ is coherent.

As $R[X]_n$ is a coherent R-module, the R-module $R^\ell \cap ker\ \varphi$ is finitely generated. Let K be the $R[X]$-submodule of $R[X]^\ell$ generated by these elements together with the elements $Xe_i - e_{k+i}$ for $i = 1,\ldots,k$. We shall show that $K = ker\ \varphi$. Clearly $K \subseteq ker\ \varphi$

Suppose $\Sigma_{i=1}^{\ell} r_i g_i = 0$; we shall show that $\Sigma_{i=1}^{\ell} r_i e_i \in K$. Choose m so that $deg\ r_i \leq m$ for each i. We proceed by induction on m. If $m = 0$, then each r_i is in R, so we are done as $R^\ell \cap ker\ \varphi \subseteq K$. If $m > 0$, write $r_i = s_i + a_i X^m$, where $a_i \in R$ and $deg\ s_i < m$. From $\Sigma\, s_i g_i + X^m \Sigma\, a_i g_i = 0$ we see that $\sum a_i g_i \in M \cap R[X]_{n-1}$, so $\Sigma_{i=1}^{\ell} a_i g_i = \Sigma_{i=1}^{k} b_i g_i$ where $b_i \in R$. Thus $\Sigma_{i=1}^{\ell} a_i e_i - \Sigma_{i=1}^{k} b_i e_i \in K$, so it suffices to show that

$$\sum_{i=1}^{\ell} s_i e_i + X^m \sum_{i=1}^{k} b_i e_i$$

is in K. Because $Xe_i - e_{k+i}$ is in K, for $i = 1,\ldots,k$, it suffices to show that

$$\sum_{i=1}^{\ell} s_i e_i - X^{m-1} \sum_{i=1}^{k} b_i e_{k+i}$$

is in K, but this is true by induction on m. Thus I is finitely presented. □

If R is a ring, and I is a left ideal of $R[X]$, define

$$L(I) = \{a_n \in R : a_n X^n + a_{n-1} X^{n-1} + \cdots + a_0 \in I\}$$

to be the set of formal leading coefficients of the polynomials in I. Note that a polynomial may have different formal leading coefficients (one of them zero) depending on how it is written, and note that $L(I)$ is a left ideal of R.

1.4 LEMMA. *Let R be a coherent Noetherian ring, and I a finitely generated left ideal of $R[X]$. Then the left ideal $L(I)$ of R is finitely generated. Let $J \supseteq I$ be a left ideal of $R[X]$ such that $L(I) = L(J)$, and let m be a positive integer. If $I \cap R[X]_m$ generates I, then $J \cap R[X]_m$ generates J.*

PROOF. Let n and $M = I \cap R[X]_n$ be as in Lemma 1.1, and let $L_n(I) \subseteq L(I)$ be the set of formal leading coefficients of polynomials in M. As $XM \cap R[X]_n \subseteq M$, the set $L_n(I)$ is the image of the map taking each polynomial in M to its coefficient of X^{n-1}. As M is finitely generated, so is $L_n(I)$. We complete the proof of the first claim by showing that $L_n(I) = L(I)$. Suppose $f = f_{m-1}X^{m-1} + \cdots + f_1 X + f_0 \in I$. If $m \leq n$, then $f_{m-1} \in L_n(I)$. If $m > n$, then $f \in \Sigma_{i=0}^{m-n} X^i M$, so we can find $g \in M$ with formal leading coefficient f_{m-1}, whence $f_{m-1} \in L_n(I)$.

Now suppose J and m are as in the second claim. We may assume that $m = n$, rechoosing n and M if necessary. Let $f = f_{d-1}X^{d-1} + \cdots + f_1 X + f_0$ be in J; we shall show that f is in the left ideal generated by $J \cap R[X]_n$. If $d \leq n$, then $f \in J \cap R[X]_n$. If $d > n$, then, since $L_n(I) = L(I) = L(J)$, we can find $g = g_{n-1}X^{n-1} + \cdots + g_1 X + g_0 \in M$ with $g_{n-1} = f_{d-1}$. Then $f - X^{d-n}g \in J \cap R[X]_{d-1}$, so is in the left ideal generated by $J \cap R[X]_n$ by induction on d. Therefore f is in the left ideal generated by $J \cap R[X]_n$. □

1.5 THEOREM (Hilbert basis theorem). *If R is a coherent Noetherian ring, then so is $R[X]$. If, in addition, R has detachable left ideals, then so does $R[X]$.*

PROOF. Lemma 1.3 says that $R[X]$ is a coherent ring, and has detachable left ideals if R does. It remains to show that $R[X]$ is Noetherian. Let $I_1 \subseteq I_2 \subseteq \cdots$ be a chain of finitely generated left ideals of $R[X]$. We construct positive integers $\upsilon(1) < \upsilon(2) < \cdots$ and nonnegative integers $n(1), n(2), \ldots$ as follows. Let $\upsilon(1) = 1$. If $\upsilon(m)$ has been constructed, choose $n(m)$ such that $I_{\upsilon(m)} \cap R[X]_{n(m)}$ generates the left ideal $I_{\upsilon(m)}$. If $\upsilon(m-1)$ and $n(m-1)$ have been constructed, use the fact that $R[X]_{n(m-1)}$ is Noetherian, and $I_i \cap R[X]_{n(m-1)}$ is finitely generated for each i (Theorem 1.2) to choose $\upsilon(m)$ greater than $\upsilon(m-1)$ such that

$$I_{\upsilon(m)} \cap R[X]_{n(m-1)} = I_{\upsilon(m)+1} \cap R[X]_{n(m-1)}.$$

The left ideals $L(I_{\upsilon(1)}) \subseteq L(I_{\upsilon(3)}) \subseteq L(I_{\upsilon(5)}) \subseteq \cdots$ are finitely generated by (1.4) so we can find m such that $L(I_{\upsilon(m-1)}) = L(I_{\upsilon(m+1)})$. Then also $L(I_{\upsilon(m-1)}) = L(I_{\upsilon(m)})$. But $I_{\upsilon(m-1)} \cap R[X]_{n(m-1)}$ generates $I_{\upsilon(m-1)}$, so Lemma 1.4 tells us that $I_{\upsilon(m)} \cap R[X]_{n(m-1)}$ generates $I_{\upsilon(m)}$. Similarly $I_{\upsilon(m)+1} \cap R[X]_{n(m-1)}$ generates $I_{\upsilon(m)+1}$. Thus $I_{\upsilon(m)} = I_{\upsilon(m)+1}$. □

In particular, $\mathbb{Z}[X_1,\ldots,X_n]$ is a coherent Noetherian ring with detachable ideals, as is $k[X_1,\ldots,X_n]$ for k a discrete field.

EXERCISES

1. Let G be the abelian group $\mathbb{Z} \oplus \mathbb{Z}$, and let P be the submonoid of G consisting of those pairs (m,n) such that $m > 0$, or $m = 0$ and $n \geq 0$. Consider the monoid ring $k^{(P)}$ where k is a discrete field. As P is additive we write elements of $k^{(P)}$ as formal sums $\Sigma\, a_p Y^p$ where $p \in P$ and $a_p \in k$. Let S be the multiplicative submonoid of $k^{(P)}$ of elements $\Sigma a_p Y^p$ with $a_{(0,0)} \neq 0$, and let R be $S^{-1}k^{(P)}$.

    (i) Show that R is a Bézout domain with recognizable units, in fact a valuation ring, hence a coherent ring with detachable ideals.

(ii) Let I be the ideal in $R[X]$ generated by the elements $1 + Y^{(0,1)}X$ and $Y^{(1,0)}$. Show that neither $I \cap R$ nor $I \cap R[X]_2$ is a finitely generated R-module.

2. Show that every ideal in $\mathbb{Z}[X]$ that is generated by a finite number of polynomials of degree at most 1 is either principal or uniquely of the form $a(X + b, c)$ with a and c positive, and $0 \leq b < c$. Develop a similar theorem for ideals generated by a finite number of polynomials of degree at most 2.

2. NOETHER NORMALIZATION AND THE ARTIN-REES LEMMA.

A homomorphism $\varphi : R \to S$ of commutative rings is said to **reflect finitely generated ideals** if $\varphi^{-1}I$ is a finitely generated ideal of R for each finitely generated ideal I of S. Corollary 1.2 says that if R is a coherent commutative Noetherian ring, then the map $R \to R[X]$ reflects finitely generated ideals; and by induction, so does the map $R \to R[X_1,\ldots,X_n]$. Classically this is not a very exciting result because *every* ideal of R is finitely generated; but it is a very useful tool for constructing generators for ideals. In this section we give a few important applications.

First we note that the remainder theorem admits the following generalization to polynomials in several variables.

2.1 LEMMA. *Let R be a commutative ring, and φ a ring homomorphism from $R[X_1,\ldots,X_n]$ to R that is the identity on R and takes X_i to a_i. Then $\ker \varphi = (X_1-a_1,X_2-a_2,\ldots,X_n-a_n)$.*

PROOF. Consider the ring endomorphism θ of $R[X_1,\ldots,X_n]$ that is the identity on R and takes X_i to $X_i - a_i$. The map $\varphi\theta$ takes each X_i to 0, so its kernel consists of those polynomials with zero constant term, that is, the ideal $(X_1,\ldots X_n)$. But θ is an automorphism, with $\theta^{-1}X_i = X_i + a_i$. Thus $\ker \varphi = \theta(\ker \varphi\theta) = (X_1-a_1,\ldots,X_n-a_n)$. □

We use (1.2) to compute relations in polynomial rings modulo finitely generated ideals; that is, given polynomials $p_1,\ldots,p_n$, and an ideal I, we construct generators for the set of polynomials f such that $f(p_1,\ldots,p_n) \in I$.

2.2 THEOREM. *Let R be a coherent commutative Noetherian ring. Let $\varphi : R[X_1,\ldots,X_m] \to R[Y_1,\ldots,Y_n]$ be a ring homomorphism that is the identity on R. Then φ reflects finitely generated ideals.*

PROOF. Extend φ to a map φ^* from $R[X_1,\ldots,X_m,Y_1,\ldots,Y_n]$ by defining $\varphi(Y_i) = Y_i$. Then φ^* is the identity on $R[Y_1,\ldots,Y_n]$, so $\ker \varphi^* = (X_1-\varphi(X_1),\ldots,X_m-\varphi(X_m))$ is a finitely generated ideal. If I is a finitely generated ideal of $R[Y_1,\ldots,Y_n]$, then $J = (\varphi^*)^{-1}I$ is a finitely generated ideal, so $J \cap R[X_1,\ldots,X_m] = \varphi^{-1}I$ is a finitely generated ideal of $R[X_1,\ldots,X_m]$ by (1.2). □

Let k be a discrete field. A commutative ring R containing k is said to be **finitely presented over k** if R is isomorphic to $k[X_1,\ldots,X_n]/I$ for some finitely generated ideal I. If R and S are two commutative rings containing the discrete field k, then a **k-algebra map** is a ring map from R to S that is the identity on k.

2.3 COROLLARY. *If R is a finitely presented commutative ring over k, then any k-algebra map from $k[X_1,\ldots,X_n]$ into R has a finitely generated kernel.*

PROOF. Let $R = k[Y_1,\ldots,Y_m]/I$, and let φ be a k-algebra map from $k[X_1,\ldots,X_n]$ to R. Then φ lifts to a map ψ into $k[Y_1,\ldots,Y_m]$, which reflects finitely generated ideals by (2.2). Thus $\psi^{-1}I$, which is the kernel of φ, is finitely generated. □

Any finitely presented commutative ring over a discrete field k can be viewed as an integral extension of a polynomial ring over k.

2.4 THEOREM (Noether normalization). *Let $R = k[x_1,\ldots,x_n]$ be a finitely presented commutative ring over a discrete field k. Then there exist $z_1,\ldots,z_n \in R$, and $m \leq n$, such that $R = k[z_1,\ldots,z_n]$ is integral over $k[z_1,\ldots,z_m]$, and $z_1,\ldots,z_m$ are algebraically independent over k.*

PROOF. We proceed by induction n. If $x_1,\ldots,x_n$ are algebraically independent, which is decidable by (2.3), then we are done. So we may assume there is a nontrivial relation

$$\Sigma\, a_j x_1^{j_1}\cdots x_n^{j_n} = 0,$$

where j is the n-tuple $j_1,\ldots,j_n$ and $a_j = 0$ for all but finitely many j. Let d be an integer such that $d > j_i$ for all i and j with $a_j \neq 0$. Let

$y_i = x_i - x_n^{d^{n-i}}$, for $i < n$, and substitute $y_i + x_n^{d^{n-i}}$ for x_i in the above equation. Expanding the result, we get

$$\Sigma\, a_j x_n{}^{j*} + f(y_1,\ldots,y_{n-1},x_n) = 0,$$

where $j* = j_1 d^{n-1} + j_2 d^{n-2} + \cdots + j_n$, and the degree of x_n in each term of f is less than $j*$ for some j. Since the $j*$ are all distinct, this is an integral equation for x_n over $k[y_1,\ldots,y_{n-1}]$. So $R = k[y_1,\ldots,y_{n-1},x_n]$ is integral over $k[y_1,\ldots,y_{n-1}]$, which is a finitely presented ring by (2.3). By induction $k[y_1,\ldots,y_{n-1}] = k[z_1,\ldots,z_{n-1}]$ is integral over $k[z_1,\ldots,z_m]$ with $z_1,\ldots,z_m$ algebraically independent over k. Set $z_n = x_n$. □

A discrete field K containing k is said to be a **finitely presented extension field** of k if K is the field of quotients of a ring that is finitely presented over k. Noether normalization gives us a way to construct a particularly nice transcendence basis for a finitely presented field extension.

2.5 COROLLARY. *Let $k \subseteq K$ be discrete fields such that K is finitely presented over k. Then there is a finite transcendence basis B of K over k such that K is finite dimensional over $k(B)$.*

PROOF. Let K be the field of quotients of the finitely presented commutative ring R. By Noether normalization we can write R as $k[X_1,\ldots,X_r,Y_1,\ldots,Y_s]/P = k[X,Y]/P$ where $k[X] \cap P = 0$ and R is integral over $k[X]$. Then $K = k(X)[Y]/(k(X)P) = k(X)[y_1,\ldots,y_s]$ is finitely presented, and algebraic, over the field $k(X)$. It suffices to show that K is finite dimensional over $k(X)$, that is, to prove the corollary when the transcendence degree r is 0.

Suppose $r = 0$, so $K = k[Y_1,\ldots,Y_s]/P = k[y_1,\ldots,y_s]$. The subfield $k[y_s]$ is a finite dimensional vector space over k by (2.3). But $K = k[Y_1,\ldots,Y_{s-1},y_s]/P'$, where P' is the image of P in $k[Y_1,\ldots,Y_{s-1},y_s]$. Thus K is finitely presented over the field $k[y_s]$, and so is finite dimensional over $k[y_s]$ by induction on s. Therefore K is finite dimensional over k. □

We also get the following result about fully factorial fields.

2.6 THEOREM. *Let K be a finitely presented field extension of the discrete field k. If k is fully factorial, then so is K.*

PROOF. Let B be a transcendence basis for K over k. Then $k(B)$ is fully factorial by (VII.3.4), so K is finite dimensional over $k(B)$, hence fully factorial. □

As another application of (2.2) we prove the **Artin-Rees lemma.**

2.7 THEOREM (Artin-Rees). *Let I be a finitely generated ideal of a coherent commutative Noetherian ring R. Let N be a finitely generated submodule of a finitely presented R-module M. Then there is k such that for all $n \geq k$ we have*

$$I^{n-k}(I^k M \cap N) = I^n M \cap N.$$

PROOF. By passing to the ring $R \oplus M$, where $m_1 m_2 = 0$ for $m_1, m_2 \in M$, we may assume that M is a finitely generated ideal of R. Let $I = (a_1, \ldots, a_m)$ and $\varphi : R[X_1, \ldots, X_m] \to R[Y]$ be the identity on R and take X_i to $a_i Y$. The kernel of φ is finitely generated, by (2.2), so the image $R[IY]$ of φ is a finitely presented $R[X_1, \ldots, X_m]$-module, hence a coherent Noetherian ring. Now $N[Y]$ is a finitely generated $R[Y]$-ideal, so $R[IY] \cap N[Y] = \varphi\varphi^{-1}(N[Y])$ is a finitely generated $R[IY]$-ideal by (2.2), whence $M[IY] \cap N[Y]$ is a finitely generated $R[IY]$-ideal as $M[IY]$ is a finitely generated $R[IY]$-ideal and $R[IY]$ is coherent. But $M[IY] \cap N[Y] = \Sigma_{i=0}^{\infty} (I^i M \cap N)Y^i$. Choose k so that $M[IY] \cap N[Y]$ is generated, as an $R[IY]$-ideal, by $\Sigma_{i=0}^{k} (I^i M \cap N)Y^i$. □

The ring $R[IY]$ in the preceding proof is known as the **Rees Ring.**

2.8 THEOREM (Krull intersection theorem). *Let M be a finitely presented module over a coherent commutative Noetherian ring R, and let I be a finitely generated ideal of R. Let $A = \cap_n I^n M$. Then $a \in Ia$ for each $a \in A$, so $IA = A$.*

PROOF. Let $N = Ra$. By Artin-Rees there is k such that for all $n \geq k$ we have $I^n M \cap N = I^{n-k}(I^k M \cap N)$. But $N \subseteq I^n M$ so, taking $n = k+1$, we get $N = IN$. □

2.9 COROLLARY. *Let M be a finitely presented module over a coherent commutative Noetherian ring R, and let I be a finitely generated quasi-regular ideal of R. Then $\cap_n I^n M = 0$.*

PROOF. If $a \in \bigcap_n I^n M$, then $a \in Ia$ by (2.8). Thus $a = \lambda a$ for some $\lambda \in I$, whence $(1 - \lambda)a = 0$, so $a = 0$. □

EXERCISES

1. Let k be a discrete field and $R = k[X_1, X_2]/(X_1X_2)$. Find z_1, z_2 in R as in the Noether normalization theorem.
2. **Krull intersection à la Herstein.** Let R be a commutative ring, I a finitely generated ideal of R, and $r \in I$. Let M be a finitely presented Noetherian R-module, and K and N finitely generated submodules of M such that $K \cap N = IN$.
   (i) Show that $L_n(r) = \{x \in M : r^n x \in K\}$ is an ascending sequence of finitely generated submodules.
   (ii) Show that if $L_n(r) = L_{n+1}(r)$, then $(r^n M + K) \cap N = IN$.
   (iii) Show that K is contained in a finitely generated submodule K' of M such that $K' \cap N = IN$ and $I^n M \subseteq K'$ for some n.
   (iv) Use (iii) to prove the Krull intersection theorem.

3. THE NULLSTELLENSATZ

Let k be a discrete field and K an algebraic extension field of k. If $\alpha \in K^n$, then the natural map of $k[X] = k[X_1,\ldots,X_n]$ onto the field $k[\alpha_1,\ldots,\alpha_n]$ has as its kernel

$$M = \{f \in k[X] : f(\alpha) = 0\},$$

so M is a detachable maximal ideal. In this section we will show that, under suitable restrictions on k, every proper finitely generated ideal of $k[X]$ is contained in such an M, and M is finitely generated if K is finite dimensional.

3.1 THEOREM. *Let $k \subseteq K \subseteq E$ be discrete fields such that E is finitely presented over K, and K is finitely presented over k. Then E is finitely presented over k.*

PROOF. Let $X = (X_1,\ldots,X_m)$ and $Y = (Y_1,\ldots,Y_n)$ be sequences of indeterminates, and let $x \in K^m$ and $y \in E^n$ be such that the maps $k[X] \to k(x) = K$ and $K[Y] \to K(y) = E$ have finitely generated kernels. Then each of the maps

$$k[X,Y] \to k[x,Y] \to k(x)[Y] \to k(x)[y]$$

reflects finitely generated ideals; the first map is onto and its kernel is generated by the kernel of $k[X] \to k[x]$, the second map reflects finitely generated ideals by (2.2) because the localization $k[x] \to k(x)$ does, and the third map is onto with a finitely generated kernel. Thus the composite map reflects finitely generated ideals; in particular, its kernel is finitely generated. □

We have seen that finitely presented extension fields are purely transcendental extensions followed by finite-dimensional extensions (2.5). The converse follows from (3.1).

3.2 COROLLARY. *Let $k \subseteq K$ be discrete fields. Then K is finitely presented over k if there is a finite transcendence basis B of K over k such that K is finite dimensional over $k(B)$.*

PROOF. Suppose such a transcendence basis B exists. Clearly $k(B)$ is finitely presented over k, so by (3.1) we may assume that K is finite dimensional over k. If $K = k$ we are done; otherwise choose $x \in K \setminus k$ and let $f \in k[X]$ be an irreducible polynomial satisfied by x. Then $k(x) \cong k[X]/(f)$ is finitely presented over k, and K is finitely presented over $k(x)$ by induction on dimension, so K is finitely presented over k by (3.1). □

From (3.1) and (2.3) it follows that if K is finite-dimensional over k, and $\alpha \in K^n$, then $\{f \in k[X] : f(\alpha) = 0\}$ is a finitely generated ideal.

Let I be a proper finitely generated ideal of the polynomial ring $k[X_1,\ldots,X_n]$. If we want to construct a *finitely generated* maximal ideal containing I, we need k to be factorial, even if $n = 1$. However, we can get the following weaker result without being able to factor.

3.3 LEMMA. *Let k be a discrete field, and I a proper finitely generated ideal of $k[X_1,\ldots,X_n] = k[X]$. Then there is a finitely generated proper ideal $J \supseteq I$ such that $k[X]/J$ is integral over k.*

PROOF. Let $k[x_1,\ldots,x_n] = k[x] = k[X]/I$. By Noether normalization there are $y_1,\ldots,y_r \in k[x]$, algebraically independent over k, such that $k[x]$ is integral over $k[y_1,\ldots,y_r]$. We shall show that y_1 is not a unit in $k[x]$, so we can replace I by $I + (Y_1)$, where $Y_1 \in k[X]$ maps onto y_1, and we are done by induction on r.

Suppose $zy_1 = 1$ for some z in $k[x]$. As $k[x]$ is integral over $k[y_1,\ldots,y_r]$, there exist $a_i \in k[y_1,\ldots,y_r]$ such that

$$z^m + a_{m-1}z^{m-1} + \cdots + a_0 = 0$$

hence

$$1 + a_{m-1}y_1 + \cdots + a_0 y_1^m = 0,$$

but this says that y_1 is a unit in $k[y_1,\ldots,y_r]$, contradicting the algebraic independence of $y_1,\ldots,y_r$. □

A discrete field k **admits splitting fields** if for each polynomial $f \in k[X]$ there is a discrete splitting field for f over k. Fully factorial fields and countable fields admit splitting fields.

3.4 LEMMA. *Let k be a discrete field admitting splitting fields, and let I be a finitely generated proper ideal of $k[X_1,\ldots,X_n] = k[X]$. Then there is an algebraic extension field K of k such that the polynomials in I have a common zero in K^n.*

PROOF. By Lemma 3.3 we may assume that $k[x] = k[X]/I$ is integral over k. For each x_i there is a monic polynomial $f_i \in k[Y]$ such that $f_i(x_i) = 0$. Let K be a splitting field of $f = \Pi_i f_i$, and let J be the ideal generated by I in $K[X]$. As $I \cap k = 0$, the ideal J is proper by (III.3.3). Let $f_1(Y) = \Pi_i(Y - \alpha_i)$ in $K[Y]$. As $f_1(X_1) \in I$,

$$J = J + \Pi_i(X_1-\alpha_i) \supseteq \Pi_i(J + (X_1-\alpha_i)),$$

so $J + (X_1-\beta_1) \neq K[X]$ for some root β_1 of f_1. Replace J by $J + (X_1-\beta_1)$, and repeat the above procedure for $f_2,\ldots,f_m$ constructing a proper ideal $J + N$, where $N = (X_1-\beta_1,\ldots,X_n-\beta_n)$. As N is a detachable maximal ideal, $N \supseteq J$, and $\beta = (\beta_1,\ldots,\beta_n)$ is a common root for the polynomials in J, thus also for the polynomials in I. □

3.5. THEOREM (Nullstellensatz). *Let k be a discrete field that admits splitting fields, I a finitely generated ideal of $k[X] = k[X_1,\ldots,X_n]$, and $f \in k[X]$. Then either $f \in \sqrt{I}$, or there exists an algebraic extension field K of k, and an element $\alpha \in K^n$, such that $f(\alpha) \neq 0$ but $g(\alpha) = 0$ for each $g \in I$.*

PROOF. Consider the ideal I' of $k[X,Y]$ generated by I and $1 - Yf$. If I' is proper, then by Lemma 3.4, there is an algebraic extension field K of k, and elements $\alpha \in K^n$ and $\beta \in K$ such that $g(\alpha) = 0$ for each $g \in I$, and $1 - \beta f(\alpha) = 0$. If $I' = k[X,Y]$, then there are polynomials $h_i \in k[X,Y]$ and

$g_i \in I$ such that

$$1 = h_0(1 - Yf) + \Sigma_{i>0} h_i g_i .$$

Substitute $Y = 1/f$ and choose r greater than the degree of Y in any h_i. Then $f^r \in I$, so $f \in \sqrt{I}$. □

EXERCISES

1. In the proof of (3.3), use Exercise VI.1.1 to show directly that $J = I + (Y_1,\ldots,Y_r)$ is the desired proper ideal, where Y_i maps to y_i.
2. Show that if k is a discrete field admitting splitting fields, and I is an ideal in $k[X_1,\ldots,X_n]$, then $\sqrt{I}$ is the intersection of the detachable maximal ideals containing I.
3. Let $k \subseteq K$ be discrete fields with K algebraically closed. A **k-variety** in K^n is a set of the form $\{x \in K^n : f_i(x) = 0$ for $i = 1,\ldots,n\}$ where $f_1,\ldots,f_n$ are in $k[X] = k[X_1,\ldots,X_n]$. The **ideal** of a k-variety V is $\{f \in k[X] : f(v) = 0$ for each $v \in V\}$. A k-variety V is **irreducible** if whenever V is the union of two k-varieties A and B, then either V = A or V = B.

   (i) Show that the ideal of the k-variety corresponding to $f_1,\ldots,f_n$ is the radical of the ideal generated by $f_1,\ldots,f_n$, and is detachable.

   (ii) Show that a k-variety is irreducible if and only if its ideal is prime

4. TENNENBAUM'S APPROACH TO THE HILBERT BASIS THEOREM

Let R be a ring and M a discrete R-module. A **Noetherian basis function** for M is a function φ taking M^n to R^{n-1}, for $n = 2,3,\ldots$, such that if $x_1,x_2,\ldots$ is an infinite sequence of elements of M, then there exist arbitrarily large n such that $x_n = \Sigma_{i=1}^{n-1} r_i x_i$, where $(r_1,\ldots,r_{n-1}) = \varphi(x_1,\ldots,x_n)$.

We often refer to the function $\rho(x_1,\ldots,x_n) = x_n - \Sigma_{i=1}^{n-1} r_i x_i$ as the basis function, suppressing reference to φ which appears implicitly when we write $x_n = \Sigma_{i=1}^{n-1} r_i x_i$ as a consequence of $\rho(x_1,\ldots,x_n) = 0$. The term *basis function* derives from the fact that $\rho(x_1)$, $\rho(x_1,x_2)$, $\ldots$, $\rho(x_1,\ldots,x_n)$ generate the same submodule as $x_1,\ldots,x_n$, so ρ effects a change of basis.

We call a basis function ρ **consistent** if $\rho(x_1,\ldots,x_n) = 0$ whenever $\rho(x_{i(1)},\ldots,x_{i(m)},x_n) = 0$ for some sequence $1 \le i(1) < i(2) < \cdots < i(m) < n$. Clearly any discrete module admitting a Noetherian basis function admits a consistent one.

4.1 THEOREM. *If a discrete R-module M admits a Noetherian basis function, then M is Noetherian.*

PROOF. Let ρ be a consistent Noetherian basis function for M, and let $I_1 \subseteq I_2 \subseteq \cdots$ be a chain of finitely generated submodules of M. Construct a sequence $x_1, x_2, \ldots$ of elements of M, and a sequence $\alpha(1) < \alpha(2) < \cdots$ of positive integers such that I_j is generated by $x_{\alpha(j)},\ldots,x_{\alpha(j+1)-1}$. Now construct a sequence $\beta(1), \beta(2), \ldots$ of positive integers such that

(i) $\alpha(j) \le \beta(j) < \alpha(j+1)$,

(ii) if $\rho(x_1,\ldots,x_{\alpha(j)-1},x_{\beta(j)}) = 0$, then $\rho(x_1,\ldots,x_{\alpha(j)-1},x_i) = 0$ for $\alpha(j) \le i < \alpha(j+1)$.

There exists n such that $\rho(x_{\beta(1)},\ldots,x_{\beta(n)}) = 0$, so

$$\rho(x_1,\ldots,x_{\alpha(n)-1},x_{\beta(n)}) = 0$$

as ρ is consistent. Therefore $\rho(x_1,\ldots,x_{\alpha(n)-1},x_i) = 0$ if $\alpha(n) \le i < \alpha(n+1)$, so $I_n = I_{n-1}$. □

4.2 THEOREM. *The ring $\mathbb{Z}$ of integers, as a module over itself, admits a Noetherian basis function.*

PROOF. Let d be the nonnegative greatest common divisor of $x_1, x_2, \ldots, x_{n-1}$. If $d = 0$, set $\rho(x_1,\ldots,x_n) = x_n$; otherwise let $\rho(x_1,\ldots,x_n)$ be the (nonnegative) remainder when x_n is divided by d. □

4.3 THEOREM. *Let B be a discrete R-module, and A a detachable submodule of B. If A and B/A admit Noetherian basis functions, then so does B. Moreover the basis function for B can be chosen to extend the basis function for A.*

PROOF. Let π denote the natural map from B to B/A, and let $b_1,\ldots,b_n$ be elements of B. Let $J = \{j_1,\ldots,j_m\}$ consist of those indices j such that $\rho_{B/A}(\pi b_1,\ldots,\pi b_j) = 0$. For $j \in J$, let

$$(r_1^j,\ldots,r_{j-1}^j) = \varphi_{B/A}(\pi b_1,\ldots,\pi b_j)$$

and set

$$a_j = b_j - \sum_{i=1}^{j-1} r_i^j b_i \in A$$

Set $\rho_B(b_1,\ldots,b_n) = b_n$ unless $n = j_m \in J$ in which case we have

$$b_n = a_n + \sum_{i=1}^{n-1} r_i^n b_i$$

so, letting $(s_1,s_2,\ldots,s_{m-1}) = \varphi_A(a_{j_1},\ldots,a_{j_m})$ we have

$$b_n = \rho_A(a_{j_1},\ldots,a_{j_m}) + \sum_{i=1}^{m-1} s_i a_{j_i} + \sum_{i=1}^{n-1} r_i^n b_i .$$

Substituting for each a_j the expression $b_j - \Sigma_{i=1}^{j-1} r_i^j b_i$ we compute elements $t_i \in R$ such that

$$b_n = \rho_A(a_{j_1},\ldots,a_{j_m}) + \sum_{i=1}^{n-1} t_i b_i .$$

Set $\varphi_B(b_1,\ldots,b_n) = (t_1,\ldots,t_{n-1})$ so $\rho_B(b_1,\ldots,b_n) = \rho_A(a_{j_1},\ldots,a_{j_m})$.

We may assume that $\varphi_{B/A}(x_1,\ldots,x_{n-1},0) = (0,\ldots,0)$, so if all the b_i are in A, then $J = \{1,\ldots,n\}$ and $a_j = b_j$ for each j in J. Thus ρ_B extends ρ_A. Given an infinite sequence $b_1,b_2,\ldots$ in B we get an infinite sequence $a_{j_1},a_{j_2},\ldots$ in A and $\rho_A(a_{j_1},\ldots,a_{j_m}) = 0$ for infinitely many m. □

For M an R-module, let $M[X]$ be the set of polynomials in X with coefficients in M. Then $M[X]$ is an $R[X]$-module in a natural way. For $M = R$ the following is the Hilbert basis theorem for rings that admit Noetherian basis functions.

4.4 THEOREM. *If the R-module M admits a Noetherian basis function, then so does the R[X]-module M[X].*

PROOF. In what follows we consider the degree of the zero polynomial to be zero. Let $M[X]_N$ be the set of polynomials in $M[X]$ of degree less than N. Let ρ be a consistent Noetherian basis function for M and let ρ_N be the basis function on the R-module $M[X]_N$ defined inductively by viewing $M[X]_N$ as an extension of $M[X]_{N-1}$ by M. For $f_1,\ldots,f_n \in M[X]$ define $g_1,\ldots,g_n$ as follows. Let $g_1 = f_1$ and suppose we have defined g_i for $i < n$. Let $N = max\{1 + deg\ g_i : 1 \le i < n\}$, and define g_n by the following iterative procedure.

If $deg\ f_n < N$, then set $g_n = f_n$. If $deg\ f_n \ge N$, let c_j be the leading coefficient of g_j, and let c be the leading coefficient of f_n. If $\rho(c_1,\ldots,c_{n-1},c) \ne 0$, set $g_n = f_n$. Otherwise let $e(i) = deg\ f_n - \deg g_i$ and $(r_1,\ldots,r_{n-1}) = \varphi(c_1,\ldots,c_{n-1},c)$, and replace f_n by $f_n - \Sigma_{i=1}^{n-1} r_i x^{e(i)} g_i$. This lowers the degree of f_n, and we start again at the top of the paragraph. Note that this construction guarantees that $\rho(c_1,\ldots,c_n) \neq 0$ if $deg\ g_n \geq N$.

Given $f_1,\ldots,f_n \in M[X]$, construct $g_1,\ldots,g_n$ as above, let $N = max\{1 + deg\ g_i : 1 \leq i < n\}$, and set $\rho(f_1,\ldots,f_n) = \rho_N(g_1,\ldots,g_n)$.

If $f_1,f_2,\ldots$ is an infinite sequence in $M[X]$, construct the sequence $g_1,g_2,\ldots$ as above and let $N(k) = max\{1 + deg\ g_i : 1 \leq i \leq k\}$. For each k there exists $n > k$ such that either $N(k) < N(n)$ or $\rho(f_1,\ldots,f_n) = 0$; indeed if $\pi_{N(k)}$ denotes the projection of $M[X]$ on $M[X]_{N(k)}$, then there exists $n > k$ such that

$$\rho_{N(k)}(\pi_{N(k)}g_1,\ldots,\pi_{N(k)}g_n) = 0.$$

Thus either $N(k) < N(n)$ or $N(k) = N(n)$ and $\rho(f_1,\ldots,f_n) = \rho_{N(n)}(g_1,\ldots,g_n) = 0$.

We can construct a sequence $\alpha(1) < \alpha(2) < \cdots$ of positive integers such that for each k

(i) either $N(\alpha(k)) < N(\alpha(k+1))$ or $\rho(f_1,\ldots,f_{\alpha(k+1)}) = 0$,

(ii) $N(\alpha(k)) = N(\alpha(k+1)) = \cdots = N(\alpha(k+1)-1)$.

If c_j is the leading coefficient of g_j, then there exist arbitrarily large values of n such that $\rho(c_{\alpha(1)},\ldots,c_{\alpha(n)}) = 0$. But this can only happen if $N(\alpha(n)) = N(\alpha(n)-1)$, so $\rho(f_1,\ldots,f_n) = 0$. □

EXERCISES

1. Show that any finite module admits a Noetherian basis function.

2. Show that any discrete module admitting a Noetherian basis function admits a consistent one.

3. Strike the phrase *arbitrarily large* from the definition of a Noetherian basis function. Show that if M admits a Noetherian basis function in this new sense, then it admits one in the old sense. Why require arbitrarily large n? It's used in the proofs of (4.3) and (4.4); can these be easily circumvented?

4. Let M be a Noetherian R-module with detachable submodules. Must M admit a Noetherian basis function? (probably not)

5. Call an R-module M a **Tennenbaum module** if there exists a discrete R'-module M' that admits a Noetherian basis function, a map $\varphi : R' \to R$ of rings, and an epimorphism $\psi : M' \to M$ of abelian groups

such that $\psi(rx) = \varphi(r)\psi(x)$ for each r in R' and x in M'. Show that every Tennenbaum module is Noetherian. Show that the Hilbert basis theorem holds for Tennenbaum rings.

6. Let a be a binary sequence, and let I be the ideal in the ring of integers $\mathbb{Z}$ generated by the elements $a_n n!$. Show that $\mathbb{Z}/I$ is a Brouwerian example of Tennenbaum ring that is not coherent. Is every Noetherian ring a Tennenbaum ring?

5. PRIMARY IDEALS

A prime ideal in a commutative ring is analogous to a prime number--more precisely, to the set of all multiples of a fixed prime number. A *primary* ideal is like a power of a prime number, and the theorem that every number is the product of powers of distinct prime numbers has an analogy in more general rings, such as polynomial rings in several variables over the integers, in the Lasker-Noether decomposition theorem: every finitely generated ideal is a finite intersection of finitely generated primary ideals. In Sections 7 and 8 we study these *Lasker-Noether* rings. In this section we present some basic properties of primary ideals in a commutative ring.

Let R be a commutative ring. An ideal Q of R is **primary** if $xy \in Q$ implies $x \in Q$ or $y^n \in Q$ for some n. Thus a detachable ideal Q is primary if and only if R/Q is a discrete ring in which every zero-divisor is nilpotent.

5.1 PROPOSITION. *If Q is a primary ideal of a commutative ring, then $\sqrt{Q}$ is prime.*

PROOF. If $xy \in \sqrt{Q}$, then $(xy)^n \in Q$, so $x^n \in Q$ or $y^{nm} \in Q$ for some m. Thus either $x \in \sqrt{Q}$ or $y \in \sqrt{Q}$. □

If Q is a primary ideal of a commutative ring R, and $\sqrt{Q} = P$, then we say that Q **belongs to** P, or P **belongs to** Q.

5.2 THEOREM. *Let P and Q be ideals of a commutative ring. Then Q is a primary ideal belonging to P, if and only if the following three conditions hold.*

(i) $Q \subseteq P$

(ii) *If $r \in P$, then $r^m \in Q$ for some m*

(iii) *If* $rs \in Q$, *then* $r \in Q$ *or* $s \in P$.

PROOF. Clearly conditions (i) and (ii) are equivalent to $Q \subseteq P \subseteq \sqrt{Q}$. Suppose (iii) holds. If $rs \in Q$, then $r \in Q$ or $s \in P$. As $P \subseteq \sqrt{Q}$, this implies that $r \in Q$ or $s^n \in Q$ for some n; thus Q is primary. To show that $\sqrt{Q} \subseteq P$, let $q \in \sqrt{Q}$. Then $q^n \in Q \subseteq P$ for some n. By (iii) we have $q^{n-1} \in Q$ or $q \in P$; so by induction on n we obtain $q \in P$. Thus $\sqrt{Q} = P$. Conversely, suppose Q is primary and $\sqrt{Q} = P$. If $rs \in Q$, then $r \in Q$ or $s^n \in Q$ for some n; but $s^n \in Q$ implies $s \in \sqrt{Q} = P$. Hence condition (iii) holds. □

5.3 COROLLARY. *Let* Q_1 *and* Q_2 *be primary ideals belonging to* P. *Then* $Q_1 \cap Q_2$ *is a primary ideal belonging to* P. □

5.4 THEOREM. *Let* P *be a detachable maximal ideal of a commutative ring* R. *Let* $Q \subseteq P$ *be an ideal such that* $p \in P$ *implies* $p^n \in Q$ *for some* n. *Then* Q *is a primary ideal belonging to* P.

PROOF. It suffices to verify condition (iii) of Theorem 5.2. Suppose $rs \in Q$. If $s \in P$ we are done, so we may assume that $s \notin P$. As P is maximal there exists x in R and p in P such that $p + xs = 1$. Choose n so that $p^n \in Q$. Then $(p+xs)^n = p^n + ys = 1$ for some y in R. Hence $r = rp^n + yrs \in Q$. □

5.5 COROLLARY. *Let* P *be a detachable maximal ideal of a commutative ring* R, *and let* Q *be an ideal such that* $P^n \subseteq Q \subseteq P$ *for some* n. *Then* Q *is a primary ideal belonging to* P. □

The following proposition extends the characteristic property of a primary ideal from elements to finitely generated ideals.

5.6 LEMMA. *Let* Q *be a primary ideal of a commutative ring* R, *and* $P = \sqrt{Q}$. *Let* I *and* J *be finitely generated ideals such that* $IJ \subseteq Q$. *Then either* $I \subseteq Q$ *or* $J \subseteq P$.

PROOF. Let $I = (a_1,\ldots,a_m)$ and $J = (b_1,\ldots,b_n)$. As $a_ib_j \in Q$, either $a_i \in Q$ or $b_j \in P$. If $b_j \in P$ for each j, then $J \subseteq P$; otherwise $a_i \in Q$ for each i, so $I \subseteq Q$. □

5.7 THEOREM. *Let* Q *be a finitely generated primary ideal of a commutative ring* R, *and let* $P = \sqrt{Q}$. *Let* I *be a finitely generated ideal of* R. *If* $Q{:}I$ *is finitely generated, then either* $I \subseteq Q$ *or* $Q{:}I$ *is a primary* *ideal belonging to P.*

PROOF. As $I(Q:I) \subseteq Q$, Lemma 5.6 tells us that either $I \subseteq Q$ or $Q:I \subseteq P$. We shall show that in the latter case $Q:I$ is a primary ideal belonging to P. We use the characterization of Theorem 5.2. We have just hypothesized condition (i) of that theorem, while condition (ii) holds because $Q \subseteq Q:I$. To check (iii) suppose $rs \in Q:I$. Then $rsI \subseteq Q$, so either $rI \subseteq Q$ or $s \in P$; that is, either $r \in Q:I$ or $s \in P$. □

5.8 PROPOSITION. *Let $\varphi : R \to R'$ be a map of commutative rings, and let P and Q be ideals of R'. If Q is detachable, then so is $\varphi^{-1}(Q)$. If Q is a primary ideal belonging to P, then $\varphi^{-1}(Q)$ is a primary ideal belonging to $\varphi^{-1}(P)$.*

PROOF. Suppose Q is detachable and $x \in R$. Then $x \in \varphi^{-1}(Q)$ if and only if $\varphi(x) \in Q$; hence $\varphi^{-1}(Q)$ is detachable. Suppose Q is a primary ideal belonging to P. Clearly $\varphi^{-1}(Q) \subseteq \varphi^{-1}(P)$. If $r \in \varphi^{-1}(P)$, then $\varphi(r) \in P$, so $\varphi(r)^n \in Q$, so $r^n \in \varphi^{-1}(Q)$. Finally if $rs \in \varphi^{-1}(Q)$, then $\varphi(rs) \in Q$, so $\varphi(r) \in Q$ or $\varphi(s) \in P$. Thus $r \in \varphi^{-1}(Q)$ or $s \in \varphi^{-1}(P)$, so $\varphi^{-1}(Q)$ is a primary ideal belonging to $\varphi^{-1}(P)$. □

EXERCISES

1. Let k be a discrete field, and let $R = k[X,Y]$. Show that R is a coherent Noetherian ring with detachable ideals. Let $P = (X,Y)$ and $Q = (X,Y^2)$. Show that the inclusions $P^2 \subseteq Q \subseteq P$ are proper. Show that Q is a primary ideal belonging to P. Conclude that a primary ideal need not be a power of a prime ideal.

2. Let k be a discrete field, and let $R = k[X,Y,Z]/(XY - Z^2)$. Let $x,y,z \in R$ be the images of X,Y,Z in R. Show that $P = (x,z)$ is a detachable prime ideal of R, but that P^2 is not primary. Compare with Corollary 5.5.

3. Let Q_1 and Q_2 be primary ideals belonging to a detachable maximal ideal P. Show that $Q_1 + Q_2$ and Q_1Q_2 are primary ideals belonging to P. Show that in the ring $\mathbb{Q}[X,Y,Z_1,Z_2]$ the ideals

$$Q_1 = ((X,Y)^3,\ Z_1X + Z_2Y) \quad \text{and} \quad Q_2 = ((X,Y)^3,\ Z_1Y + Z_2X)$$

are primary belonging to (X,Y), but that $Q_1 + Q_2$ is not primary because it contains $Z_1(X^2 - Y^2)$ (Seidenberg).

4. Let φ be a homomorphism from a ring R onto a ring R', and Q an ideal containing the kernel of φ. Show that Q is a primary ideal belonging to P if and only if $\varphi(Q)$ is a primary ideal belonging to $\varphi(P)$.

6. LOCALIZATION.

Let S a multiplicative submonoid of a commutative ring. For each R-module M we define the **S-torsion submodule** of M to be

$$\tau_S(M) \quad = \quad \{x \in M : sx = 0 \text{ for some } s \text{ in } S\}.$$

It is readily checked that $\tau_S(M)$ is indeed a submodule of M. If $M = \tau_S(M)$, we say that M is **S-torsion.**

The following theorem shows that the map $R \to S^{-1}R$ reflects finitely generated ideals if and only if $\tau_S(R/I)$ is finitely generated for each finitely generated ideal I of R.

6.1 THEOREM. *Let S be a multiplicative submonoid of a commutative ring R, let I be an ideal of R, and let $x \in R$. Then the following conditions are equivalent.*

(i) $x/1 \in S^{-1}I$,

(ii) $sx \in I$ *for some* s *in* S,

(iii) x *represents an element of* $\tau_S(R/I)$.

PROOF. Clearly (ii) and (iii) are equivalent. If (i) holds, then $x/1 = y/s_1$ for some $y \in I$ and $s_1 \in S$. Therefore there exists s_2 in S such that $s_2(s_1x - y) = 0$, so we can take $s = s_2s_1$ in (ii). Conversely if (ii) holds, then $x/1 = (sx)/s \in S^{-1}I$. □

We say that a module M is **S-bounded** if there exists s in S such that $sM = 0$. It is readily seen that any finitely generated S-torsion module is S-bounded. For S-torsion submodules of finitely presented modules over a coherent ring, the converse is true.

6.2 THEOREM. *Let S be a multiplicative submonoid of a coherent commutative ring R. If M is a finitely presented R-module, and $\tau_S(M)$ is S bounded, then $\tau_S(M)$ is finitely generated.*

PROOF. Let s in S be such that $s\tau_S(M) = 0$. Then $\tau_S(M)$ is the kernel of the endomorphism of M induced by multiplication by s, so $\tau_S(M)$ is finitely generated by (III.2.2) and (III.2.6). □

6.3 LEMMA. *Let S be a multiplicative submonoid of a commutative ring R. Let M' be a submodule of the R-module M. If* $\tau_S(M')$ *and* $\tau_S(M/M')$ *are S bounded, then so is* $\tau_S(M)$.

PROOF. Choose s and t in S such that $s\tau_S(M') = 0$ and $t\tau_S(M/M') = 0$. If $x \in \tau_S(M)$, then x represents an element of $\tau_S(M/M')$, so $tx \in M'$ whence $tx \in \tau_S(M')$. Therefore $stx = 0$, so we have proved that $st\tau_S(M) = 0$. □

6.4 THEOREM. *Let S be a multiplicative submonoid of a coherent commutative ring R. If* $\tau_S(R/I)$ *is S-bounded for each finitely generated ideal I of R, (in particular, if* $R \to S^{-1}R$ *reflects finitely generated ideals) then*

(i) $\tau_S(M)$ *is finitely generated for each finitely presented R-module M.*

(ii) *If R has detachable ideals, then so does* $S^{-1}R$.

PROOF. To prove (i) let $x_1,\ldots,x_n$ generate M and let M' be the submodule of M generated by $x_1,\ldots,x_{n-1}$. By induction on n we have $\tau_S(M')$ is S-bounded. The ideal $I = \{r \in R : rx_n \in M'\}$ is finitely generated because R is coherent and M is finitely presented. By hypothesis $\tau_S(R/I)$ is S-bounded, so $\tau_S(M/M') \cong \tau_S(R/I)$ is S-bounded. Thus $\tau_S(M)$ is S-bounded by (6.3), hence finitely generated by (6.2).

To prove (ii) let J be a finitely generated ideal of $S^{-1}R$. Then $J = S^{-1}I$ for some finitely generated ideal I of R. Choose t in S so that $t\tau_S(R/I) = 0$. If $x \in R$ and $tx \in I$, then $x/s = tx/(ts) \in J$ for each s in S. Conversely, if $x/s \in J$, then $x/1 \in J$ so $s_1x \in I$ for some s_1 in S, by (6.1), whereupon $tx \in I$. Thus we can tell whether $x/s \in J$ by checking to see if $tx \in I$. □

6.5 THEOREM. *Let P be a finitely generated detachable proper prime ideal of a coherent commutative ring R, and let M be a finitely presented R-module such that* $P^nM = 0$ *for some positive integer n. Then* $\tau_{R\setminus P}(M)$ *is finitely generated.*

PROOF. Let $S = R\setminus P$. As M is a module over R/P^n, we may assume that $P^n = 0$, and proceed by induction on n. If $n = 1$, then $P = 0$ so R is a discrete integral domain and S consists of the nonzero elements of R. If I is any finitely generated ideal of R, then $\tau_S(R/I) = 0$ if $I = 0$, and

$s\tau_S(R/I) = 0$ for any nonzero element s in I; we may decide which of these alternatives holds because R is discrete and I is finitely generated. Thus $\tau_S(R/I)$ is S-bounded for any finitely generated ideal I, so $\tau_S(M)$ is bounded by (6.4), and hence is finitely generated by (6.2).

If $n > 1$, then the modules PM and M/PM are finitely presented and annihilated by P^{n-1}. Thus $\tau_S(PM)$ and $\tau_S(M/PM)$ are finitely generated by induction on n, and hence S-bounded. So $\tau_S(M)$ is S-bounded by (6.3) and therefore finitely generated by (6.2). □

6.6 THEOREM. *Let S be a multiplicative submonoid of the commutative Noetherian ring R. Then $S^{-1}R$ is Noetherian.*

PROOF. Let $J_1 \subseteq J_2 \subseteq \cdots$ be a chain of finitely generated ideals of $S^{-1}R$. Then we can construct a chain $I_1 \subseteq I_2 \subseteq \cdots$ of finitely generated ideals of R such that $J_j = S^{-1}I_j$ for each j. There exists n such that $I_n = I_{n+1}$, so $J_n = J_{n+1}$. □

6.7 COROLLARY. *Let R be a coherent Noetherian ring with detachable ideals. Let P be a finitely generated prime ideal of R such that $P^n = 0$ for some n. Then R_P is a coherent Noetherian ring with detachable ideals.*

PROOF. Theorem 6.6 shows that R_P is Noetherian. Coherence follows from Exercise III.3.4. As $P^n = 0$, Theorem 6.5 says that $\tau_{R\setminus P}(M)$ is finitely generated for each finitely presented R-module M, so R_P has detachable ideals by (6.4). □

See Exercise 8.5 for a strengthening of Corollary 6.7. We turn our attention to the behavior of primary ideals under localization.

6.8 LEMMA. *Let S be a multiplicative submonoid of a commutative ring R, and let Q be a primary ideal of R such that $Q \cap S$ is empty. Then $x/1 \in S^{-1}Q$ if and only if $x \in Q$. If Q belongs to the prime ideal P, then $S^{-1}Q$ is a primary ideal belonging to the prime ideal $S^{-1}P$.*

PROOF. Obviously $x/1 \in S^{-1}Q$ if $x \in Q$. Conversely if $x/1 \in S^{-1}Q$, then $sx \in Q$ for some $s \in S$ by (6.1). Thus $x \in Q$ or $s^n \in Q$ for some n, but the latter is impossible as $Q \cap S$ is empty.

Clearly $S^{-1}Q \subseteq S^{-1}P$ and some power of each element of $S^{-1}P$ lies in $S^{-1}Q$. Suppose $(x/s_1)(y/s_2) \in S^{-1}Q$. Then $sxy \in Q$ for some s in S, so either $x \in P$, whence $x/s_1 \in S^{-1}P$, or $sy \in Q$, whence $y/s_2 \in S^{-1}Q$. □

6.9 THEOREM. *Let S be a multiplicative submonoid of a commutative ring R. Let $Q_1,\ldots,Q_n$ be detachable primary ideals of R such that $Q_i \cap S$ is empty for $i = 1,\ldots,m$, and $Q_i \cap S$ is nonempty for $i = m+1,\ldots,n$. If $I = Q_1 \cap \cdots \cap Q_n$, then $S^{-1}I = \bigcap_{i=1}^{m} S^{-1}Q_i$.*

PROOF. If $Q \cap S$ is nonempty, then $S^{-1}Q = S^{-1}r$. Clearly $S^{-1}I \subseteq \bigcap_{i=1}^{n} S^{-1}Q_i = \bigcap_{i=1}^{m} S^{-1}Q_i$. Conversely, suppose $x/s \in \bigcap_{i=1}^{m} S^{-1}Q_i$. By (5.8) we have $x \in I$, so $x/s \in S^{-1}I$. □

From Lemma 6.8 and Proposition 6.9 it follows that if R is a commutative ring with detachable ideals such that each finitely generated ideal is a finite intersection of finitely generated primary ideals, and P is a finitely generated prime ideal of R, then R_P is a commutative ring with detachable ideals (but see Exercise 8).

Let P be a detachable prime ideal of a commutative ring R, and let n be a positive integer. If P^n is a primary ideal, then P^n belongs to P. Although P^n need not be primary if P is not maximal (see Corollary 5.5 and Exercise 5.2), there is a closely related ideal that is primary. The **symbolic power** $P^{(n)}$ of P is the ideal

$$P^{(n)} = \{x \in R : sx \in P^n \text{ for some } s \text{ in } R\backslash P\}.$$

Observe that $P^{(n+1)} \subseteq P^{(n)} \subseteq P^{(1)} = P$, that $P^{(n)}/P^n = \tau_S(R/P^n)$, and that $P^{(n)}$ is the preimage of $S^{-1}P^n$ in R.

6.10 THEOREM. *Let P be a detachable prime ideal of a commutative ring R, and let n be a positive integer. Then $P^{(n)}$ is a primary ideal belonging to P. If P^n is primary, then $P^{(n)} = P^n$.*

PROOF. If $xy \in P^{(n)}$, then $sxy \in P^n$ for some s in $R\backslash P$. If $x \in P$, then $sx \in R\backslash P$ so $y \in P^{(n)}$. If P^n is primary, and $x \in P^{(n)}$, then $sx \in P^n$ for some $s \in R\backslash P$, so $x \in P^n$ or $s^m \in P^n$ for some m. Thus $x \in P^n$. □

Let P be a prime ideal of a commutative ring R. Then P is a **minimal prime ideal over** an ideal I if $P \supseteq I$, and for each prime ideal P' such that $P \supseteq P' \supseteq I$ we have $P = P'$. The ideal P is a **minimal prime ideal of** R if P is a minimal prime over 0.

6.11 THEOREM. *Let P be a finitely generated proper prime ideal of a commutative ring R. If $P^{(n)} = P^{(n+1)}$ for some n, then P is a minimal prime ideal over 0.*

PROOF. Let Q be a prime ideal of R such that $P \supseteq Q$. From $P^{(n)} = P^{(n+1)}$ it follows that $(P_P)^n = (P_P)^{n+1} = P_P(P_P)^n$. The R_P-module $(P_P)^n$ is finitely generated and P_P is a quasi-regular ideal of R_P, hence by the Nakayama Lemma (III.1.4) we have $(P_P)^n = 0_P \subseteq Q_P$. So $P^n \subseteq P^{(n)} \subseteq Q$, by (6.8), and thus $P \subseteq Q$. □

Under the additional assumptions of coherence and detachable ideals, $P^{(n)}$ is a finitely generated ideal.

6.12 THEOREM. *Let P be a finitely generated proper prime ideal of a coherent commutative ring R with detachable ideals, and let n be a positive integer. Then $P^{(n)}$ is a finitely generated primary ideal belonging to P.*

PROOF. By Proposition 6.10 it suffices to show that $P^{(n)}$ is finitely generated. But $P^{(n)}/P^n = \tau_{R\setminus P}(R/P^n)$ is finitely generated by Theorem 6.5, so $P^{(n)}$ is finitely generated. □

EXERCISES

1. Consider the rings $\mathbb{Z} \subseteq \mathbb{Z}[X]/(2X-4)$. Show that the ideal P generated by 2, in either one of these rings, is prime. Show that $2 \in P^{(2)}$ in one ring and $2 \notin P^{(2)}$ in the other. Construct a Brouwerian example of a ring R with detachable ideals, and a finitely generated prime ideal P such that $P^{(2)}$ is not detachable.

2. Let S be a finitely generated multiplicative submonoid of a coherent commutative Noetherian ring R. Show that if M is a finitely presented R-module, then $\tau_S(M)$ is finitely generated. (Hint: Let s be the product of the generators of S and consider $M_n = \{x \in M : s^m x = 0\}$)

3. Let S be a finitely generated multiplicative submonoid of a coherent commutative Noetherian ring R with detachable ideals. Show that $S^{-1}R$ is a coherent Noetherian ring with detachable ideals.

4. Let R be the polynomial ring over the integers in the indeterminates $s, x_1, x_2, \ldots$ modulo the ideal generated by the elements sx_i, and let S be the multiplicative submonoid of R

generated by s. Show that $\tau_S(R)$ is S-bounded but not finitely generated. Why doesn't (6.2) apply.

5. Let α be a binary sequence and let S be the multiplicative submonoid of the ring of integers $\mathbb{Z}$ generated by $\{1+\alpha_n : n = 1,2,\ldots\}$. Show that $S^{-1}\mathbb{Z}$ is a Brouwerian example of a ring that does not have detachable ideals. Why doesn't (6.4) apply?

6. For p a prime, let A_p denote the ring of pairs (x,y) where $x \in \mathbb{Z}$ and $y \in \mathbb{Z}_p$ (the ring of integers modulo p), and multiplication is defined by $(x_1,y_1)(x_2,y_2) = (x_1y_1, x_1y_2+x_2y_1+y_1y_2)$. Let $p(n)$ be the nth odd prime, and let α be a binary sequence. If $\alpha_i = 0$ for each $i \leq n$, set $R_n = \mathbb{Z}$; otherwise let R_n be the ring $A_{p(i)}$ where i is the first index not exceeding n such that $\alpha_i \neq 0$. Let R be the union (direct limit) of the rings R_n. Show that R is a coherent Noetherian ring with detachable ideals. Let P be the ideal of R generated by 2. Show that $\tau_{R\setminus P}(R)$ is a Brouwerian example of an R-module that is not finitely generated. Why doesn't (6.5) apply?

7. Let S be a multiplicative submonoid of a commutative ring R, and let φ be the natural map from R to $S^{-1}R$. Show that $\tau_S(R/I)$ is finitely generated for each finitely generated ideal I of R if and only if $\varphi^{-1}(J)$ is finitely generated for each finitely generated ideal J of $S^{-1}R$.

8. Use the rings $\mathbb{Z} \subseteq \mathbb{Z}[X]/(2X)$ to construct a Brouwerian example of a ring R with a prime ideal P generated by the element 2 such that R has detachable ideals but R_P is not discrete.

9. Let K be a discrete field and $R = K[S,X,Y,Z]/(SXY-Z^2)$. Use the two prime ideals $I = (x,z)$ and $J = (x,y,z)$ of R to construct a Brouwerian example of a prime ideal P of R such that $xy \in P^{(2)}$ but it is not the case that $x \in P$ or $y \in P^{(2)}$. Why doesn't (6.10) apply?

7. PRIMARY DECOMPOSITIONS

An ideal I of a commutative ring has a **primary decomposition** if there are finitely generated primary ideals $Q_1,\ldots,Q_n$, belonging to finitely generated prime ideals, such that $I = \cap_i Q_i$. Classically every ideal in a

Noetherian ring has a primary decomposition (see Exercise 4).

A primary decomposition is **irredundant** if no primary ideal of the decomposition contains the intersection of the other primary ideals, and no two primary ideals belong to the same prime ideal. In a coherent ring with detachable ideals we can replace primary ideals belonging to the same prime ideal by their intersection (5.3), and delete primary ideals that contain the intersection of the other primary ideals, so that any ideal that has a primary decomposition has an irredundant one.

Let I be an ideal of a commutative ring R, and P a proper finitely generated prime ideal of R. We say that P is an **associated prime ideal** of I if $P = \sqrt{I:a}$ for some a in R.

7.1 THEOREM. *Let R be a coherent commutative ring with detachable ideals. Let $I = \cap_i Q_i$ be an irredundant primary decomposition of a proper ideal I of R. Then the set of associated prime ideals of I consists of the ideals $\sqrt{Q_i}$.*

PROOF. To see that $\sqrt{Q_i}$ is an associated prime ideal of I, choose $a \in \cap_{j \neq i} Q_j$ such that $a \notin Q_i$. Then $I:a = \cap_j (Q_j:a)$, and $Q_j:a = R$ if $j \neq i$. Theorem 5.7 tells us that $I:a = Q_i:a$ is a primary ideal belonging to $\sqrt{Q_i}$.

Conversely $\sqrt{I:a} = \cap \sqrt{Q_i:a}$ which, by (5.7), is equal to the intersection of those ideals $\sqrt{Q_i}$ such that $a \notin Q_i$; so if $\sqrt{I:a}$ is a prime ideal, then $\sqrt{I:a} = \sqrt{Q_i}$ for some i by (II.2.4). □

7.2 THEOREM. *Let R be a coherent commutative ring with detachable ideals, and let I be an ideal of R having a primary decomposition. Then each minimal prime over I is an associated prime of I.*

PROOF. Let $P_1, \ldots, P_n$ be the associated prime ideals of I, and let P be a minimal prime over I. Then $P \supseteq \sqrt{I} = \cap_i P_i$, hence $P \supseteq P_i$ for some i. By minimality $P = P_i$. □

In the situation of (7.2) the associated prime ideals of I that are not minimal over I are called **embedded prime ideals**. The primary ideals belonging to embedded prime ideals need not be unique (see Exercise 1), but the primary ideals belonging to minimal prime ideals are unique.

7.3 THEOREM. *Let R be a coherent commutative ring with detachable ideals. Let $I = \cap_i Q_i$ be an irredundant primary decomposition of an ideal I with associated prime ideals P_i. For each i define $Q'_i = \{x \in R : sx \in I$*

for some $s \in R \setminus P_i\}$. *Then* Q_i' *is a detachable ideal contained in* Q_i, *and* $Q_i' = Q_i$ *if* P_i *is a minimal prime ideal over* I.

PROOF. We easily verify that Q_i' is an ideal of R. As R is coherent, the ideal $I:x$ is finitely generated, so we can decide whether $I:x \subseteq P_i$. Thus Q_i' is detachable. If $x \in Q_i'$, then $sx \in I \subseteq Q_i$ for some $s \notin P_i$, so $x \in Q_i$. Thus $Q_i' \subseteq Q_i$. If P_i is a minimal prime over I, then P_i does not contain $\cap_{j \neq i} P_j$ (the ideals P_j are distinct because the primary decomposition is irredundant), so using coherence we can find $a \in \cap_{j \neq i} P_j \setminus P_i$. Then $a^m \in \cap_{j \neq i} Q_j \setminus P_i$ for some m, and $a^m Q_i \subseteq I$. Thus $Q_i \subseteq Q_i'$. □

7.4 COROLLARY. *Let* R *be a coherent commutative ring with detachable ideals. Let* I *be an ideal of* R *having a primary decomposition, and let* P *be a minimal prime ideal over* I. *Then the primary ideal that belongs to* P, *in an irredundant primary decomposition of* I, *is* $\{x \in R : sx \in I$ *for some* $s \in R \setminus P\}$. *Hence the primary ideals belonging to the minimal primes are the same for all irredundant primary decompositions.* □

7.5 LEMMA. *Let* I *be an ideal of a commutative ring* R.

(i) *If* I *has a primary decomposition, then so does* $\sqrt{I}$.

(ii) $\sqrt{I}$ *has a primary decomposition if and only if* $\sqrt{I}$ *is the intersection of finitely many finitely generated prime ideals.*

(iii) *If* R *is a coherent ring with detachable ideals, and* $\sqrt{I}$ *has a primary decomposition, then the minimal primes over* I *are precisely the associated primes of* $\sqrt{I}$, *and* $\sqrt{I}$ *is their intersection.*

PROOF. If I, or $\sqrt{I}$, is equal to $\cap Q_i$, then $\sqrt{I} = \cap \sqrt{Q_i}$. This, and the observation that prime ideals are primary and belong to themselves, proves (i) and (ii). To prove (iii) suppose that $\sqrt{I}$ has a primary decomposition. From (ii) we can write $\sqrt{I}$ as an intersection of finitely many finitely generated prime ideals, hence we can get an irredundant primary representation of $\sqrt{I}$ from among these prime ideals. By (7.2) these are associated primes of $\sqrt{I}$ as they are clearly all minimal. As the minimal primes of $\sqrt{I}$ are the same as the minimal primes of I, we have established (iii). □

Let R be a coherent commutative ring with detachable ideals, and let I and P be finitely generated proper ideals of R such that P is a minimal prime over I. Consider $Q = \{x \in R : sx \in I$ for some $s \in R\backslash P\}$. We easily verify that Q is a detachable ideal containing I, and that $Q/I = \tau_{R\backslash P}(R/I)$. If Q is finitely generated, then $Q = I:s$ for some $s \notin P$, so $I:Q$ is not contained in P. If I has a primary decomposition, then Theorem 7.3 tells us that Q is the primary ideal of I belonging to P. Under certain other circumstances it is also true that the ideal Q is a finitely generated primary ideal.

7.6 LEMMA. *Let I and P be finitely generated ideals of a coherent commutative ring R with detachable ideals. Suppose P is prime and $P^n \subseteq I \subseteq P$ for some n. Then P is the unique minimal prime over I, and $Q = \{x \in R : sx \in I$ for some $s \in R\backslash P\}$ is a finitely generated primary ideal belonging to P.*

PROOF. From $P^n \subseteq I$ it follows that P is contained in each prime ideal containing I. It remains to show that Q is a finitely generated primary ideal belonging to P. As I is finitely generated, we may assume $I = 0$. Then $Q = P^{(n)}$ is a finitely generated primary ideal belonging to P by Proposition 6.12. □

7.7 THEOREM. *Let R be a coherent commutative ring with detachable ideals. Let I be a finitely generated ideal such that $\sqrt{I}$ has a primary decomposition. If P is a minimal prime ideal over I, then $Q^* = \{x \in R : sx \in I$ for some $s \in R\backslash P\}$ is a finitely generated primary ideal belonging to P.*

PROOF. Let K be the product of the minimal prime ideals over I different from P. There is n such that $(PK)^n \subseteq I$. Let $J = I:K^n$. Then $P^n \subseteq J \subseteq P$. By Lemma 7.6 the ideal

$$Q = \{x \in R : sx \in J \text{ for some } s \in R\backslash P\}$$

is a finitely generated primary ideal belonging to P. As I is contained in J, the ideal Q^* is contained in Q; we shall show that they are equal. If $r \in Q$, then $srK^n \subseteq I$ for some s in $R\backslash P$. If $t \in K^n\backslash P$, then $str \in I$ whence $r \in Q^*$. □

The ideal Q^* in (7.7) is called the **isolated primary ideal of I belonging to P.**

EXERCISES

1. Let $R = \mathbb{Z}[X]$. Show that (2) is a prime ideal in R, and that $(4,X)$ and $(4,X-2)$ are primary ideals in R belonging to the prime $(2,X)$. Conclude that $(4,X) \cap (2)$ and $(4,X-2) \cap (2)$ are irredundant primary decompositions of $(2X,4)$; thus the primary ideals belonging to embedded prime ideals need not be unique.

2. Let k be the ring of integers modulo 2 and consider the pair of rings $k \subseteq k[X]/(X^2)$. Construct a Brouwerian example of a coherent, Noetherian ring R with detachable ideals, such that every finitely generated ideal of R is primary, R has a detachable proper prime ideal, but R has no finitely generated proper prime ideals; thus 0 is a primary ideal with no primary decomposition.

3. Let R be a coherent Noetherian ring, I a finitely generated ideal of R, and $a,b \in R$ with $ab \in I$. Show that there exists n such that $I = (I + (a)) \cap (I + (b^n))$. Hint: look at the ascending chain of ideals $I:(b^n)$.

4. Call an ideal I **irreducible** if whenever I is written as the intersection of two ideals, then one of the ideals is equal to I. Use Exercise 3 to show that, if R is a coherent Noetherian ring, and I is a finitely generated irreducible ideal, then I is primary. Give a classical proof that every ideal in a Noetherian ring is an intersection of primary ideals using the principle that every ideal is either primary or it isn't, and that every set of ideals in a Noetherian ring contains a maximal element.

8. LASKER-NOETHER RINGS

A **Lasker-Noether** ring is a coherent Noetherian ring with detachable ideals such that the radical of each finitely generated ideal is the intersection of a finite number of finitely generated prime ideals. Classically every Noetherian ring is a Lasker-Noether ring (see Exercise 7.4). Discrete fields are Lasker-Noether rings, as is the ring of integers. The name *Lasker-Noether* refers to the Lasker-Noether decomposition of Theorem 8.5.

If k is a discrete field, then $k[X]$ is a coherent Noetherian ring with detachable ideals by the Hilbert basis theorem (1.5). Each finitely generated ideal of $k[X]$ is principal, so if the radical of the principal ideal (f) is the intersection of a finite set of finitely generated prime ideals, then every prime factor of f is an associate of a generator of one of those (principal) prime ideals. Thus if $k[X]$ is Lasker-Noether, then k is a factorial field.

Theorem 7.5.iii guarantees that in a Lasker-Noether ring we can find lots of minimal prime ideals over a finitely generated ideal. The class of Lasker-Noether rings is closed under localizing with respect to finitely generated prime ideals, and under passing to quotients modulo finitely generated ideals.

8.1 THEOREM. *Let S be a multiplicative submonoid of a Lasker-Noether ring R such that $I \cap S$ is either empty or nonempty for each finitely generated ideal I of R. Then $S^{-1}R$ is a Lasker-Noether ring.*

PROOF. The ring $S^{-1}R$ is Noetherian by (6.6) and coherent by Exercise III.3.4. Let J be a finitely generated ideal of $S^{-1}R$. We must show that $\sqrt{J}$ is the intersection of a finite number of finitely generated prime ideals of $S^{-1}R$. Write $J = S^{-1}I$ for some finitely generated ideal I of R. Then $\sqrt{J} = S^{-1}\sqrt{I} = S^{-1}(P_1 \cap \cdots \cap P_n)$ where P_i is a finitely generated prime ideal of R. We may assume that P_i has empty intersection with S for $i \leq m$, and nonempty intersection for $i > m$. Then (6.9) says that we can write $\sqrt{J} = \cap_{i=1}^{m} S^{-1}P_i$. But $S^{-1}P_i$ is a (finitely generated) prime ideal by (6.8). □

8.2 THEOREM. *Let R be a Lasker-Noether ring, and let I be a finitely generated ideal of R. Then R/I is a Lasker-Noether ring.*

PROOF. Trivial. □

A **composition series** for a coherent finitely generated module M with detachable submodules is a maximal finite chain in the lattice of finitely generated submodules of M. It is easy to see that a finite-dimensional vector space over a field has a composition series. As the lattice of finitely generated submodules of M is modular, the Jordan-Hölder-Dedekind theorem applies, so a module with a composition series is Noetherian and also satisfies the descending chain condition on finitely generated submodules.

8.3 THEOREM. *Let R be a Lasker-Noether ring and P a minimal prime ideal of R such that every element of $R \setminus P$ is a unit. Then the left R-module R has a composition series.*

PROOF. The ring $F = R/P$ is a discrete field. As P is the unique minimal prime ideal over 0, there exists n such that $P^n = 0$. The modules P^i/P^{i+1} are vector spaces over F, and are finite dimensional because R is coherent. Thus R admits a composition series of finitely generated detachable ideals. □

8.4 LEMMA. *Let R be a commutative ring, I an ideal of R, and $P_1,\ldots,P_n,Q_1,\ldots,Q_n$ detachable ideals of R such that $I \subseteq \cap_i Q_i$, and Q_i is a primary ideal belonging to the prime ideal P_i for each i. Let $f \in (I : \cap_i Q_i) \setminus (\cup_i P_i)$. Then*

(i) $I:f = \cap_i Q_i$.

(ii) $I:f = I:f^2$.

(iii) $I = (I:f) \cap (I,f)$.

PROOF. If $x \in \cap_i Q_i$, then $xf \in I$, so $\cap_i Q_i \subseteq I:f$. Conversely, if $xf \in I$, then $xf \in Q_i$, so $x \in Q_i$ since $f \notin P_i$. Thus $I:f \subseteq \cap_i Q_i$.

As $f^2 \in (I:\cap_i Q_i)\setminus(\cup_i P_i)$, part (i) says $I:f = \cap_i Q_i = I:f^2$.

Obviously $I \subseteq (I:f) \cap (I,f)$. If $x \in (I:f) \cap (I,f)$, then there is $a \in I$ and $r \in R$ such that $x = a + rf \in I:f$. Thus $af + rf^2 \in I$, hence $rf^2 \in I$. By (ii) this implies $rf \in I$, so $x = a + rf \in I$. □

8.5 THEOREM (Primary decomposition theorem). *Let R be a Lasker-Noether ring. Then each finitely generated ideal of R has a primary decomposition.*

PROOF. Let I be a finitely generated proper ideal of R. We shall construct finitely generated ideals J and K such that J has a primary decomposition, $I = J \cap K$, and K properly contains I.

By Theorem 7.7 the isolated primary ideals $Q_1,\ldots,Q_k$ of I belonging to the minimal prime ideals $P_1,\ldots,P_k$ over I are finitely generated. Let $J = Q_1 \cap \cdots \cap Q_k$. As $I:Q_i$ is not contained in P_i, the ideal $I:J$ is not contained in any P_i, so there is $f \in (I:J)\setminus(\cup_i P_i)$ by II.2.3. If $K = (I,f)$, then $I = J \cap K$ by Lemma 8.4.

We construct an ascending chain of finitely generated ideals H_n of R as follows. Let $H_1 = I$. Let $H_{n+1} = R$ if $H_n = R$; otherwise choose H_{n+1} so that $H_n = J \cap H_{n+1}$, where J has a primary decomposition and H_{n+1} properly contains H_n. Note that if H_n has a primary decompo-sition, then so does H_{n-1}, and so on down to I. As R is Noetherian there exists n such that $H_n = H_{n+1}$; but that can only happen if $H_n = R$ in which case H_n, and hence I, has a primary decomposition. □

8.6 THEOREM. *Let P be a detachable proper prime ideal of a Lasker-Noether ring R, and let $\varphi : R \to R_P$ be the natural map. Then R_P is a Lasker-Noether ring, and φ reflects finitely generated ideals.*

PROOF. That R_P is a Lasker-Noether ring follows from (8.1). If I is a finitely generated ideal of R_P, then there is a finitely generated ideal $J \subseteq R$ such that $I = J_P$. By Theorem 8.5 the ideal J has a primary decomposition $J = Q_1 \cap \cdots \cap Q_n$. If $Q_1,\ldots,Q_s$ are the primary ideals of this decomposition that are contained in P, then by (6.8) and (6.9) the ideal $\varphi^{-1}(I) = \varphi^{-1}(J_P) = Q_1 \cap \cdots \cap Q_s$ is finitely generated. □

EXERCISES

1. Let R be a Lasker-Noether ring that is a principal ideal domain. Show that R is a unique factorization domain.

2. Let F be the Brouwerian example in Exercise VII.1.5 of a field that is factorial but not fully factorial. Show that $F[X,Y]$ is a unique factorization domain and a coherent Noetherian ring with detachable ideals. Show that each principal ideal of $F[X,Y]$ has a primary decomposition. Show that $F[X,Y]$ is not a Lasker-Noether ring.

3. Let k be a discrete field. Show that k is factorial if and only if $k[X]$ is a Lasker-Noether ring.

4. Let I be a finitely generated ideal of a Lasker-Noether ring. Show that $\sqrt{I}$ is finitely generated, and that I contains a power of $\sqrt{I}$.

5. Let R be a coherent Noetherian ring with detachable ideals. Let P be a finitely generated prime ideal of R such that $P^n = 0$. Show that R_P is a Lasker-Noether ring with a composition series.

6. Let I be a finitely generated ideal of a Lasker-Noether ring R. Let $P \supseteq I$ be a detachable proper prime ideal of R. Show that P is a minimal prime ideal of I if and only if there is n such that

$(P_P)^n \subseteq I_P$ in R_P.

7. Let R be a Lasker-Noether ring and I an ideal of R consisting of zero-divisors. Show that $rI = 0$ for some nonzero $r \in I$. (Hint: use (II.2.3). One should be able to weaken this hypothesis to coherent Noetherian, or just Noetherian.)

9. FULLY LASKER-NOETHER RINGS

The property of being a Lasker-Noether ring is not inherited by polynomial rings: any discrete field k is a Lasker-Noether ring, but $k[X]$ is a Lasker-Noether ring only if k is factorial.

Let R be a ring such that $R[X_1,\ldots,X_n]$ is a Lasker-Noether ring for every n. Let P be a finitely generated proper prime ideal of R, and K the field of quotients of R/P. Let E be a finite-dimensional algebraic extension field of K. We can write $E = K[\alpha_1,\ldots,\alpha_n]$ with α_i integral over R/P for each i. Then E is isomorphic to the field of quotients of $R[X_1,\ldots,X_n]/I$ where the prime ideal I is generated by P and, for each i, the preimage in $R[X_1,\ldots,X_i]$ of the minimum polynomial of α_i over $K[\alpha_1,\ldots,\alpha_{i-1}]$. So each finite-dimensional algebraic extension field of K is factorial, hence K is fully factorial. This suggests the following definition.

Call R a **fully Lasker-Noether** ring if it is a Lasker-Noether ring and if for each finitely generated prime ideal P of R, the field of quotients of R/P is fully factorial. Note that the ring of integers $\mathbb{Z}$ is a fully Lasker-Noether ring, as is any fully factorial field.

9.1 THEOREM. *Let I be a finitely generated ideal of a fully Lasker-Noether ring R. Then R/I is a fully Lasker-Noether ring.*

PROOF. By Theorem 8.2 the ring R/I is a Lasker-Noether ring. Let P be a finitely generated prime ideal of R/I. The preimage P' of P in R is a finitely generated prime ideal of R, so the field of quotients of $(R/I)/P \cong R/P'$ is fully factorial. □

9.2 THEOREM. *If P is a detachable prime ideal of a fully Lasker-Noether ring R, then R_P is a fully Lasker-Noether ring.*

PROOF. By Theorem 8.1 the ring R_P is a Lasker-Noether ring. Let Q be a finitely generated prime ideal of R_P. The preimage Q' of Q in R is a finitely generated prime ideal by (8.6), so the field of quotients of

R_P/Q, which is isomorphic to the field of quotients of R/Q', is fully factorial. □

If $R[X_1,\ldots,X_n]$ is a Lasker-Noether ring for each n, then R is a fully Lasker-Noether ring. We shall show that the converse holds.

9.3 LEMMA. *Let S be a multiplicative subset of a coherent, Noetherian commutative ring R. If $R \to S^{-1}R$ reflects finitely generated ideals, then so does $R[X] \to S^{-1}R[X]$.*

PROOF. As $R \to S^{-1}R$ reflects finitely generated ideals, the kernel of $R \to S^{-1}R$ is finitely generated, so we may assume that $R \subseteq S^{-1}R$. The ring $S^{-1}R$ is coherent and Noetherian. Let I be a finitely generated ideal of $S^{-1}R[X]$, and let $R[X]_m = \{f \in R[X] : deg\ f < m\}$. By (1.1) there exists m such that $M = I \cap S^{-1}R[X]_m$ is a finitely generated $S^{-1}R$-module, and $I \cap S^{-1}R[X]_n = \Sigma_{i=0}^{n-m} MX^i$ for each $n \geq m$. Let M' be the R-module generated by a finite set of generators of the $S^{-1}R$-module M. Then $\tau_S(R[X]_m/M') = (R[X]_m \cap I)/M'$ is finitely generated by Theorem 6.4, so the R-module $R[X]_m \cap I = M \cap R[X]$ is finitely generated.

Let J be the ideal of $R[X]$ generated by $M \cap R[X]$. Then J is a finitely generated ideal of $R[X]$ contained in $I \cap R[X]$. Let A be the (finitely generated) ideal in $S^{-1}R$ consisting of the coefficients of X^{m-1} of elements of M. As $R \to S^{-1}R$ reflects finitely generated ideals, $A \cap R$ is finitely generated, so there exist $f_1,\ldots,f_k$ in M whose X^{m-1} coefficients generate $A \cap R$. Choose $s \in S$ so that $sf_i \in R[X]$ for each i. As $R[X]$ is coherent and Noetherian, there is p such that $J:s^p = J:s^{p+1}$. We shall show that $I \cap R[X] = J:s^p$, so $I \cap R[X]$ is finitely generated.

Suppose $s^p h \in J$ for $h \in R[X]$. Then $s^p h \in I$, hence $h \in I$, so $h \in I \cap R[X]$. Conversely suppose $h \in I \cap R[X]$. Then $h \in R[X]_n$ for some n, and we proceed by induction on n. If $n \leq m$, then $h \in J \subseteq J:s^p$. Suppose $n > m$. Then $h = \Sigma_i r_i f_i X^{n-m} + g$, where $g \in S^{-1}R[X]_{n-1}$ and $r_i \in R$. So $sh = \Sigma_i r_i sf_i X^{n-m} + sg$, whence $sg \in I \cap R[X]_{n-1}$. By induction on n we have $sg \in J:s^p$. Thus $h \in J:s^{p+1} = J:s^p$. □

9.4 LEMMA. *Let R be a fully Lasker-Noether ring and I a proper finitely generated ideal of $R[X]$. Let M be a minimal prime ideal over $I \cap R$, let K the field of quotients of R/M, and let J be the ideal generated by I in $K[X]$. If J is proper, then the preimage P of $\sqrt{J}$ in $R[X]$ is a finite intersection of finitely generated prime ideals, and $I:P \neq I$.*

PROOF. As K is a factorial field, $\sqrt{J}$ is the intersection of finitely many principal prime ideals in $K[X]$. From Lemma 9.3 we see that the intersection of each of those prime ideals with $(R/M)[X]$ is finitely generated, so P is the intersection of finitely many finitely generated prime ideals.

To prove that $I:P \neq I$, consider first the special case when $M \subseteq \sqrt{R \cap I}$ and $K = R/M$. As $\sqrt{J}$ is a principal ideal of $K[X]$, there exist g in $R[X]$ such that the image of g generates $\sqrt{J}$. As $M \subseteq \sqrt{R \cap I}$, there exists $n \geq 1$ such that $M^n \subseteq I$ and $M^{n-1} \backslash I$ is nonempty. Choose $t \in M^{n-1} \backslash I$ and note that $tM \subseteq I$. As $g \in \sqrt{I + M[X]}$, there exists $m \in \mathbb{N}$ such that $tg^m \notin I$ and $tg^{m+1} \in I$. Then $tg^m \in (I:P) \backslash I$. Notice that we did not need (9.3) for this case as P is an intersection of prime ideals of the form $(p) + M[X]$ which are clearly finitely generated.

For the general case set $S = R \backslash M$. Then $S^{-1}R$ is a Lasker-Noether ring, $S^{-1}M$ is the radical of $S^{-1}I \cap S^{-1}R$, and $S^{-1}R/S^{-1}M = K$. The preimage of $\sqrt{J}$ in $S^{-1}R[X]$ is $S^{-1}P$. The special case gives us $S^{-1}I:S^{-1}P \neq S^{-1}I$, so $I:P \neq I$. □

9.5 LEMMA. *If R is a fully Lasker-Noether ring, then $R[X]$ is a coherent Noetherian ring with detachable ideals such that for each finitely generated prime ideal P of $R[X]$, the field of quotients of $R[X]/P$ is fully factorial.*

PROOF. From the Hilbert basis theorem (1.5) it follows that $R[X]$ is a coherent Noetherian ring with detachable ideals. If $P' = P \cap R$, then P' is a finitely generated prime ideal of R by (1.2), so the field of quotients of R/P' is fully factorial. The field of quotients of $R[X]/P \cong (R/P')[X]/(P/P'[X])$ is a finitely presented field extension of a fully factorial field , so by Theorem 2.5 is fully factorial. □

9.6 THEOREM. *If R is a fully Lasker-Noether ring, then so is $R[X]$.*

PROOF. By (9.5) it suffices to show that if I is a proper finitely generated ideal of $R[X]$, then there exist a finite number of finitely generated prime ideals containing I so that some power of the intersection of those prime ideals is contained in I. As some product of minimal prime ideals over $R \cap I$ is contained in $R \cap I$, there exists a minimal prime ideal M over $R \cap I$ such that $I + M[X] \neq R[X]$. If K is the field of quotients of R/M, then the ideal J generated by the image of I in $K[X]$ is proper. By (9.4) the preimage P of $\sqrt{J}$ in $R[X]$ is a finite intersection of finitely generated prime ideals, and $I:P \neq I$. Let L be the intersection of all those ideals P as M ranges over the minimal prime ideals of R such that J is proper. Replace I by $I_2 = I:L \supseteq I:P \neq I$ and start again. This provides us with an ascending chain of ideals $I = I_1 \subseteq I_2 \subseteq \cdots$ and finite intersections of finitely generated prime ideals $L = L_1, L_2, \ldots$ such that $I_{n+1} = I_n:L_n$, and $I_n \neq I_{n+1}$ unless $I_n = R[X]$. There is n such that $I_n = I_{n+1}$ so $R[X] = I_n \subseteq I:L_1L_2\cdots L_{n-1}$ whence $L_1L_2\cdots L_{n-1} \subseteq I$. □

EXERCISES

1. Show that a discrete field is fully factorial if and only if it is a fully Lasker-Noether ring.
2. Show that the ring of integers is fully Lasker-Noether. Conclude that the field of rational numbers is fully factorial.
3. By considering the rings $\mathbb{Q} \subseteq \mathbb{Q}[X]$, construct a Brouwerian example of a fully Lasker-Noether ring that does not have a finitely generated maximal ideal.
4. **A bad Noetherian ring (Nagata).** Let k be a discrete field and $X_1, X_2, \ldots$ a countable set of indeterminates. Let $m_1, m_2, \ldots$ be a sequence of positive integers such that $0 < m_i - m_{i-1} < m_{i+1} - m_i$ for each i. Let P_i be the (prime) ideal in $K = k[X_1, X_2, \ldots]$ generated by $\{X_j : m_i \leq j < m_{i+1}\}$, let S be those polynomials in K that are not in any P_i, and let $R = S^{-1}K$. Show that
   (i) K, and hence R, is coherent.
   (ii) If I is a proper, nonzero, finitely generated ideal of R, then $\{i : I \subseteq S^{-1}P_i\}$ is nonempty and finite.
   (iii) The localization R_i of K at P_i is Noetherian.
   (iv) R is a (fully) Lasker-Noether ring if k is (fully) factorial.
   (v) R has arbitrarily long chains of finitely generated prime ideals.
5. Let S be a finitely generated multiplicative subset of a fully Lasker-Noether ring R. Show that $S^{-1}R$ is fully Lasker-Noether.
6. Let K be a fully factorial field. Let $R = K[X,Y,Z]$. Let $I = (X^2 - YZ, Y^2 - XZ)$. Determine the minimal primes of I.

10. THE PRINCIPAL IDEAL THEOREM

By the **height** of a finitely generated detachable prime ideal of a Noetherian ring R, we mean its height in the set of finitely generated detachable prime ideals of R ordered by inclusion (see the definition of depth and height on pages 24 and 25).

10.1 THEOREM (Principal ideal theorem). *Let a be an element of a Lasker-Noether ring R, and P a finitely generated proper prime ideal that is minimal over (a). Then P has height at most 1.*

PROOF. Let Q be a finitely generated prime ideal of R such that $P \supseteq Q$. We shall show that $Q = P$ or Q is minimal over 0. By Lemma 6.8 we may localize at P and so assume that every element of $R \backslash P$ is a unit. If $a \in Q$, then $Q = P$ by the minimality of P, so we may assume $a \notin Q$. We shall show that Q is minimal over 0. Consider the decreasing sequence of symbolic powers $Q^{(1)} \supseteq Q^{(2)} \supseteq \cdots$ of Q. The ideals $Q^{(i)} + (a)$ form a decreasing sequence of ideals containing (a). Since $P/(a)$ is a minimal prime of $R/(a)$, the ring $R/(a)$ is a Lasker-Noether ring with a composition series by (8.3). By Proposition 6.12 the ideals $Q^{(i)}$ are finitely generated, so there is n such that $Q^{(n)} + (a) = Q^{(n+1)} + (a)$. So if $q \in Q^{(n)}$, then there exist $r \in Q^{(n+1)}$ and $x \in R$ such that $q = r + xa$. Now $a \notin Q$, while $xa = q - r \in Q^{(n)}$, so $x \in Q^{(n)}$. Thus $Q^{(n)} = Q^{(n+1)} + Q^{(n)}a$. By the Nakayama lemma $Q^{(n)} = Q^{(n+1)}$, so Q is a minimal prime over 0 by Theorem 6.11. □

10.2 LEMMA. *Let R be a Lasker-Noether ring. Let $P_0 \supseteq \cdots \supseteq P_{n-1} \supseteq P_n$ be a strictly decreasing sequence of finitely generated proper prime ideals, and let $x \in P_0$. Then there are finitely generated prime ideals $P_1^*, \ldots, P_{n-1}^*$ such that $P_0 \supseteq P_1^* \supseteq \cdots \supseteq P_{n-1}^* \supseteq P_n$ is strictly decreasing, and $x \in P_{n-1}^*$.*

PROOF. We may assume that $n = 2$, that $x \notin P_1$, and that $P_2 = 0$. Let P_1^* be a minimal prime ideal over (x) that is contained in P_0. Obviously P_1^* is nonzero. The prime P_0 has height at least 2, while the prime P_1^* is a minimal prime over the principal ideal (x), so by the principal ideal theorem (10.1) has height at most 1. Thus P_0 properly contains P_1^*. □

10.3 COROLLARY. *Let R be a Lasker-Noether ring. Let $P_0 \supseteq \cdots \supseteq P_{n-1} \supseteq P_n$ be a strictly decreasing sequence of finitely generated proper prime* *ideals, and let* $Q_1,\ldots,Q_m$ *be finitely generated proper prime ideals not containing* P_0. *Then there are finitely generated prime ideals* $P_1^*,\ldots,P_{n-1}^*$ *such that* $P_0 \supsetneq P_1^* \supsetneq \cdots \supsetneq P_{n-1}^* \supsetneq P_n$ *is strictly decreasing, and such that no* Q_i *contains* P_{n-1}^*.

PROOF. By Proposition II.2.3 the ideal P_0 is not contained in the union $Q_1 \cup \cdots \cup Q_m$. Take $x \in P_0 \setminus (Q_1 \cup \cdots \cup Q_m)$ and apply Lemma 10.2. □

10.4 THEOREM (Generalized principal ideal theorem). *Let R be a Lasker-Noether ring. Let* $I = (a_1,\ldots,a_n)$. *Then every minimal prime ideal over I has height at most n.*

PROOF. Let P be a minimal prime ideal over I, and proceed by induction on n. If $n = 0$, then $I = 0$, and P is a minimal prime ideal of R, so P has height 0.

Let $J = (a_1,\ldots,a_{n-1})$. If P is a minimal prime ideal over J, then P has height at most $n - 1$ by induction. Thus we may assume that P is not contained in any of the minimal prime ideals over J.

We must show that there cannot exist a strictly decreasing sequence $P = P_0 \supsetneq P_1 \supsetneq \cdots \supsetneq P_n \supsetneq P_{n+1}$ of finitely generated prime ideals. Given such a sequence, by Corollary 10.3 we could construct a strictly decreasing sequence of finitely generated prime ideals $P = P_0^* \supsetneq P_1^* \supsetneq \cdots \supsetneq P_n^* \supsetneq P_{n+1}$ with P_n^* not contained in any of the minimal primes over J. The prime ideal P/J of R/J is a minimal prime ideal over the principal ideal I/J, so has height 1 by the principal ideal theorem (10.1). The ideal $(P_n^* + J)/J$ is not contained in any minimal prime ideal of R/J, but $P/J \supseteq (P_n^* + J)/J$. Thus P/J is a minimal prime over $(P_n^* + J)/J$, so P is a minimal prime over $P_n^* + J$, so P/P_n^* is a minimal prime over $(P_n^* + J)/P_n^*$ in R/P_n^*. In R/P_n^* the ideal $(P_n^* + J)/P_n^*$ is generated by $n - 1$ elements. By induction P/P_n^* has height at most $n - 1$. Thus $P_i^*/P_n^* = P_{i+1}^*/P_n^*$ for some $i < n$, so $P_i^* = P_{i+1}^*$, a contradiction. □

The converse of (10.4) is not true: a prime ideal of height at most 1 need not be principal (Exercise 3). However the following is true.

10.5 THEOREM. *Let P be a finitely generated proper prime ideal of a Lasker-Noether ring R. Then there is m such that P has height m, and P is a minimal prime over some ideal generated by m elements.*

PROOF. As P is minimal over P, by (10.4) there is n such that P has height at most n. We proceed by induction on n. There are finitely many minimal prime ideals $Q_1,\ldots,Q_k$ of R contained in P. If $P = Q_i$ for some i, then P has height 0 and is minimal over the ideal 0, which is generated by 0 elements. Otherwise for each i we can choose $x_i \in P \backslash Q_i$. There are finitely many prime ideals that are minimal over $Q_i + (x_i)$, and if $\hat{Q}$ is one of them, then $P/\hat{Q}$ has height at most $n - 1$, so by induction $P/\hat{Q}$ has a height. Let $m - 1$ be the maximum of the heights of $P/\hat{Q}$ where $\hat{Q} \subseteq P$ is a minimal prime ideal over some $Q_i + (x_i)$ for some i. We shall show that P has height m.

Clearly P has height at least m because each $\hat{Q}$ has height at least 1. Let $P = P_0 \supseteq \cdots \supseteq P_n$ be a strictly decreasing sequence of finitely generated prime ideals. Then $P_n \supseteq Q_i$ for some i. By Lemma 10.2 there is a strictly decreasing sequence $P = P_0 \supseteq P_1^* \supseteq \cdots \supseteq P_{n-1}^* \supseteq P_n$ such that $x_i \in P_{n-1}^*$. Thus P_{n-1}^* contains one of the primes $\hat{Q}$ minimal over $Q_i + (x_i)$, so $n \leq m + 1$ and P has height at most m.

Let P have height m. We shall show that P is a minimal prime ideal over an ideal I generated by m elements. We proceed by induction on m. We may assume that $m > 0$. Let $Q_1,\ldots,Q_n$ be the minimal primes of R contained in P. As P properly contains each Q_i, by Theorem II.2.3 there is $x \in P \backslash (Q_1 \cup \cdots \cup Q_n)$. Then $P/(x)$ has height at most $m - 1$ in $R/(x)$. By induction there is an ideal $J \subseteq R$ such that J is generated by $m - 1$ elements and $P/(x)$ is minimal prime over $(J + (x))/(x)$. Hence P is a minimal prime over $I = J + (x)$. □

10.6 COROLLARY. *Each finitely-generated proper prime ideal in a Lasker Noether ring has a height.* □

EXERCISES

1. Let R be a coherent Noetherian ring with detachable ideals. Show that each finitely-generated prime ideal of R of height 1 is a minimal prime ideal over some principal ideal.

2. Let R be a Lasker-Noether domain. Show that R is a unique factorization domain if and only if each prime ideal of height 1 is principal.

3. Let R be $\mathbb{Z}[\sqrt{-5}]$, and P the ideal generated by 2 and $1+\sqrt{-5}$. Show that R is a Lasker-Noether domain. Show that R/P is a discrete

field with 2 elements, so P is a finitely-generated prime ideal. Show that P has height 1 but is not principal.

NOTES

A constructive version of the Hilbert basis theorem was proved by Jonathan Tennenbaum in his 1973 dissertation directed by Errett Bishop at the University of California, San Diego. Tennenbaum employed a basis *operation* rather than a basis *function*. This seems to be one of the few instances where Bishop's notion of an operation from A to B cannot be interpreted as a function from A to the set of nonempty subsets of B. The original scope and spirit of Tennenbaum's result can be retrieved by ignoring the function φ in Section 3 and considering the function ρ to be an operation. Exercises 4.5 and 4.6 on Tennenbaum rings are an attempt to restore the scope of Tennenbaum's result without relying on the notion of an operation.

The Hilbert basis theorem for coherent Noetherian rings with detachable ideals was proved by Richman (1974), who relied on Tennenbaum's result. Seidenberg (1974a) showed how to remove the dependence on Tennenbaum's theorem, and at the same time showed that the Hilbert basis theorem also holds for coherent Noetherian rings (with no reference to detachability of ideals). Note that this latter Hilbert basis theorem is neither more nor less general than the one requiring detachable ideals: both the hypothesis and the conclusion are weaker.

The main theorem on primary decomposition of ideals in Noetherian rings — that $R[X]$ is a fully Lasker-Noether ring if R is — was first proved by Seidenberg (1984) who cannot, however, be blamed for the terminology.

The bad Noetherian ring in Exercise 9.4 is from [Nagata 1962]. It provides an example of a fully Lasker-Noether ring that is not built up from a discrete field by adjoining finitely many indeterminates, taking quotients, and localizing.

Chapter IX. Finite Dimensional Algebras

1. REPRESENTATIONS.

Let k be a discrete field. A **k-algebra** is a ring A that is also a vector space over k, satisfying $\lambda(ab) = (\lambda a)b = a(\lambda b)$ for each λ in k and a,b in A. If A and B are k-algebras, then a **homomorphism** from A to B is a ring homomorphism that is also a k-linear transformation. The term **finite dimensional**, when applied to a structure S that is a vector space over k, like a k-algebra, signifies that S is a finite-dimensional vector space over k.

If M is an n-dimensional vector space over k, then the k-linear transformations from M into M form a finite-dimensional k-algebra $E(M)$ which may be identified with the algebra of $n \times n$ matrices over k. Any finite dimensional k-algebra A is isomorphic to a finite-dimensional subalgebra of $E(A)$ by associating with each element a of A the linear transformation T_a from A to A given by $T_a(x) = ax$.

A **representation** of a finite-dimensional k-algebra A is a finite-dimensional vector space M over k, together with a k-algebra homomorphism $\varphi : A \to E(M)$. If we drop explicit reference to φ, and write am for $\varphi(a)m$, then we have the notion of a finite-dimensional A-module. If the kernel K of the representation φ is zero, then we say that φ (or M) is **faithful**. If M is a faithful A-module, then A may be identified with a subalgebra of $E(M)$. The following theorem gives a number of important constructions that yield finite-dimensional vector spaces.

1.1 THEOREM. *Let A be a finite-dimensional k-algebra, and M a finite-dimensional A-module. Let I be a finite-dimensional subspace of A, and N a finite-dimensional subspace of M. Then the following are finite-dimensional vector spaces over k.*

(i) *The subspace IN of M generated by $\{ax : a \in I \text{ and } x \in N\}$.*

(ii) *The subalgebra $I + I^2 + I^3 + \cdots$ of A generated by I.*

(iii) *The* **center** *$\{a \in A : ax = xa \text{ for all } x \text{ in } A\}$ of A.*

(iv) *The representation kernel* $K = \{a \in A : aM = 0\}$ *of* M.

(v) *The* ***centralizer*** $\{b \in E(M) : ab = ba$ *for all* $a \in A\}$ *of* A/K.

PROOF. The subspace IN is clearly finitely generated, hence finite dimensional. For each n the subspace $S_n = I + I^2 + \cdots + I^n$ is finite dimensional, so we can find n such that $S_n = S_{n+1}$, and so $S_m = S_n$ for each $m \geq n$. The center of A is the intersection of the kernels of the maps $f_x : A \to A$ given by $f_x(a) = ax - xa$ where x ranges over a basis for A. The representation kernel K of M is the intersection of the kernels of the maps $g_m : A \to M$ given by $g_m(a) = am$ where m ranges over a basis for M. The centralizer of A/K is the intersection of the kernels of the maps $h_a : E(M) \to E(M)$ given by $h_a(b) = ab - ba$ where a ranges over a basis for A. □

By (1.1.iv) we can decide whether or not M is faithful; in either case we may view M as a faithful module over A/K. The endomorphism ring of M is the centralizer of the algebra A/K viewed as a subalgebra of $E(M)$.

If a and b are nonzero elements of a finite-dimensional k-algebra A, and $ab = 0$, then we say that a is a **left zero-divisor**, and that b is a **right zero-divisor.**

1.2 COROLLARY. *Let* A *be a finite-dimensional* k*-algebra. Then each nonzero element of* A *is either a unit, or a left and right zero-divisor.*

PROOF. For $a \in A$, let $\rho_a : A \to A$ be defined by $\rho_a(x) = xa$. By (II.6.2) either ρ_a has a nonzero kernel, in which case a is a right zero-divisor, or ρ_a is onto, in which case a has a left inverse. In the latter case there is b such that $ba = 1$, so ρ_a is onto, hence one-to-one. But $\rho_a(ab) = \rho_a(1)$, so $ab = 1$, whence a is a unit. Thus a is either a unit or a right zero-divisor. Similarly, a is either a unit or a left zero-divisor. □

As $Aa = A$ if and only if a is a unit, a finite-dimensional algebra A has a nontrivial finite-dimensional left ideal if and only if A has a zero divisor. We cannot always decide whether *every* nonzero element of A has an inverse, that is, whether A is a **division algebra**.

1.3 EXAMPLE. Let a be a binary sequence and let $k = \cup_n \mathbb{Q}(ia_n)$. Let X be an indeterminate and A the two-dimensional algebra $k[X]/(X^2+1)$ over k. It is easy to verify that A is a division algebra precisely when $a_n = 0$ for all n. □

A finite-dimensional k-algebra is a division algebra if and only if it has no zero-divisors. It is a classical theorem that if A is a division algebra, then any A-module is free. Thus, classically, either A has a zero-divisor, or every finite-dimensional left A-module is free. Although we may not be able to determine which of these alternatives holds, we can, given an A-module M, either construct a basis for M as a free A-module, or construct a zero-divisor of A.

1.4 THEOREM. *Let A be a finite-dimensional k-algebra, M a finite-dimensional A-module, and u a nonzero element of M. Then either A has a zero-divisor, or M is a free A-module with a basis containing u.*

PROOF. The set $\{a \in A : au = 0\}$ is a finite-dimensional left ideal of A. If it is nonzero we are done by (1.2), so we may assume that Au is a rank-1 free A-module. Then $N = M/Au$ is a finite-dimensional A-module of dimension less than the dimension of M. If $N = 0$, then $\{u\}$ is basis for M. Otherwise, by induction on the dimension of M, either A has a zero-divisor, or N is a free A-module; in the latter case M is a free A-module with a basis containing u. □

Let K be a finite-dimensional k-algebra. Many constructions that could be effected if K were a division ring can be attempted whether K is a division ring or not, the result being either the desired construction or the construction of a zero-divisor of K. Theorem 1.4 is an example of this. A related example is the construction of minimal polynomials.

Let A be a finite dimensional k-algebra, K a subalgebra of the center of A, and $a \in A$. A monic polynomial $f \in K[X]$ is called the **minimal polynomial** of a over K if for each $g \in K[X]$ we have $g(a) = 0$ if and only if f divides g. It is easily seen that the minimal polynomial is unique, if it exists, and that $K[a] \cong K[X]/(f)$.

1.5 THEOREM. *Let A be a finite-dimensional k-algebra, K a finite-dimensional subalgebra of the center of A, and $a \in A$. Then we can construct either a zero-divisor in K, or the minimal polynomial of a over K.*

PROOF. For each nonnegative integer n let $M_n = K + Ka + \cdots + Ka^n$. Consider the map $\varphi_n : K \to M_n/M_{n-1}$ that takes 1 to a^n, and choose the least n such that $ker\ \varphi_n \neq 0$; we will find such an $n \leq 1 + dim\ A$, because we must have $M_n = M_{n-1}$ at some point. If $ker\ \varphi_n \neq K$, then K has a zero-divisor. Otherwise, it is readily seen that $1, a, \ldots, a^{n-1}$ is a basis for

$M_n = M_{n-1}$ over K, and if $a^n = c_0 + c_1 a + \cdots + c_{n-1}a^{n-1}$, then $p(X) = X^n - c_{n-1}X^{n-1} - \cdots - c_1 X - c_0$ is the minimal polynomial for a over K. □

EXERCISES

1. Let A be a finite-dimensional k-algebra. Show that A is a coherent Noetherian ring with detachable ideals.
2. Show that the algebra A of Example 1.3 is a division algebra precisely when $a_n = 0$ for all n.
3. Construct a Brouwerian counterexample to the theorem that if A is a finite-dimensional k-algebra, then either A has a zero-divisor, or every left A-module is free.
4. The **quaternion algebra** H over k is the four-dimensional k-algebra with basis $\{1,i,j,ij\}$ satisfying

$$i^2 = j^2 = -1$$
$$ij = -ji.$$

   For what k is H a division algebra? For what k does H have a zero-divisor?
5. Show that if k is algebraically closed, and A is a finite-dimensional k-algebra, then either A is a division algebra or A has a zero-divisor.

2. THE DENSITY THEOREM.

A module M is **reducible** if it has a nontrivial submodule—otherwise it is **irreducible** (or **simple**). The classical **density theorem** asserts that if M is a faithful irreducible A-module, then A is a dense ring of endomorphisms of M, when M is viewed as a vector space over the centralizer B of A (which is a division ring). The terminology refers to the situation where M is infinite dimensional over B, and we conclude that A is dense in the sense that any finite B-independent set can be mapped anywhere by multiplying by some element of A. In the finite-dimensional case, which is what we are dealing with here, we conclude that A is the full ring of B-module endomorphisms of M, that is, A is the centralizer of B. From a constructive point of view, the density theorem is much more useful if we can apply it to a possibly reducible M.

2.1 THEOREM (density). *Let* A *be a finite-dimensional* k*-algebra of linear transformations on a finite-dimensional* k*-space* M*, and let* u *be a nonzero element of* M*. Let* B *be the centralizer of* A*. Then either*

(i) M *has a nontrivial* A*-submodule.*

(ii) M *is a free* B*-module with a basis containing* u*, and* A *is the centralizer of* B*.*

PROOF. By Theorem 1.4, either B has a zero-divisor or M is a free B-module with a basis containing u. If $b \in B$ is a zero-divisor, then bM is a nontrivial A-submodule of M. Thus we may assume that M is a free B-module with a basis $u_1,\ldots,u_n$ containing u. For $j = 0,\ldots,n$ set

$$L_j = \{a \in A : au_i = 0 \text{ for each } i \leq j\}.$$

Then L_j is a finite-dimensional left ideal of A. We may assume that the finite-dimensional A-submodule L_ju_i of M is either 0 or M for each j and i. Let

$$M_j = \{x \in M : L_jx = 0\}$$

Then M_j is a B-submodule of M containing $\Sigma_{i=1}^{j} Bu_i$; we shall show, by induction on j, that $M_j = \Sigma_{i=1}^{j} Bu_i$. As $L_0 = A$, both sides are 0 for $j = 0$. Suppose the equality holds for some $j < n$. Then $L_ju_{j+1} = M$ because $u_{j+1} \notin M_j$. Let $x \in M_{j+1}$. As $L_{j+1}x = 0$, setting $b(\ell u_{j+1}) = \ell x$ for $\ell \in L_j$ defines $b \in B$, with $L_j(x - bu_{j+1}) = 0$. Thus $x - bu_{j+1} \in M_j = \Sigma_{i=1}^{j} Bu_i$, so $x \in \Sigma_{i=1}^{j+1} Bu_i$.

To show that A is the centralizer of B we construct, given $v_1,\ldots,v_n$ in M, an element a in A such that $au_i = v_i$. As $L_{i-1}u_i = M$, we can construct $a_i \in L_{i-1}$ inductively such that

$$a_iu_i = v_i - (a_1 + \cdots + a_{j-1})u_j.$$

Let $a = a_1 + a_2 + \cdots + a_n$. □

We use the density theorem to reduce the question of whether an algebra is simple to the commutative case.

2.2 THEOREM. *Let* A *be a finite-dimensional* k*-algebra with center* C*. Then either* A *has a nontrivial ideal, or there is a 1-1 correspondence between ideals* I *of* A *and ideals* J *of* C *given by*

$$J = I \cap C$$
$$I = AJ.$$

PROOF. Let R be the (finite-dimensional) subalgebra of $E(A)$ generated by left and right multiplication by elements of A. The centralizer of R is the center C of A, and the R-submodules of A are precisely the ideals of A. By the density theorem, either A has a nontrivial ideal, or A is a free C-module with basis containing 1, and R is the centralizer of C. Assuming the latter alternative holds, let I be an ideal of A and let J be an ideal of C. That $J = (AJ) \cap C$ follows immediately from the fact that A is a free C-module with basis containing 1. We must show that $I = (I \cap C)A$. Let $1 = u_0, \ldots, u_n$ be a basis for A over C, and let $r_i \in R$ be such that $r_i u_i = 1$ and $r_i u_j = 0$ if $j \neq i$. If $x \in I$, then $x = \Sigma\, c_i u_i$ so $r_i x = c_i$ whence $x \in (I \cap C)A$. □

We shall improve on this theorem in the next section, replacing the condition "A has a nontrivial ideal" by "rad A $\neq 0$".

EXERCISES

1. **Schur's lemma.** Let M be a finite-dimensional faithful irreducible module over the finite-dimensional k-algebra A. Show that the centralizer of A is a division ring.
2. **Burnside's theorem.** Let M be a finite-dimensional vector space over an algebraically closed field k, and let A be a subalgebra of $E(M)$. Show that either M has a nontrivial A-submodule, or $A = E(M)$.
3. Let A be a finite-dimensional k-algebra. Show that either the center of A has dimension greater than one, or A has a nontrivial ideal, or every ideal of A is trivial.
4. Refer to the first two sentences of the proof of Theorem 2.2. Show that the centralizer of R is the center C of A.

3. THE RADICAL AND SUMMANDS.

Let A be a finite dimensional algebra. A left ideal L of A is **nilpotent** if $L^n = 0$ for some n. The **radical** of A is defined by

$$rad\ A = \{x \in A : Ax \text{ is nilpotent}\}.$$

It is readily verified that rad A is an ideal of A. Moreover rad A is detachable from A, for if L is a left ideal of dimension n, then L^n is finite dimensional, and $L^n = 0$ if and only if L is nilpotent.

The classical Wedderburn structure theorem for finite-dimensional algebras A with *rad* A = 0 hinges on the construction of idempotents in A. We want the additional information provided by an algorithm that constructs either an idempotent or a nonzero element of *rad* A.

3.1 LEMMA. *Let L be a nonzero left ideal of a finite-dimensional k-algebra A. Then either L contains a nonzero nilpotent left ideal or a nonzero idempotent.*

PROOF. Passing to $Ax \subseteq L$ we may assume that L is finite-dimensional and proceed by induction on the dimension of L. If $L^2 = 0$ we have found our nonzero nilpotent left ideal; if $0 \neq L^2 \neq L$, we are done by induction. Thus we may assume that $L^2 = L$.

As $L^2 = L \neq 0$, we can find x in L such that $Lx \neq 0$. Define $f : L \to L$ by $f(y) = yx$. By (II.6.2) either *ker* f is a nonzero left ideal properly contained in L, or f is an isomorphism. In the former case we are done by induction, so we may assume the latter. Hence $yx = x$ for some y in L, so $y^2x = yx = x$ whereupon $(y^2 - y)x = 0$ so $y = y^2$ is a nonzero idempotent in L. □

The following is the standard result relating idempotents to summands.

3.2 LEMMA. *Let L be a finite-dimensional left ideal of a finite dimensional k-algebra A. Then L = Ae for some idempotent e ∈ A if and only if A = L ⊕ K for some left ideal K.*

PROOF. If e is an idempotent, then $A = Ae \oplus A(1-e)$. If $A = L \oplus K$, write $1 = e + f$ where $e \in L$ and $f \in K$. If $x \in L$, then $x = xe + xf$ so $xf = 0$ and $x = xe$. Conversely if $x = xe$, then $x \in L$. Thus $L = Ae$ and $e^2 = e$. □

3.3 THEOREM. *Let A be a finite-dimensional k-algebra and L a finite dimensional (left) ideal of A. Then either L ∩ rad A ≠ 0 or A = L ⊕ N for some (left) ideal N.*

PROOF. We may assume $L \neq 0$. By Lemma 3.1 either $L \cap$ *rad* $A \neq 0$ or L contains a nonzero idempotent and hence, by Lemma 3.2, a nonzero left ideal summand K of A. So $A = K \oplus M$ and $L = K \oplus (M \cap L)$. By induction on *dim* L we may assume that $M \cap L$ is a summand of A, hence of M, so $A = K \oplus (M \cap L) \oplus N = L \oplus N$. If L is an ideal, and $R = \{r \in A : Lr = 0\}$, then $N \subseteq R$. As $R \cap L$ is a finite-dimensional nilpotent right ideal contained in the ideal L, either L contains the nonzero nilpotent ideal $A(R \cap L)$, or $R \cap L = 0$ so $N = R$ is an ideal. □

If A_1 and A_2 are k-algebras, then $A_1 \times A_2$ is a k-algebra in a natural way, called the **product** of A_1 and A_2. The algebras A_1 and A_2 may be identified with the ideals in $A_1 \times A_2$ generated by the idempotents (1,0) and (0,1) respectively. Conversely, if an algebra A can be written as the direct sum of two-sided ideals L and N, as in (3.3), then L and N are algebras in their own right, and A is isomorphic to their product.

If L in Theorem 3.3 is an ideal, then we either get a nonzero element of $L \cap \text{rad } A$, or A decomposes into a product of k-algebras. This latter decomposition will be used frequently for proofs by induction on the dimension of A. An example is the following stronger version of Theorem 2.2.

3.4 THEOREM. *Let A be a finite-dimensional k-algebra with center C. Then either rad A $\neq 0$ or there is a 1-1 correspondence between ideals I of A and ideals J of C given by*

$$J = I \cap C$$
$$I = AJ$$

PROOF. By Theorem 2.2 we may assume that A has a nontrivial finite-dimensional ideal L. By Theorem 3.3 either rad A $\neq 0$ or A is a product of algebras, in which case we are done by induction on the dimension of A. □

3.5 LEMMA. *Let K be a finite-dimensional commutative k-algebra with rad $K = 0$. If f is a separable polynomial in $K[X]$, then rad $K[X]/(f) = 0$.*

PROOF. If K has a zero-divisor, then K has nontrivial ideal, so K is a product by Theorem 3.3, and we are done by induction on the dimension of K. We shall show that if $g \in K[X]$ and f divides g^2, then either f divides g or K has a zero-divisor. The Euclidean algorithm constructs either a zero-divisor of K, or a generator d of the ideal (f,g). Let $f = da$ and $g = db$. As f is separable, a is separable and strongly relatively prime to d. As $f = da$ divides $g^2 = d^2b^2$, either a divides db^2 or K has a zero-divisor. But a and d are strongly relatively prime, so a divides b^2, whence either K has a zero-divisor or a divides b by induction on the degree of f. Thus f divides g or K has a zero-divisor. □

3.6 LEMMA. *Let p be a prime and r an element of a commutative ring R such that $pR = 0$. Let q be a power of p, and f and g be monic polynomials in $R[X]$ with $fg = X^q - r$. If f and g are strongly relatively prime, then either f or g has degree 0.*

PROOF. If $q = 1$, then the theorem is clearly true, so we may assume that $q > 1$. Let $sf + tg = 1$. Differentiating $fg = X^q - r$ we get $f'g + fg' = 0$. As $sf + tg = 1$ we conclude that f divides f', so $f' = 0$. Similarly $g' = 0$. Thus we can write $f(X) = f_0(X^p)$ and $g(X) = g_0(X^p)$. Let $s(X) = \Sigma_{i=0}^{p-1} X^i s_i(X^p)$ and $t(X) = \Sigma_{i=0}^{p-1} X^i t_i(X^p)$. Then $s_0 f_0 + t_0 g_0 = 1$ and $f_0 g_0 = X^{q/p} - r$, so we are done by induction on q. □

The next lemma generalizes (VI.6.6) from fields to finite-dimensional commutative algebras with zero radical. Classically the latter are products of fields and the generalization is routine. As we may not know how to write the algebra as a product of fields, the constructive proof is more delicate.

3.7 LEMMA. *Let k be a discrete field of characteristic p, let K be a finite-dimensional commutative k-algebra, and let a be an element of K. If $X^p - a = fg$ where f and g are monic in $K[X]$ and have degrees greater than zero, then either*

(i) $a \in K^p$

(ii) $rad\ K \neq 0$.

PROOF. If we encounter a nontrivial ideal L of K, then by Theorem 3.3 either $rad\ K \neq 0$ or $K = L \oplus N$ for nontrivial ideals L and N. In the latter case the theorem holds for L and N by induction on dimension. Hence either $rad\ L$ or $rad\ N$ is nonzero, or the components of a in L and N are in L^p and N^p respectively, so $a \in K^p$. Thus we may assume that every nonzero element of K that we construct is invertible.

We want to write $X^q - a$ as a nontrivial power. Suppose $X^q - a = fg^i$ with f and g monic polynomials of positive degree. Let d be the monic GCD of f and g. If $d = 1$, then Lemma 3.6 is violated, so we may assume that $deg\ d > 0$. Let f_1 and g_1 be monic polynomials such that $f = f_1 d$ and $g = g_1 d$. Then $X^q - a = f_1 d(g_1 d)^i = (f_1 g_1^i)d^{i+1}$. If $f_1 g_1^i = 1$, then we have written $X^q - a$ in the desired form; otherwise we continue with $f_1 g_1^i$ as the new f, and d as the new g. After at most q steps we will have written $X^q - a = h(X)^m$ for some monic $h(X)$ and $m > 0$. Clearly m divides q, so m

is a power of p. Therefore $a = (-h(0)^{m/p})^p \in K^p$. □

3.8 LEMMA. *Let k be a discrete field of characteristic p satisfying condition P. Let K be a finite-dimensional commutative k-algebra, $a \in K$, and $L = K[X]/(X^p - a)$. Then either*

(i) *rad $L \neq 0$ or*

(ii) *rad $L \neq 0$ implies rad $K \neq 0$.*

PROOF. Let $e_1,\ldots,e_n$ be a basis for K over k. As k satisfies condition P, either $e_1^p,\ldots,e_n^p$ are independent over k^p or they are dependent over k^p. If $\Sigma\, a_j^p e_j^p = 0$, then $\Sigma\, a_j e_j \in$ *rad* K, so we may assume that they are independent, in which case the natural map from K to K^p is an isomorphism of rings.

As k satisfies condition P, we can decide whether elements of K are linearly dependent over k^p, so the $K^p[a]$ is a finite-dimensional k^p-algebra. By Corollary 1.5 we can either construct a zero-divisor of K^p, hence of K, or the minimal polynomial of a over K^p. In the former case (3.1) either gives us an idempotent of K, and we are done by induction on *dim* K, or an element of *rad* K, and we have trivially established (ii). Otherwise, let g be the minimal polynomial of a over K^p. We may assume that g is monic. Either $g = X^p - a^p$ or $X^p - a^p$ factors over K^p. In the latter case either *rad* $K \neq 0$ or $a \in K^p$, by Lemma 3.7, so *rad* $L \neq 0$ and (i) holds. If $X^p - a^p$ is the minimal polynomial of a over K^p, then $K^p[a]$ is isomorphic to $K^p[X]/(X^p - a^p)$ which is isomorphic, as a ring, to L. Since $K^p[a] \subseteq K$ we have (ii). □

EXERCISES

1. Show that if L is an n-dimensional left ideal of a finite-dimensional algebra, then L^n is finite dimensional, and L is nilpotent if and only if $L^n = 0$.

2. Show that the radical of a finite-dimensional algebra A is the Jacobson radical of A. Show that $rad(A_1 \times A_2) = rad\, A_1 \times rad\, A_2$.

3. Let A be a finite-dimensional algebra with a finite-dimensional radical J. The **left socle** of A is $\{x \in A : Jx = 0\}$. Show that the left socle is a two-sided ideal that has nonzero intersection with every nonzero left ideal. Identify the radical, and the left and right socles, of the algebra of lower triangular $n \times n$

matrices over a discrete field.

4. Show that the following four conditions on a finite-dimensional algebra A are equivalent.

   (i) *rad* A = 0

   (ii) each finite-dimensional left ideal of A is a summand

   (iii) each finite-dimensional submodule of a finite-dimensional A-module is a summand.

   (iv) each finite-dimensional A-module is projective.

5. **Maschke's theorem.** Let G be a finite group, k a discrete field such that $n = \#G$ is not zero in k, and A the group algebra of G over k. Let M be a finite-dimensional A-module and $\varphi : M \to M$ a k-space map such that $\varphi^2 = \varphi$ and $\varphi(M)$ is a left A-module. Let $\psi(x) = \frac{1}{n}\Sigma_{g\in G}\, g\varphi(g^{-1}x)$. Show that ψ is a map of A-modules, $\psi^2 = \psi$, and $\varphi(M) = M$. Conclude that *rad* A = 0 (see Exercise 4).

4. WEDDERBURN'S THEOREM, PART ONE.

In order to prove Wedderburn's theorem about finite-dimensional algebras with zero radical, we must put some restriction on the field k.

4.1 LEMMA. *Let A be a finite-dimensional k-algebra, and $K \neq A$ a finite dimensional subalgebra of the center of A. Then either K has a nontrivial idempotent or a nontrivial nilpotent, or there is an element in $A \backslash K$ whose minimal polynomial g over K is either separable, or $g(X) = X^p - r$ where $p = char\ k$, or $g(X) = X^m$.*

PROOF. We may assume that $A = K[a]$ and construct, by Corollary 1.5, either a zero-divisor in K or the minimal polynomial $g(X)$ of a over K. If K has a zero-divisor, then K has a nontrivial idempotent or a nontrivial nilpotent by Lemma 3.1. By (VI.6.3) either K has a nontrivial ideal or we can factor g into pairwise strongly relatively prime polynomials of the form $f^m(X^q)$ where m is a positive integer, q is either 1 or a power of the finite characteristic p of k, and f separable. This induces a decomposition of A into a product of algebras of the form $K[X]/f^m(X^q)$, so we may assume that $g(X) = f^m(X^q)$. If $a^q \in K$, then $g(X) = X^q - r$. If $q = p$ we are done; otherwise $a^p \notin K$ while $(a^p)^{q/p} \in K$, so $K[a^p]$ is a nontrivial subalgebra of A and we are done by induction on dim A. If $a^q \notin K$ and $m = 1$, then f is the minimal polynomial of a^q over K and is separable. If $a^q \notin K$ and $m > 1$, then the minimal polynomial of $f(a^q)$ over K is X^m. □

4.2 THEOREM. *A discrete field k is separably factorial if and only if each finite-dimensional k-algebra A with rad A = 0 is either simple or has a nontrivial ideal.*

PROOF. By Theorem 2.2 we can restrict our attention to commutative k-algebras. Suppose A is a finite-dimensional commutative k-algebra. If $A = k$ we are done; otherwise we can find $a \in A \backslash k$ with minimal polynomial g.

First suppose k is separably factorial and $rad\ A = 0$. Then g cannot be of the form X^m. If $g(X) = X^p - r$, then g is irreducible by (VI.6.6) lest $r \in k^p$ in which case $rad\ A \neq 0$. If g is irreducible, then we can replace k by $k[X]/(g)$ and we are through by induction on $dim\ A$. So, by Lemma 4.1, we may assume that g is separable. Factor g into irreducible polynomials. If a satisfies one of these factors, then $k(a)$ is a field and, as A is commutative, A is a finite-dimensional $k(a)$-algebra. Thus, as $k(a)$ is also separably factorial, we are done by induction on $dim\ A$. If a does not satisfy any of the factors f of g, then each $f(a)$ generates a nontrivial ideal of A.

Conversely let K be a finite-dimensional extension field of k. If $f \in K[X]$ is a separable polynomial, construct the finite-dimensional commutative k-algebra $A = K[X]/(f)$. By Lemma 3.5 we have $rad\ A = 0$. The finitely generated ideals of A are in 1-1 correspondence with the monic factors of f, so f is either irreducible or has a nontrivial factorization. □

We now characterize separably factorial fields in terms of decomposing algebras into products of simple algebras. This is the first part of Wedderburn's theorem.

4.3 THEOREM. *A discrete field k is separably factorial if and only if every finite-dimensional k-algebra with zero radical is a product of simple algebras.*

PROOF. Let A be a finite-dimensional k-algebra. By Theorem 4.2, A is either simple or has a nontrivial ideal L. In the former case we are done. In the latter we may apply Theorem 3.3 and write A as a nontrivial product of algebras, and we are done by induction on the dimension of A.

Conversely, if each A with *rad* A = 0 is a product of simple algebras, then A is either simple or has a nontrivial ideal, so k is separably factorial by Theorem 4.2. □

4.4 THEOREM. *A discrete field k satisfies condition P if and only if each finite-dimensional k-algebra has a finite-dimensional radical.*

PROOF. Suppose k satisfies condition P. If we construct a nonzero element a of *rad* A, then we can pass to $A/(AaA)$ and we are done by induction on *dim* A. Let C be the center of A. By Theorem 3.4 either *rad* A ≠ 0, or *rad* A = A·*rad* C. Thus we may assume that A is commutative. If we construct a nontrivial idempotent of A, then we can factor A into two algebras of smaller dimension and we are done by induction on *dim* A. Let K be a finite-dimensional k-subalgebra of A with *rad* K = 0. Initially we take $K = k$, and we proceed by induction on *dim* A − *dim* K.

Either $K = A$, in which case we are done, or we can apply Lemma 4.1 to construct an element a in $A\backslash K$ whose minimal polynomial g is either separable, or $g(X) = X^p - r$, where p = *char* k, or $g(X) = X^m$. If $g(X) = X^m$, then a is a nonzero element of *rad* A. If $g(X) = X^p - r$, then, as k satisfies condition P, Lemma 3.8 says that either *rad* $K[a] \neq 0$ or *rad* $K[a] = 0$. In the latter case we replace K by $K[a]$ and we are done by induction on *dim* A − *dim* K. If g is separable, then *rad* $K[a] = 0$ by Lemma 3.5, and we replace K by $K[a]$ as before.

Conversely suppose K is a finite-dimensional extension field of k, and $a \in K$. Consider the k-algebra $L = K[X]/(X^p - a)$. If *rad* $L = 0$, then $a \notin K^p$ for if $a = r^p$, then $0 \neq X - r \in$ *rad* L. If *rad* $L \neq 0$, then there is a polynomial $f \in K[X]$ such that $X^p - a$ divides f^2 but does not divide f. Then the greatest common divisor of $X^p - a$ and f is a proper factor of $X^p - a$, so $a \in K^p$ by (VI.6.6). □

4.5 COROLLARY. *A discrete field k is fully factorial if and only if every finite-dimensional algebra A over k has a finite-dimensional nilpotent ideal I such that A/I is a product of simple k-algebras.*

PROOF. As k is fully factorial if and only if k is separably factorial and satisfies condition P, this follows immediately from Theorems 4.3 and 4.4. □

EXERCISES

1. Let K be a finite-dimensional commutative k-algebra and $g \in K[X]$. Suppose $g = f_1 f_2 \cdots f_n$ where the f_i are pairwise strongly relatively prime. Show that $K[X]/(g)$ is isomorphic to the product of the k-algebras $K[X]/(f_i)$.

2. Let k be a factorial field and $f \in k[X]$. Identify the radical I of $A = k[X]/(f)$, and describe A/I as a product of simple k-algebras.

3. Let G be the symmetric group on {1,2,3} and A the group algebra of G over $\mathbb{Q}$. Decompose A into a product of simple $\mathbb{Q}$-algebras (see Exercise 3.5).

5. MATRIX RINGS AND DIVISION ALGEBRAS.

The second part of Wedderburn's structure theorem for semi-simple algebras says that a finite-dimensional simple algebra is isomorphic to a full ring of matrices over a division algebra. This theorem hinges on being able to construct nontrivial left ideals. Once we have a nontrivial left ideal we can use it to decompose the original algebra.

5.1 THEOREM. *Let* A *be a finite-dimensional k-algebra, and L a nontrivial left ideal of* A. *Then either*

(i) A *has a nonzero radical*

(ii) A *is a product of finite dimensional k-algebras*

(iii) A *is isomorphic to a full matrix ring over some k-algebra of dimension less than* A.

PROOF. We may assume that L is finite dimensional by passing to the principal left ideal generated by a nonzero element of L. If L is not a faithful A-module, then the representation kernel of L is a nonzero ideal of A; whence by Theorem 3.3, either *rad* A $\neq 0$ or A is a product. Thus we may assume that L is faithful.

Let B be the centralizer of A on L (that is, the A-endomorphism ring of L). By the density theorem (2.1), either L is reducible (and we are done by induction on the dimension of L) or A is a full matrix ring over the opposite ring of B. It remains to show that the dimension of B is less than that of A. By Theorem 3.3 either rad A $\neq 0$ or L is a summand of A. In the latter case the algebra B is a proper subalgebra of the centralizer of A on A, which is the opposite ring of A. □

The fundamental problem is to be able to recognize whether a given finite-dimensional algebra is a division algebra or not, in the sense of being able either to assert that it is a division algebra or to construct a nontrivial left ideal. If we could do that, then Theorem 5.1 would imply that every finite-dimensional k-algebra has a finite dimensional radical, and modulo its radical it is a product of full matrix rings over division algebras. This condition is equivalent to being able to recognize whether an arbitrary finite-dimensional representation of a finite-dimensional k-algebra is reducible.

5.2 THEOREM. *The following conditions on a discrete field k are equivalent.*

(i) *Each finite-dimensional k-algebra is either a division algebra or has a nontrivial left ideal.*

(ii) *Each finite-dimensional left module M over a finite dimensional k-algebra A is either reducible or irreducible.*

(iii) *Each finite-dimensional k-algebra A has a finite-dimensional radical, and A/rad A is a product of full matrix rings over division algebras.*

PROOF. Clearly both (ii) and (iii) imply (i). Suppose (i) holds and we wish to establish (ii). We may assume that M is a faithful A-module. Let B be the centralizer of A on M. If B has a zero-divisor b, then bM is a nontrivial A-submodule of M. If B is a division algebra, then by the density theorem either M is reducible or A is the centralizer of B so M is irreducible.

Now suppose (i) holds and we wish to establish (iii). If A is a division algebra we are done. If A has a nontrivial left ideal, then by Theorem 5.1 either A has a nonzero finite-dimensional nilpotent ideal I, in which case we pass to A/I and are done by induction on *dim* A, or A is a product of finite-dimensional k-algebras and we are done by induction on *dim* A, or A is isomorphic to a full matrix ring and we are done by induction on *dim* A. □

For what fields k do the conditions of Theorem 5.2 hold? Finite fields and algebraically closed fields provide trivial examples. The field of algebraic real numbers admits only three finite-dimensional division algebras, and a constructive proof of this statement shows that this field satisfies the conditions of Theorem 5.2.

5.3 THEOREM. *Let k be a discrete subfield of $\mathbb{R}$ that is algebraically closed in $\mathbb{R}$, and $H = k(i,j)$ the quaternion algebra over k. If A is a finite-dimensional algebra over k, then either A has a zero-divisor, or A is isomorphic to k, to $k(i)$, or to H.*

PROOF. If $A = k$ we are done; otherwise let $\alpha \in A\backslash k$. As A is finite dimensional we can construct a nontrivial polynomial satisfied by α. The field $k(i) \subseteq \mathbb{C}$ is discrete and algebraically closed, so k is factorial, and no irreducible polynomial over k has degree greater than 2. Thus either A has a zero-divisor or α is of degree 2, so we may assume that $\alpha \in k(i)$ where $i \in A$ and $i^2 = -1$. The centralizer of i in A is a finite-dimensional algebra over the algebraically closed field $k(i)$, hence either has a zero-divisor or is $k(i)$ itself. We may assume the latter. If $A = k(i)$ we are done. Otherwise we can construct $\beta \in A\backslash k(i)$ with $\beta^2 = -1$ just like we constructed $i \in A$. As $i\beta + \beta i$ commutes with i and β, we may assume that $i\beta + \beta i \in k(i) \cap k(\beta) = k$. Let $i\beta + \beta i = r \in k$, and set $j = \beta + ri/2$. Then $ij + ji = 0$ and $j^2 \in k$. If $j^2 = s^2$ for some $s \in k$, then $j - s$ is a nonzero zero-divisor of A. Otherwise $j^2 < 0$ so we can normalize j to get $j^2 = -1$. If j' is another such j, then jj' commutes with i, so $jj' \in k(i)$ whereupon $j' \in k(i,j)$. □

Does the field $\mathbb{Q}$ of rational numbers satisfy the conditions of Theorem 5.2? Certainly we are not going to produce a Brouwerian counterexample when $k = \mathbb{Q}$. Probably a close analysis of the classical theory of division algebras over $\mathbb{Q}$, in analogy with Theorem 5.3, will yield a proof.

EXERCISES

1. Show that every finite-dimensional commutative algebra over a fully factorial field is either a division algebra or has a nontrivial left ideal. Show that over a factorial field every algebra of prime dimension is either a division algebra or has a nontrivial left ideal.

NOTES

Most of the material in this chapter appeared in [Richman 1982]. It might be interesting to see how the theory develops for rings with composition series (Artinian rings). Frobenius algebras, quasi-Frobenius algebras, and Lie algebras are also natural topics for investigation.

Chapter X. Free Groups

1. EXISTENCE AND UNIQUENESS.

A group F is a **free group** on a subset S of F if for any group H and any function f from S to H, there is a unique homomorphism from F to H extending f. If F is a free group on S, then S is called a (**free**) **basis** for F. Before showing how to construct free groups, we show that there is, up to isomorphism, only one free group on a set.

1.1 THEOREM. *Let F_1 and F_2 be free groups on S_1 and S_2 respectively. If f is an isomorphism from S_1 to S_2, then there is a unique isomorphism from F_1 to F_2 that extends f.*

PROOF. Let g be the inverse of f. As F_1 is a free group on S_1, there is a unique homomorphism f_* from F_1 to F_2 extending f. Similarly there is a unique homomorphism g_* from F_2 to F_1 extending g. The homomorphism f_*g_* from F_2 to F_2 extends the identity map on S_2. Since F_2 is a free group on S_2, and the identity map on F_2 also extends the identity map on S_2, the identity map on F_2 must equal f_*g_*. Similarly g_*f_* is the identity map on F_1. Thus f_* is the required isomorphism; f_* is unique because F_1 is free on S_1. □

To construct a free group on an arbitrary set S, define the set $S \cup S^{-1}$ to be $S \times \{1,-1\}$, with S identified with $S \times \{1\}$, and $(s,-\epsilon)$ denoted by $(s,\epsilon)^{-1}$. Let $F(S)$ be the free monoid on $S \cup S^{-1}$. We define two equalities on $F(S)$. The first, denoted by the symbol '$\equiv$', is the usual equality on a free monoid. To define the second equality, we take the notation x^{-1} seriously: Call two words in $F(S)$ **adjacent** if one can be written as vw and the other as $vxx^{-1}w$ with v and w in $F(S)$, and x in $S \cup S^{-1}$. Two words v and w are **equal** in $F(S)$, written $v = w$, if there exists a sequence of words $v \equiv w_1, w_2, \ldots, w_n \equiv w$, such that w_i is adjacent to w_{i+1} for each $i < n$.

We must show that the natural map of S into $F(S)$ embeds S as a subset of $F(S)$: that is, if s and s' are in S, and $s = s'$ as elements of $F(S)$, then $s = s'$ as elements of S. To this end, and for its general utility, we establish the **Church-Rosser property** for $F(S)$. If $x_i \in S \cup S^{-1}$ for $i = 1,\ldots,n$, then the **length** of $z \equiv x_1x_2\cdots x_n$ is defined to be $\ell(z) = n$. The length is a function on $F(S)$ with respect to '$\equiv$' but not with respect to '$=$', as the words x and $xx^{-1}x$ are equal but their lengths are 1 and 3 respectively.

1.2 THEOREM (**Church-Rosser property**). *Let $u = v$ be words in $F(S)$. Then we can find an integer n, and a chain $u \equiv w_1,\ldots,w_n \equiv v$ of adjacent words such that if $\ell(w_{i-1}) < \ell(w_i)$ for $i < n$, then $\ell(w_i) < \ell(w_{i+1})$.*

PROOF. As $u = v$ there is a sequence of adjacent words $u \equiv w_1,\ldots,w_n \equiv v$. We shall prove the result by induction on the total length $N = \Sigma\, \ell(w_i)$ of the sequence. We may assume that $n > 1$. If for some i we have $\ell(w_{i-1}) < \ell(w_i)$ and $\ell(w_i) > \ell(w_{i+1})$, then w_{i-1} is obtained from w_i by deleting a part xx^{-1}, and w_{i+1} is obtained from w_i by deleting a part yy^{-1}. If these two parts coincide, then w_i and w_{i+1} may be omitted from the sequence and we are done by induction. If these two parts overlap without coinciding, then w_i has a part $zz^{-1}z$ and w_{i-1} and w_{i+1} are both obtained by replacing this part by z, so $w_{i-1} \equiv w_{i+1}$ once again, and w_i and w_{i+1} may be omitted. In the remaining case, where the two parts are disjoint, we can delete both parts from w_i to obtain a new sequence with total length $N-4$ and we are done by induction. □

It is an easy consequence of (1.2) that S is a subset of $F(S)$.

The (associative) multiplication on $F(S)$ respects the equality '$=$'; that is, if $v = v'$ and $w = w'$, then $vw = v'w'$. If $v \equiv x_1x_2\cdots x_n$, set $v^{-1} \equiv x_n^{-1}\cdots x_2^{-1}x_1^{-1}$; then $vv^{-1} = v^{-1}v = 1$, where 1 is the empty word. Thus $F(S)$ is a group.

To show that $F(S)$ is a free group on S, let f be a function from S to a group H. If $\overline{f}$ is a homomorphism from $F(S)$ to H extending f, then $\overline{f}(x_1\cdots x_n)$ must be equal to $f(x_1)\cdots f(x_n)$, so $\overline{f}$ is unique. Moreover, setting $\overline{f}(x_1\cdots x_n) = f(x_1)\cdots f(x_n)$ defines a homomorphism because $\overline{f}(w) = \overline{f}(w')$ if $w = w'$. Thus $F(S)$ is a free group on S, which we will refer to as **the free group on** S. We summarize the above in the following.

1.3 THEOREM. *If S is a set, then $F(S)$ is a free group on S.* □

A word $w \equiv x_1x_2\cdots x_n$ in $F(S)$, with each x_i in $S \cup S^{-1}$, is **reducible** if $x_ix_{i+1} = 1$ for some $i = 1,\ldots,n-1$. If w is not reducible we say that w is **reduced.** Note that if u and v are in $F(S)$, and if uv is reduced, then u and v are reduced. If S is a discrete set, and u and v are reduced words in $F(S)$, then uv is either reduced or $u \equiv ax$ and $v \equiv x^{-1}b$ with ab reduced.

If $w = w'$, and w' is reduced, then w' is called the **reduced form** of w; the Church-Rosser property of $F(S)$ implies that if $w = w'$, and w and w' are both reduced, then $w \equiv w'$, so the reduced form is unique. If S is discrete, then every element of $F(S)$ has a unique reduced form. From this we immediately get

1.4 THEOREM. *If S is a discrete set, then $F(S)$ is a discrete group.* □

1.5 THEOREM. *If $w \in F(S)$ and $w^n = v$ with $\ell(v) \leq \ell(w)$ for some $n > 1$, then $w = 1$ or w is reducible. Thus a free group is torsion free; that is, if $w^n = 1$ for some $n > 0$, then $w = 1$.*

PROOF. Write $w \equiv u^{-1}w_1u$ (for example: $u \equiv 1$ and $w_1 \equiv w$) and induct on $\ell(w_1)$. If $w^n = v$ with $\ell(v) \leq \ell(w)$ for some $n > 1$, then by (1.2) either $\ell(u^{-1}w_1^nu) \leq \ell(v)$, or w is reducible, or $w_1 \equiv x^{-1}w_2x$ for some x in $S \cup S^{-1}$. In the first case $w_1 \equiv 1$ so $w = 1$, in the second we are done, and in the third we are done by induction on $\ell(w_1)$. □

1.6 THEOREM. *Let F and F' be free groups on finite sets S and S' respectively. Then F and F' are isomorphic if and only if $\#S = \#S'$.*

PROOF. If $\#S = \#S'$, then F and F' are isomorphic by Theorem 1.1. To prove the converse we show how to recover the number $\#S$ from F. Let N be the subgroup of F generated by the elements v^2 with v in F. Clearly N is a normal subgroup; the quotient group F/N is abelian because the square of every element is 1, so $xyx^{-1}y^{-1} = xyxy = 1$. We will show that F/N is a finite set with $2^{\#S}$ elements.

If $w \equiv x_1x_2\cdots x_n \in F$, and $s \in S$, let $v_s(w)$ denote the number of indices i such that $x_i = s$ or $x_i = s^{-1}$. If $w = w'$, then $v_s(w)$ is congruent to $v_s(w')$ modulo 2. Let $D = \{w \in F : v_s(w)$ is even for each $s \in S\}$. Clearly $N \subseteq D$ and D is a detachable subset of F. Conversely, as F/N is abelian, and the square of any element of F/N is 1, we have $D \subseteq N$, and each element of F/N can be written uniquely as a product of distinct elements of S. □

If $F(S)$ is the free group on a finite set S, then $\#S$ is an invariant of

$F(S)$, called the **rank** of $F(S)$. If S is a countably infinite set then $F(S)$ is said to be of **countable rank**.

1.7 THEOREM. *If G is a group, then there is a free group F and a epimorphism $f : F \to G$. If G is discrete, then F can be taken to be discrete.*

PROOF. Let $F = F(G)$ be the free group on the set G. Using the identity function from G as a set to G as a group we obtain, by the definition of a free group, a unique homomorphism f from $F(G)$ to G which is the identity on G. □

1.8 LEMMA. *Let U be a subset of a group G such that $U \cap U^{-1} = \phi$. Then U is a free basis for a subgroup of G if and only if whenever $u_1u_2\cdots u_n = 1$ with $n \geq 1$ and each $u_i \in U \cup U^{-1}$, then $u_iu_{i+1} = 1$ for some i.*

PROOF. If U is a basis for a free subgroup of G, then U satisfies the conditions of the lemma by (1.2). Conversely suppose that U satisfies the conditions of the lemma, and let $F(U)$ be the free group on U. The inclusion map from U to G extends uniquely to a group homomorphism f from $F(U)$ to G whose image is the subgroup generated by U. Suppose

$$1 = f(u_1\cdots u_n) = f(u_1)\cdots f(u_n),$$

where each $u_i \in U \cup U^{-1}$. As f is the inclusion map on U, we have $f(u_i) \in U \cup U^{-1}$ in G, so there is i with $f(u_i)f(u_{i+1}) = 1$, and therefore $u_iu_{i+1} = 1$. Thus the kernel of f is trivial, so $F(U)$ is isomorphic to the subgroup generated by U. □

EXERCISES

1. Show that if $F(S)$ is abelian, and $a,b \in S$, then $a = b$.
2. Show that if $w \in F(S)$ and $s \in S$, and $sw = ws$, then $w = s^n$ for some integer n.
3. Show that every word in $F(S)$ has a reduced form if and only if S is discrete.
4. Show that $F(S) \cong F(T)$ implies S is isomorphic to T if:
   (i) S is finite,
   (ii) S is $\mathbb{N}$,
   (iii) S is a detachable initial (no gaps) subset of $\mathbb{N}$.

5 Let S be a finite set of cardinality m. Show that if T is a finite subset of $F(S)$ of cardinality greater than m, then there exists a nonempty product of distinct elements of T that is equal to a product of squares.

6. Let $S = \{s_1,\ldots,s_m\}$ and $T = \{t_1,\ldots,t_n\}$ (not necessarily discrete). Use Exercise 5 to show that if $F(S) \cong F(T)$, and $m < n$, then $t_i = t_j$ for some $i < j$.

2. NIELSEN SETS.

Let S be a discrete set. We study conditions on a subset U of $F(S)$ that ensure that U is a free basis for the subgroup $\langle U\rangle$ generated by U. The **reduced length** of an element w in $F(S)$ is the length of its reduced form, and is denoted by $|w|$.

A subset U of the free group $F(S)$ is called a **Nielsen set** if

(N0) $U \cap U^{-1} = \phi$,

and for all $x,y,z \in U \cup U^{-1}$ we have

(N1) If $xy \neq 1$, then $|xy| \geq \max\{|x|, |y|\}$

(N2) If $xy \neq 1$ and $yz \neq 1$, then $|xyz| > |x| - |y| + |z|$.

Note that N0 ensures that $1 \notin U$. We will show that if U satisfies N0 and N2, then U is a free basis for $\langle U\rangle$.

2.1 EXAMPLES. *Conditions* N1 *and* N2 *are independent*. Let $S = \{s,t,u,v\}$. Then the set $U = \{s^2,st,ts\}$ satisfies N1 but, as $(st)^{-1}(s^2)(ts)^{-1} = t^{-2}$, it does not satisfy N2. The set $V = \{tuv, suv\}$ satisfies N2 but not N1. □

If u and v are reduced words, then there exist unique reduced reduced words a, b and c such that $u \equiv ab^{-1}$ and $v \equiv bc$, and ac is reduced. We call b **the part of** v **that cancels in the product** uv. Similarly b^{-1} is called **the part of** u **that cancels in the product** uv. The following lemma helps to explain the meaning of N1.

2.2 LEMMA. *Let U be a finite set of reduced words, satisfying* N1, *in the free group $F(S)$. If u and v are words in U, and b is the part of v that cancels in the product uv, then $2|b| \leq \min(|u|,|v|)$.*

PROOF. This is an easy consequence of N1. □

The following lemma explains the significance of N2, and sets the stage for invoking Lemma 1.8.

2.3 LEMMA. *Let S be a discrete set, and let U be a set of reduced words in $F(S)$ satisfying* N0 *and* N2. *Let $w \equiv u_1u_2\cdots u_n$ with each u_i in $U \cup U^{-1}$, and $u_iu_{i+1} \neq 1$ for each $i < n$. Then $u_i \equiv a_ib_ic_i$, with $b_i \neq 1$, and $b_1b_2\cdots b_n$ is the reduced form of w.*

PROOF. Set $a_0 \equiv c_n \equiv 1$. Let c_i be the part of u_i, and $a_{i+1} \equiv c_i^{-1}$ the part of u_{i+1}, that cancels in the product u_iu_{i+1} for $i < n$. By N2 we have

$$|u_{i-1}u_iu_{i+1}| > |u_{i-1}| - |u_i| + |u_{i+1}|$$

so $u_i \equiv a_ib_ic_i$, with $b_i \neq 1$, if $1 < i < n$. From (1.5) we know that $u_1u_1 \neq 1$ and $u_nu_n \neq 1$, so N2 says that $|u_1u_1u_2| > |u_2|$ and $|u_{n-1}u_nu_n| > |u_{n-1}|$; therefore $b_1 \neq 1$ and $b_n \neq 1$. As c_i is the part of u_i that cancels in the product u_iu_{i+1}, we have $a_ib_ib_{i+1}c_{i+1}$ is the reduced form of u_iu_{i+1}, so $b_1b_2\cdots b_n$ is reduced. □

2.4 COROLLARY. *If U is a set of words satisfying* N0 *and* N2 *in a free group on a discrete set, then U is a free basis for $\langle U\rangle$.*

PROOF. This follows from Lemma 2.3 and Lemma 1.8. □

As the set V of Example 2.1 satisfies N0 and N2, it follows that V is a free basis for $\langle V\rangle$.

Unlike in the abelian case, a free group of small rank can have a subgroup that is free of large rank. In fact there are infinite-rank free subgroups of a finite-rank free group.

2.5 THEOREM. *A free group on a finite set of cardinality two contains a countable-rank free subgroup.*

PROOF. Let $\{x,y\}$ be the basis for the free group, and consider the set

$$U = \{y,\ xyx^{-1},\ x^2yx^{-2},\ x^3yx^{-3},\ \ldots\ \}.$$

It is readily verified that U is a Nielsen set, hence a free basis for the subgroup it generates. □

EXERCISES

1. Let F be the free group on the two-element set $\{x,y\}$. Show that $\{x^my^nx^{-m}y^{-n} : m,n \in \mathbb{Z}\setminus\{0\}\}$ is free basis for the commutator

subgroup of F.

2. Show that if U is a Nielsen set of words in a discrete free group, and if $w \equiv u_1 \cdots u_n$ with u_i in $U \cup U^{-1}$ and each $u_i u_{i+1} \neq 1$, then $|w| \geq max\ \{|u_1|, \ldots, |u_n|\}$.
3. Show that the subgroup constructed in (2.5) is detachable.

3. FINITELY GENERATED SUBGROUPS OF FREE GROUPS.

In this section we show how to transform a finite set of generators for a subgroup of a free group on a finite set S into a Nielsen set.

3.1 DEFINITION. *Let U and V be finite sets of words. We say that V is obtained from U by a* ***Nielsen transformation*** *if either*

(T0) $V = U \setminus \{1\}$

(T1) $V = (U \setminus \{u\}) \cup \{u^{-1}\}$ *where* $u \in U$

(T2) $V = (U \setminus \{u\}) \cup \{v\}$ *where v is either uu' or $u'u$, for some u' in $U \cup U^{-1}$ different from u and u^{-1}.*

A transformation of type T2 is referred to as **replacing u by v in U.** Note that if V is obtained from U by a Nielsen transformation, then $\langle U\rangle = \langle V\rangle$ and $\#U \geq \#V$. (Transformations of types T1 and T2 may decrease the cardinality of the set U. For example if $U = \{a,ab,b\}$, the transformation of type T2 replacing b by ab decreases the cardinality.) Also, if V is obtained from U by a transformation of type T1 or T2, and $\#U = \#V$, then V is obtained from U by a transformation of the same type.

If S is a finite set, we can linearly order the words of $F(S)$, with respect to '$\equiv$', as follows. Fix a linear ordering of $S \cup S^{-1}$ and extend lexicographically to an ordering on the words of $F(S)$. If u and v are words of $F(S)$, define $u < v$ if either

(i) $\ell(u) < \ell(v)$ or

(ii) $\ell(u) = \ell(v)$ and u comes before v in the lexicographic ordering.

Note that each word has a finite number of predecessors under this ordering.

3.2 THEOREM. *Let S be a finite set and U a finite subset of $F(S)$. Then there is a sequence of Nielsen transformations that transforms U to a Nielsen set.*

PROOF. First we set up a measure of the size of U on which to base our induction. If w is a word, then we can write the reduced representative of w uniquely as $w_L w_R$ where $|w_L|$ is the greatest integer not exceeding $(|w|+1)/2$. We define the function φ from $F(S)$ to $\mathbb{N}$, somewhat cryptically, by setting $\varphi(w)$ equal to the number of words v such that $v < w_L w_R^{-1}$. Note that if $|u_1| < |u_2|$, then $\varphi(u_1) < \varphi(u_2)$. Let $\varphi U = \Sigma_{u \in U}\varphi(u)$.

We proceed by induction on φU. By a series of transformations of type T0 and T1 we may assume that $U \cap U^{-1} = \phi$. If $|xy| < |x|$ for $x,y \in U \cup U^{-1}$ and $xy \neq 1$, then $x \neq y$ by (1.5) and we can replace x by xy (or x^{-1} by $y^{-1}x^{-1}$) and decrease φU. Thus we may assume that U satisfies N1.

Let x, y and z be reduced words in $U \cup U^{-1}$ with $xy \neq 1$ and $yz \neq 1$. Suppose $|xyz| \leq |x| - |y| + |z|$. Then we can write

$$x \equiv ap^{-1}$$
$$y \equiv pq^{-1}$$
$$z \equiv qc.$$

If $|p| > |q|$, then $|xy| < |x|$; if $|p| < |q|$, then $|yz| < |z|$; if $|p| > |a|$, then $|xy| < |y|$; if $|q| > |c|$, then $|yz| < |y|$; so none of these cases occur as U satisfies N1. Thus $|p| = |q| \leq min(|a|,|c|)$, and $|xy| = |x|$ and $|yz| = |z|$. Note that $p \neq q$ since $y \neq 1$. If $p < q$ (in the lexicographic ordering), then $\varphi(yz) < \varphi(z)$, while if $q < p$, then $\varphi(xy) < \varphi(x)$. In either case we can apply a Nielsen transformation that reduces $\varphi(u)$ for one element u in U while leaving the other elements alone, thus decreasing φU. □

3.3 COROLLARY. *Every finitely generated subgroup of a finite-rank free group is free, and has a Nielsen set as a free basis.* □

3.4 THEOREM. *If F is a free group of finite rank n, and U is a set of generators for F, then U contains at least n elements. Moreover, if U has exactly n elements, then U is a free basis for F.*

PROOF. As each element of a free basis for F is a product of finitely many words in U, we may assume that U is finite. By (3.2) we can transform U into a Nielsen set V by a sequence of Nielsen transformations. Thus $\langle U\rangle = \langle V\rangle$ and $\#V \leq \#U$. As V is a Nielsen set, and U is a generating set, V is a free basis for F, so V has n elements. If U has n elements, then no transformations of type T0 were used in transforming U to V. Therefore V can be transformed to U by transformations of types T1 and T2.

But these transformations can be used to define a function from the free basis V onto U, that when extended to a homomorphism of F is an isomorphism. □

3.5 THEOREM. *Let F be a finite-rank free group and G a finitely generated subgroup of F. Then G is detachable.*

PROOF. Since G is finitely generated, (3.3) says that G has a Nielsen set U as a finite free basis. Let $w \in F$. From (2.3) it follows that if $w \in G$, then w can be written as a product of not more than $|w|$ elements of $U \cup U^{-1}$. As F is discrete, we can check to see if w can be so written. □

EXERCISES

1. Show that the commutator subgroup of a free group on a two-element set is not finitely generated (see Exercise 2.1).
2. Give a Brouwerian example of a countable subgroup of a finite-rank free group that is not free.
3. Show that a finite-rank free group F is **Hopfian** in the sense that any map from F onto F is one-to-one.
4. Show that finitely generated subgroups of discrete free groups are free and detachable.

4. DETACHABLE SUBGROUPS OF FINITE-RANK FREE GROUPS.

In this section we prove that detachable sugroups of finite-rank free groups are free. Also, a subgroup of finite index n in a free group of finite rank r is free of rank $n(r-1) + 1$.

Let F be a group and G a subgroup of F. A function T from F to F is a **(right) transversal** for G if $T(x) \in Gx$ for each x in F, and if $T(x) = T(y)$ whenever $Gx = Gy$. In other words, T is a choice function for the set of right cosets of G in F. Note that $T(T(x)y) = T(xy)$ for all x,y in F.

Let F be the free group on a finite set S. If $w \equiv uv$ is a reduced word in F, then u is called an **initial segment** of w, and v is called a **final segment** of w. A transversal T for a subgroup G of F is a **Schreier transversal** if $T(w)$ is reduced for each w in F, and if $T(F) = \{T(w) : w \in F\}$ is closed under taking initial segments. If $T(F)$ is also closed under taking final segments, then T is a **two-sided Schreier transversal.** Note that a Schreier transversal is a function from F with the equality

'=', to F with the equality '$\equiv$'.

4.1 THEOREM. *If F is the free group on a finite set S, and G is a detachable subgroup of F, then G has a Schreier transversal. If G is also a normal subgroup, then G has a two-sided Schreier transversal.*

PROOF. Equip F with the total ordering defined just prior to (3.2). Each element in F has a finite number of predecessors, and G is detachable, so for each w in F we can define $T(w)$ to be the first word in Gw (note that Gw is the set of all words that are equal to a word of the form gw). As 1 is the first word in $G\cdot 1$, we have $1 \in T(F)$. Also each word in $T(F)$ is reduced.

If $w \equiv uv$ is the smallest element of Gw, then u is the smallest element of Gu, for if there is g in G with $gu = c < u$, then $gw = cv < uv \equiv w$. Thus T is a Schreier transversal. If G is normal, then $Gw = wG$, so v is the smallest element of $Gv = vG$, whence T is a two-sided Schreier transversal. □

4.2 LEMMA. *Let F be the free group on the finite set S, and let T be a Schreier transversal for a subgroup G of F. Let s and s' be elements of $S \cup S^{-1}$, and t and t' be elements of $T(F)$ such that neither ts nor $t's'$ is equal to an element of $T(F)$. Let u be the reduced form of $T(ts)^{-1}t'$. Then*

(i) *$tsT(ts)^{-1}$ and $t's'T(t's')^{-1}$ are reduced,*

(ii) *If $tsT(ts)^{-1} = t's'T(t's')^{-1}$, then $t \equiv t'$ and $s \equiv s'$,*

(iii) *sus' is reduced unless $u = 1$ and $s' = s^{-1}$.*

PROOF. If ts is not reduced, then $t \equiv t''s^{-1}$ with $t'' \in T(F)$, as T is a Schreier transversal. Thus $ts = t'' \in T(F)$, contrary to the hypothesis. If $sT(ts)^{-1}$ is not reduced, then $T(ts) \equiv t''s$ with $t'' \in T(F)$, as T is a Schreier transversal. But then

$$t'' = T(t'') = T(T(ts)s^{-1}) = T(tss^{-1}) = T(t) = t,$$

so $ts = t''s \in T(F)$, contrary to the hypothesis. As ts and $sT(ts)^{-1}$ are both reduced, it follows that $tsT(ts)^{-1}$ is reduced. Similarly $t's'T(t's')^{-1}$ is reduced.

If $tsT(ts)^{-1} = t's'T(t's')^{-1}$, then, as they are both reduced, and as ts is not an initial segment of t', and $t's'$ is not an initial segment of t', it follows that $t = t'$ and $s = s'$.

To show (iii) it suffices to show that su and us' are reduced. We have

$u = T(ts)^{-1}t'$, and u and $sT(ts)^{-1}$ are reduced. If su is not reduced, then $T(ts)^{-1}$ must cancel in the product $T(ts)^{-1}t'$ leaving an s^{-1} on the left, so $T(ts)s^{-1}$ is an initial segment of t'. Therefore $T(ts)s^{-1}$ is in $T(F)$ since T is a Schreier transversal. Hence $t = T(T(ts)s^{-1}) = T(ts)s^{-1}$, whereupon $ts = T(ts) \in T(F)$, contrary to the hypothesis. So su is reduced.

Similarly, if $us' = T(ts)^{-1}t's'$ is not reduced, then, as $t's'$ is reduced, $t's'$ is an initial segment of $T(ts)$, hence is in $T(F)$, contrary to the hypothesis. □

4.3 LEMMA. *Let T be a transversal for a subgroup of a group F. Let $t, t' \in T(F)$ and $s \in F$. Then the following conditions are equivalent:*

(i) $t' = T(ts)$

(ii) $t = T(t's^{-1})$.

Moreover, if these conditions hold, then

(iii) $f(ts)f(t's^{-1}) = 1$, *where* $f(w) = wT(w)^{-1}$.

PROOF. If $t' = T(ts)$, then

$$T(t's^{-1}) = T(T(ts)s^{-1}) = T(tss^{-1}) = T(t) = t.$$

so (i) is implies to (ii) whence (ii) implies (i). If (i) and (ii) hold, then

$$f(ts)^{-1} = T(ts)s^{-1}t^{-1} = T(T(t's^{-1})s)s^{-1}t^{-1} = T(t')s^{-1}t^{-1} = t's^{-1}t^{-1} = (t's^{-1})t^{-1} = f(t's^{-1}). \quad \square$$

4.4 THEOREM. *Let G be a subgroup of the free group F on a discrete set S. Let T be a Schreier transversal for G. Define the function f from F to G by $f(w) = wT(w)^{-1}$. Then the set*

$$Y = \{f(ts) : s \in S,\ t \in T(F),\ \text{and } f(ts) \neq 1\}$$

is a basis for G.

PROOF. If $w \in G$, then $T(w) = 1$, so $f(w) = w$. Thus to show that $G = \langle Y\rangle$ it is enough to show that $f(w) \in \langle Y\rangle$ for all w in F. Let $w \in F$ and $s \in S$. Then

$$f(w)f(T(w)s) = wT(w)^{-1}T(w)sT(T(w)s)^{-1} = wsT(ws)^{-1} = f(ws).$$

As $f((Tw)s) \in Y \cup \{1\}$ it follows that $f(ws) \in \langle Y\rangle$ if and only if $f(w) \in \langle Y\rangle$.

Now let w be a reduced word in F. Either $w = 1 \in \langle Y\rangle$, or $|ws^{-1}| < |w|$ for some s in S, or $|ws| < |w|$ for some s in S. In the latter two cases, by induction on length, either $f(ws^{-1}) \in \langle Y\rangle$ or $f(ws) \in \langle Y\rangle$. Hence $w \in \langle Y\rangle$.

We will show that Y is a basis by appealing to Lemma 1.8. First observe that

$$Y^{-1} = \{f(ts^{-1}) : s \in S,\ t \in T(F), \text{ and } f(ts^{-1}) \neq 1\}$$

follows from Lemma 4.3. So if $y \in Y \cup Y^{-1}$, then $y = f(ts)$ with s in $S \cup S^{-1}$, and t in $T(F)$. Next we note that $Y \cap Y^{-1} = \phi$ follows from Lemma 4.2.ii. Now suppose that $y_1y_2\cdots y_n = 1$ with each y_i in $Y \cup Y^{-1}$. Write $y_i \equiv f(t_is_i) \equiv t_is_iT(t_is_i)^{-1}$, with s_i in $S \cup S^{-1}$, and t_i in $T(F)$. Let u_i be the reduced form of $T(t_is_i)^{-1}t_{i+1}$. Then

$$(*) \qquad y_1y_2\cdots y_n = t_1s_1u_1s_2u_2\cdots u_{n-1}s_n$$

Were t_is_i equal to an element of $T(F)$, then y_i would equal 1; therefore we can apply Lemma 4.2.iii and conclude that either $s_iu_is_{i+1}$ is reduced for each $i < n$, or for some $i < n$ we have $u_i = s_is_{i+1} = 1$. In the former case the right hand side of (*) is reduced, so the left hand side cannot equal 1. In the latter case $t_{i+1} = T(t_is_i)$, so $y_iy_{i+1} = 1$ by lemma 4.3. □

Theorems 4.1 and 4.4 imply that detachable subgroups of finite-rank free groups are free. If the subgroup has finite index, then we can compute its rank as follows.

4.5 THEOREM. *Let $F = F(S)$ be a free group of rank r and G a subgroup of finite index n. Then G is a free group of rank $n(r-1) + 1$.*

PROOF. Let T be a Schreier transversal for G, and Y be the basis defined in Theorem 4.4. We only need to show that Y has $n(r - 1) + 1$ elements. Define maps λ and μ

$$\lambda : T(F)\backslash\{1\} \rightarrow T(F) \times S$$
$$\mu : Y \rightarrow T(F) \times S$$

by

$$\lambda(t) = \begin{cases} (t',s) & \text{if } t \equiv t's \quad \text{with } s \in S \\ (t,s) & \text{if } t \equiv t's^{-1} \text{ with } s \in S \end{cases}$$

and

$$\mu(y) = (t,s) \text{ such that } f(ts) = y.$$

Lemma 4.2 guarantees that the map μ is well defined; it is clearly one-to-one. The map λ maps into $T(F) \times S$ because T is a Schreier transversal; is easily seen that λ is one-to-one. The range of λ is the set of pairs

(t,s) in $T(F) \times S$ for which $ts \in T(F)$, which is the same as the set of pairs (t,s) such that $f(ts) = 1$. Thus the ranges of λ and μ partition $T(F) \times S$ into a union of two disjoint sets, the first containing $n-1$ elements and the second containing the same number of elements as Y. As $T(F) \times S$ has nr elements, we see that $\#Y = nr - (n-1) = n(r-1) + 1$. □

EXERCISES

1. Show that if G is a subgroup of a discrete free group, then G is detachable if and only if G has a Shreier transversal.
2. Construct a Brouwerian example of a free subgroup of a finite-rank free group that is not detachable.
3. Let G be a finitely generated subgroup of a finite-rank free group. Show how to determine whether or not G has finite index.
4. Use (2.5) to show that a detachable subgroup of a countable discrete free group is free.

5. CONJUGATE SUBGROUPS.

Let S be a set and $u,v \in F(S)$. Then u and v are **conjugates** in $F(S)$ if there exists c in F with $u = c^{-1}vc$. A word $w \equiv x_1 \cdots x_n$, with $x_i \in S \cup S^{-1}$ is **cyclically reduced** if it is reduced and $x_n x_1 \neq 1$. If S is discrete, then each reduced word w of $F(S)$ may be written uniquely as $w \equiv v^{-1}w'v$ where w' is cyclically reduced.

5.1 THEOREM. *If F is a free group on a discrete set S, and $u,v \in F$, then either u and v are conjugates or they are not.*

PROOF. We may assume that u and v are reduced. Write $u \equiv a^{-1}u'a$ and $v \equiv b^{-1}v'b$ with u' and v' cyclically reduced. Then u and v are conjugates if and only if u' and v' are conjugates. We will show that if $u' = c^{-1}v'c$, then u' is a cyclic permutation of v'. We may assume c is reduced. As u' is cyclically reduced, either $c^{-1}v'$ or $v'c$ is not reduced. If the first case holds, then $c \equiv sd$ and $v' \equiv sw$, with s in $S \cup S^{-1}$. Then $u' = d^{-1}s^{-1}swsd = d^{-1}wsd$ and ws is a cyclic permutation of v'. As $|d| < |c|$, we are done by induction on $|c|$. The second case is treated similarly. So u and v are conjugates if and only if u' and v' are cyclic permutations of each other. □

Let F be a group and U a subset of F. For $w \in F$ let $w^{-1}Uw = \{w^{-1}uw : u \in U\}$. If U is a subgroup, then $w^{-1}Uw$ is a subgroup. The subgroups G and H are **conjugate** if there exists w in F with $H = w^{-1}Gw$.

5.2 THEOREM. *Let F be a finite-rank free group, and let G and H be finitely generated subgroups of F. Then either G is conjugate to a subgroup of H or it is not; and either G is conjugate to H or it is not.*

PROOF. Let U be a finite free basis for G, and V a finite free basis for H. We may assume that V is a Nielsen set of generators for H and, by replacing G with a conjugate subgroup, we may assume that U contains an element u which is cyclically reduced (the case $U = \phi$ is trivial). Let $m = max\ \{|w| : w \in U \cup V\}$. We will show that, for each w in F,

(*) If $w^{-1}Gw \subseteq H$, and $|w| > m$, then we can find w' in GwH with $|w'| < |w|$.

As $w' \in GwH$ we have $w'^{-1}Gw \subseteq H$. It follows by induction that we can find w' in GwH so that $|w'| \leq m$. Once we have shown (*), the theorem is proved as follows. Since F has finite rank there are only a finite number of words w with $|w| \leq m$. For each such w we test to see if $w^{-1}Uw \subseteq H$; we can do this because U is finite and H is detachable. If there is no such w, then G is not conjugate to a subgroup of H, while if such a w exists, then $w^{-1}Gw$ is a subgroup of H conjugate to G. If $w^{-1}Uw \subseteq H$, then we can decide if $\langle w^{-1}Uw\rangle = H$ because H is finitely generated and $\langle w^{-1}Uw\rangle$ is detachable. If so, then G and H are conjugates; if not, then they are not conjugates.

To prove (*) suppose that w is reduced, $|w| > m$, and $w^{-1}Gw \subseteq H$. Let u in U be cyclically reduced. Then one of $u^{-1}w$ or uw is reduced. We may assume, replacing u by u^{-1} if necessary, that uw is reduced. If $|w| > |u^{-1}w|$, then we may take $w' = u^{-1}w$, so we may assume that $|w^{-1}u| \geq |w^{-1}|$. Therefore no more than half of u cancels in the product $w^{-1}u$. As uw is reduced, and $|w^{-1}| = |w| > m \geq |u|$, we see that the reduced form of the product $w^{-1}uw$ begins with more than half of the factor w^{-1}, and ends with the factor w. In particular, $|w^{-1}uw| > |w| > m$.

Write $w^{-1}uw$ in terms of the basis V of H; that is, $w^{-1}uw = v_1v_2\cdots v_n$ with $v_i \in V \cup V^{-1}$ and $v_iv_{i+1} \neq 1$ for all $i < n$. As $|w^{-1}uw| > m \geq |v_1|$, we must have $n > 1$. If $|wv_1| < |w|$, then we may take $w' = wv_1$, so we may assume that $|w| \leq |wv_1|$. As V is a Nielsen set, (2.2) and (2.3) say that the reduced form of $w^{-1}uw = v_1\cdots v_n$ begins with at least half of v_1, and by the above it also begins with at least half of w^{-1}. As $|w| > m \geq |v_1|$ this implies that w^{-1} begins with at least half of v_1. However $|w| \leq |wv_1|$, so w^{-1} begins with no more than half of v_1. Thus the reduced form of $w^{-1}uw$ begins with exactly half of v_1, so half of v_1 cancels in v_1v_2. Similarly, it ends with exactly half of v_n. Suppose that $|v_1| \leq |v_n|$. As w^{-1} begins with half of v_1 and w ends with half of v_n, it follows that half of v_1 cancels in the product v_nv_1, and all of it in the product $v_nv_1v_2$. As $v_1v_2 \neq 1$, and V is a Nielsen set, it follows that $v_1 = v_n^{-1}$. Similarly, if $|v_n| \leq |v_1|$, then $v_1 = v_n^{-1}$. Thus $v_1 = v_n^{-1}$. As w ends in half of $v_n = v_1^{-1}$, it follows that at least half of v_1 cancels in the product $w' = wv_1$, so $|w'| \leq |w|$. But $w'^{-1}uw' = v_2 \cdots v_{n-1}$ and we are done by induction on n. □

EXERCISES.

1. Show that (5.2) is true if F is a discrete free group.
2. Construct Brouwerian counterexamples to generalizations of (5.2), one with G countably generated and one with H countably generated.

NOTES

The **word problem** for a group G is to decide whether or not two elements of G are equal; that is, to solve the word problem for G is to show that G is discrete. The terminology comes from considering quotients F/N where F is a free group and N a normal subgroup of F; in this setting the problem becomes how to decide whether or not a word in F is in N or not. It would seem plausible that we should be able to solve the word problem when F is of finite rank, and N is finitely generated as a normal subgroup, that is, there is a finite subset A of N so that each element of N can be written as a product of conjugates of elements of A. However a famous result of Novikov and Boone constructs such F and A in such a way that the word problem cannot be solved by a Turing machine, and therefore no algorithm for deciding whether a word in F is in N can be written in any standard programming language.

The **generalized word problem** for a group G relative to a subgroup H is to decide whether or not an element of G is in H or not; we solve the generalized word problem by showing that H is detachable. The generalized word problem for a finite-rank free group relative to a finitely generated subgroup is solved by the Nielsen construction (3.5).

The Schreier construction shows, classically, that any subgroup of any free group is free. To construct a Schreier transversal, well-order the elements of $S \cup S^{-1}$ and proceed as in the finite case.

Chapter XI. Abelian Groups

1. FINITE-RANK TORSION-FREE GROUPS.

An abelian group is a module over the ring $\mathbb{Z}$ of integers, so when studying abelian groups we may appeal to the general facts about modules developed in Chapter III, and about modules over a PID developed in Chapter V. The structure theorem for finitely presented abelian groups is a special case of the structure theorem (V.2.3) for finitely presented modules over a PID. In this section we are concerned with the simplest kinds of torsion-free abelian groups that are not finitely presented.

If G is a module over a commutative ring R, and $r \in R$, then $rG = \{rx : x \in G\}$ is a submodule of G. Given $x \in G$, it is of interest to know for what $r \in R$ we have $x \in rG$. If $R = \mathbb{Z}$, this question reduces to the question of when is $x \in qG$ for q a prime power.

1.1 LEMMA. *If G is a module over a commutative ring R, and a and b are strongly relatively prime elements of R, then* $abG = aG \cap bG$.

PROOF. Clearly $abG \subseteq aG \cap bG$, so suppose $x = ay = bz$ is in $aG \cap bG$. We can write $1 = sa + tb$, so $x = sax + tbx = sabz + tbay = ab(sz + ty) \in abG$. □

An abelian group G is **torsion-free** if, for each nonzero $n \in \mathbb{N}$ and each $x \in G$, if $nx = 0$, then $x = 0$. If G is a torsion-free abelian group, then the natural map $G \to \mathbb{Q} \otimes G$, that sends x to $1 \otimes x$, is a monomorphism. Note that $\mathbb{Q} \otimes G$ is discrete if and only if G is discrete, and that, in any case, $\mathbb{Q} \otimes G$ is a vector space over $\mathbb{Q}$. A torsion-free group G is said to be of **rank** n if $\mathbb{Q} \otimes G$ is an n-dimensional discrete vector space over $\mathbb{Q}$. A group is a torsion-free group of rank one if and only if it is isomorphic to a nonzero subgroup of the additive group of $\mathbb{Q}$. Classically the rank-one torsion-free groups are classified by equivalence classes of functions from the set of primes to $\mathbb{N} \cup \{\infty\}$.

If x is an element of a torsion-free group G, then we define the

p-height of x to be $h_p x = sup\{n : x \in p^n G\}$, where the supremum is taken in $\mathbb{N} \cup \{\infty\}$. Of course there is no reason to believe that we can compute $h_p x$ in general; if we can, for each prime p, and $x \in G$, we say that the group G **has heights.** If $G = A \oplus B$, then it is easily seen that G has heights if and only if A and B have heights, in which case $h_p^G x = min(h_p^A x, h_p^B x)$.

1.2 LEMMA. *Let G be a torsion-free group with heights. If p is a prime, $x \in G$, and $m \in \mathbb{N}$, then $h_p mx \geq h_p x$ with equality holding if $(p,m) = 1$. If $h_p x \in \mathbb{N}$, then $h_p px = h_p x + 1$.*

PROOF. Clearly $h_p mx \geq h_p x$. Suppose $(p,m) = 1$ and $mx = p^n y$. Write $sp^n + tm = 1$. Then $x = sp^n x + tmx = p^n(sx + y)$; thus $h_p x \geq h_p mx$. Finally, $px = p^{n+1} y$ if and only if $x = p^n y$ as G is torsion-free. □

A **type** is a function from the set of primes to $\mathbb{N} \cup \{\infty\}$; two types are **equal** if they are equal except at a finite number of places where each is finite. The set of types admits a natural partial order by setting $\tau_1 \leq \tau_2$ if $\tau_1(p) \leq \tau_2(p)$ except at a finite number of places where each is finite. It is easily seen that the set of types forms a distributive lattice with a greatest element and a least element.

The **type of an element** x in a torsion-free group G with heights is the function $h_p x$ of p, viewed as a type. From (1.1) it follows that if $x \in G$, and m is a nonzero integer, then the type of x is equal to the type of mx. As any two nonzero elements of a rank-one torsion-free group have a common nonzero multiple, we can define the **type of a rank-one torsion-free group with heights** to be the type of any nonzero element. The next theorem shows that two such groups with the same type are isomorphic.

1.3 THEOREM. *Let G and G' be rank-one torsion-free abelian groups with heights, of types τ and τ' respectively. Then there is a nonzero map from G to G' if and only if $\tau \leq \tau'$. If $\tau = \tau'$, then G and G' are isomorphic.*

PROOF. Suppose $\varphi : G \to G'$ is a nonzero map. Clearly $h_p \varphi x \geq h_p x$ for all x in G, so $\tau \leq \tau'$.

Now suppose the $\tau \leq \tau'$, and x and x' are nonzero elements of G and G'. From (1.1) there exist a nonzero integers m and m' such that $h_p mx \leq h_p m'x'$ for all primes p, with equality if $\tau = \tau'$. Let H be the subgroup of $G \oplus G'$ generated by $(mx, m'x')$, and let

$$K = \{w \in G \oplus G' : nw \in H \text{ for some nonzero } n \in \mathbb{N}\}$$

Clearly K is a subgroup of $G \oplus G'$. If $y \in G$, then $ny = \ell m x$ for some nonzero ℓ and n as G is rank-one. Therefore $\ell m' x' \in nG'$ by (1.1) and (1.2). Thus there exists $y' \in G'$ such that $ny' = \ell m' x'$, so $(y,y') \in K$. If $\tau = \tau'$, then, by symmetry, for each $y' \in G'$, there exists $y \in G$ such that $(y,y') \in K$. If (y,y') and (y,z') are in K, then $(0,y'-z') \in K$, so $y' = z'$ as G and G' are torsion-free. Setting $\varphi(y) = y'$ gives a nonzero map which is an isomorphism if $\tau = \tau'$. □

The rank-one torsion-free abelian groups with heights form a **semirigid class** in the following sense.

1.4 COROLLARY. *Let A and B be rank-one torsion-free abelian groups with heights. If there exist nonzero maps from A to B and from B to A, then A and B are isomorphic.* □

The hypothesis that A and B have heights cannot be dropped in (1.4); an example is sketched in Exercise 7.

Classically every finitely generated torsion-free abelian group is a direct sum of cyclics. This is not the case constructively (see Exercise 2) but the following theorem holds.

1.5 THEOREM. *Finitely generated subgroups of finite-rank torsion-free abelian groups are direct sums of cyclic groups.*

PROOF. It suffices to consider finitely generated subgroups G of $\mathbb{Q}^n$. Multiplying G by a common denominator of the coordinates of the generators of G we may assume that $G \subseteq \mathbb{Z}^n$. Then G is finitely presented, being a submodule of a finitely presented module over a coherent Noetherian ring, hence G is a direct sum of cyclic groups by the structure theorem. □

The simplest kind of finite-rank torsion-free groups are finite direct sums of rank-one groups with heights. These groups may be specified, up to isomorphism, by a finite family of types. It is *a priori* possible that distinct families of types might give rise to isomorphic groups, but this turns out not to be the case. In the classical context we can compute, in an invariant manner, the number of rank-one summands of type τ, by considering the subgroup

$$G(\tau) = \{x \in G : type(x) \geq \tau\}$$

and the subgroup $G(\tau^*)$ generated by $\{G(\sigma) : \sigma > \tau\}$. The rank of the quotient $G(\tau)/G(\tau^*)$ is the number of summands of type τ. But since the set of types is not discrete, we cannot determine this rank in the constructive context. Our approach hinges on the following lemma about bases of finite-dimensional vector spaces.

1.6 LEMMA. *Let $e_1,\ldots,e_n$ and $e_{n+1},\ldots,e_{2n}$ be bases of a finite dimensional vector space over a discrete field. Let $\pi_i(x)$ be the scalar multiple of e_i in the expression for x relative to the appropriate basis. Let R be the transitive closure of the relation $\pi_i(e_j) \neq 0$ on $\{1,\ldots,2n\}$. Then each equivalence class of elements of $\{1,\ldots,2n\}$, under the equivalence $i \equiv j$ if $R(i,j)$ and $R(j,i)$, has exactly half of its elements in $\{1,\ldots,n\}$.*

PROOF. Let C be an equivalence class of elements of $\{1,\ldots,2n\}$ and suppose $C = A \cup B$ where $A = C \cap \{1,\ldots,n\}$ and $B = C \cap \{n+1,\ldots,2n\}$. For S equal to A or to B let $\pi_S = \Sigma_{i\in S}\pi_i$ and let V_S be the subspace generated by $\{e_i : i \in S\}$, that is, the image of π_S. We shall show that V_A has the same dimension as V_B. By symmetry it suffices to show that $\pi_A V_B = V_A$. We shall show that $e_i = \pi_A\pi_B e_i$ for each $i \in A$. Now $\pi_A\pi_B e_i = \Sigma_{k\in A}\Sigma_{j\in B}\pi_k\pi_j e_i$. But $\pi_k\pi_j e_i = 0$ if $i,k \in A$ and $j \in \{n+1,\ldots,2n\}\backslash B$ by the definition of the equivalence classes. So $\pi_A\pi_B e_i = \Sigma_{k\in A}\Sigma_{j=n+1}^{2n}\pi_k\pi_j e_i = \pi_A e_i = e_i$. □

1.7 THEOREM. *Let $G = H_1 \oplus H_2 \oplus \cdots \oplus H_n = K_1 \oplus K_2 \oplus \cdots \oplus K_m$ where each H_i and K_j is a rank-one torsion-free group with heights. Then $m = n$, and there is a permutation σ of $\{1,\ldots,n\}$ such that $H_i \cong K_{\sigma(i)}$.*

PROOF. Let $e_1,\ldots,e_n$ be nonzero elements of $H_1,\ldots,H_n$ respectively, and $e_{n+1},\ldots,e_{n+m}$ be nonzero elements of $K_1,\ldots,K_m$ respectively. Then $e_1,\ldots,e_n$ and $e_{n+1},\ldots,e_{n+m}$ are bases of $\mathbb{Q} \otimes G$, so $m = n$. For notational convenience, set $H_{n+i} = K_i$ for $i = 1,\ldots,n$.. Let the π_i be as in Lemma 1.6, and note that $\pi_i G = H_i$, so if $\pi_i e_j \neq 0$ then there is a nonzero map from H_j to H_i. The result now follows from (1.6) and (1.4). □

EXERCISES

1. Show that any finite abelian group is finitely presented.
2. Let a be a binary sequence with at most one 1. Let S be the subgroup of $\mathbb{Z} \oplus \mathbb{Z}$ generated by $\{(1,na_n) : n \in \mathbb{N}\}$. Show that $(\mathbb{Z} \oplus \mathbb{Z})/S$ is a Brouwerian example of a finitely generated discrete torsion-free abelian group that is not a direct sum of cyclics.

3. For each nonnegative integer i, let A_i be the subgroup of $\mathbb{Q}$ generated by the set $\{1/p^i : p \text{ is a prime}\}$. Show that A_i has heights, and that A_i and A_j are not isomorphic if $i \neq j$. For which values of i is A_i finitely generated?

4. Let a be a binary sequence, and A be the subgroup of $\mathbb{Q}$ generated by 1 and the set $\{1/2^n : a_n = 1\}$. Show that A is a Brouwerian example of a rank-one torsion-free group that does not have heights. Show that A is detachable from $\mathbb{Q}$ if a is decreasing, but not in general.

5. Given a function f from the set of primes to $\mathbb{N} \cup \{\infty\}$, construct a rank-one torsion-free group, and an element x, such that $h_p x = f(p)$ for all primes p.

6. A subgroup A of an abelian group B is **full** if B/A is torsion. Show that a torsion-free abelian group is of rank n if and only if it contains a full subgroup isomorphic to $\mathbb{Z}^n$.

7. Let a be a binary sequence, and let A be the subgroup of $\mathbb{Q}$ generated by 1 and $\{a_n/2 : n \in \mathbb{N}\}$. Construct nonzero maps from $\mathbb{Z}$ to A and from A to $\mathbb{Z}$, but show that A is a Brouwerian example of a group that is not isomorphic to $\mathbb{Z}$. Why doesn't (1.4) apply? Construct an example of this kind where A and B have the property that mA and mB are detachable subgroups for every m.

8. Construct a Brouwerian example showing that the hypothesis that the groups in (1.7) have heights is necessary.

9. An example of an indecomposable rank-two torsion-free group with heights. Let G be the subgroup of $\mathbb{Q} \oplus \mathbb{Q}$ generated by elements of the form $(2^{-m}, 0)$, $(0, 3^{-m})$, and $(5^{-m}, 5^{-m})$. To compute $h_p(x)$ for $p > 5$, compute $h_p(30^m x)$ in $\mathbb{Z} \oplus \mathbb{Z}$. In computing $h_2(x)$, we may ignore $(2^{-m}, 0)$, and similarly for h_3 and h_5. That G is indecomposable follows from the fact that it has elements of three pairwise incomparable types.

2. DIVISIBLE GROUPS.

A group is **p–divisible** if $pG = G$, **divisible** if it is p–divisible for each prime p. From (1.1) we see that G is divisible if and only if $nG = G$ for each nonzero integer n. The additive group of rational numbers $\mathbb{Q}$ is a divisible torsion-free group. The simplest example of a nontrivial divisible torsion group is the p-primary subgroup of the torsion group $\mathbb{Q}/\mathbb{Z}$, which is called $\mathbb{Z}(p^\infty)$; the cyclic group of order p^∞, so to speak.

In a torsion-free divisible group, the endomorphism induced by multiplication by a nonzero integer n is both one-to-one and onto, hence an isomorphism. Therefore a torsion-free divisible group admits a unique structure as a vector space over the field $\mathbb{Q}$. Conversely, it is clear that the additive group of a vector space over the field $\mathbb{Q}$ is torsion-free and divisible.

A **coherent abelian group** is one that is coherent as a $\mathbb{Z}$-module. It is easily seen that a coherent group has detachable (finitely generated) subgroups, and that any discrete torsion group is coherent.

2.1 THEOREM. *Let D be a divisible subgroup of a group G such that G/D is countable and coherent. Then we can construct a countable subgroup K of G such that $G = K \oplus D$.*

PROOF. Let $x_1, x_2, \ldots$ be elements of G that enumerate G/D. As G/D is coherent, we can arrange so that either

$$\langle x_{i+1}\rangle \cap (\langle x_1,\ldots,x_i\rangle + D) = 0,$$

$$\text{or} \qquad px_{i+1} \in \langle x_1,\ldots,x_i\rangle + D \text{ for some prime } p.$$

We shall construct a sequence of finitely enumerable subgroups $K_1 \subseteq K_2 \subseteq \cdots$ of G such that $K_i + D = \langle x_1,\ldots,x_i\rangle + D$, and $K_i \cap D = 0$. Then $K = \bigcup K_i$ is as desired.

Set $K_0 = 0$. Given K_i, construct K_{i+1} as follows. If $x_{i+1} \in K_i + D$ (a decidable question), then set $K_{i+1} = K_i$. If $x_{i+1} \notin K_i + D$, then either $\langle x_{i+1}\rangle \cap (K_i + D) = 0$, or $px_{i+1} \in K_i + D$. In the former case, set $K_{i+1} = \langle x_{i+1}\rangle + K_i$; in the latter case write $px_{i+1} = k_i + d$, where $k_i \in K_i$ and $d \in D$. As D is divisible, we can find $d' \in D$ such that $d = pd'$. Let $y = x_{i+1} - d'$. Then $py = k_i$, and we set $K_{i+1} = \langle y\rangle + K_i$. Certainly $K_{i+1} + D = \langle x_1,\ldots,x_{i+1}\rangle + D$; we must show that $K_{i+1} \cap D = 0$. For $w \in K_{i+1} \cap D$, write $w = ny + z$, where $z \in K_i$. So $ny = w - z \in D + K_i$. But $py \in K_i$, so if $(p,n) = 1$, then $y \in K_i + D$, so $x_{i+1} \in K_i + D$, contrary to our assumption. Therefore p divides n, so $ny \in K_i$ whereupon $w \in K_i \cap D = 0$. □

We can use (2.1) to get a structure theorem for countable coherent divisible groups.

2.2 THEOREM. *Let G be a countable coherent divisible group. Then G is a countable direct sum of subgroups isomorphic to* $\mathbb{Q}$ *and to* $\mathbb{Z}(p^\infty)$ *for various primes p.*

PROOF. Let T be the torsion subgroup of G. Then G/T is coherent because, as G is coherent, finitely generated subgroups of G are direct sums of finite and infinite cyclic groups. Thus (2.1) says that we can write $G = T \oplus F$, where F is torsion-free. Thus it suffices to prove the theorem under the assumption that G is torsion, or that G is torsion-free.

If G is torsion-free divisible, then G admits a unique structure as a vector space over the field $\mathbb{Q}$, so each nonzero element of G is contained in a unique subgroup of G that is isomorphic to the additive group $\mathbb{Q}$. Let $x_0, x_1, \ldots$ be an enumeration of G. Define a detachable subset S of $\mathbb{N}$ by putting $i \in S$ if x_i is not in the vector space generated by $x_0, \ldots, x_{i-1}$; this is decidable because G is coherent. It is readily seen that G is a direct sum of the subspaces $\mathbb{Q}x_i \cong \mathbb{Q}$ for $i \in S$.

If G is torsion, then G is a direct sum of its primary components G_p, so we may assume that G is a p-group. It suffices to show that every element of a discrete divisible p-group is contained in a subgroup that is isomorphic to $\mathbb{Z}(p^\infty)$; we then apply (2.1) repeatedly. But given such an x we can construct a sequence $x = y_0, y_1, \ldots$ such that $py_{i+1} = y_i$ for each i. The subgroup generated by the y's is the desired subgroup. □

A subgroup A of a discrete abelian group B is **essential** if for each nonzero $b \in B$ there is $n \in \mathbb{Z}$ such that nb is a nonzero element of A; in particular, B/A is torsion (but that is not enough). A **divisible hull** of a discrete abelian group A is a discrete divisible abelian group B containing A as an essential subgroup.

2.3 THEOREM. *Any countable discrete abelian group has a countable discrete divisible hull.*

PROOF. We may assume that the given group is F/K, where F is a countable-rank free abelian group, and K is a detachable subgroup of F. Let $QF = Q \otimes F$ and construct a countable subgroup N of QF as follows. First note that if A is a finitely generated subgroup of QF, then $A \cap F$ is finitely generated, as A and the relevant basis elements of F can be put inside a finite-rank free subgroup of QF. Let $a_0, a_1, \ldots$ be an enumeration of QF and set

$$\begin{aligned} N_0 &= 0 \\ N_{i+1} &= N_i + \mathbb{Z}a_i \text{ if } (N_i + \mathbb{Z}a_i) \cap F \subseteq K \\ &= N_i \qquad \text{otherwise.} \end{aligned}$$

The decision as to whether to put a_i in N_{i+1} can be made because $N_i + \mathbb{Z}a_i$ is finitely generated.

Set $N = \cup N_i$ and let $D = QF/N$. The subgroup N is detachable from QF because $a_i \in N$ if and only if $(N_i + \mathbb{Z}a_i) \cap F \subseteq K$; therefore D is discrete. Clearly $N \cap F = K$, so we may view F/K as a subgroup of $QF/N = D$. Finally, if $a_i \in QF\backslash N$ is a nonzero element of D, then there exist $x \in N_i$ and $n \in \mathbb{Z}$ such that $x + na_i \in F\backslash K$. Thus na_i is equal to a nonzero element of F/K. □

The divisible hull of a *coherent* countable discrete abelian group is coherent. In fact the following more general theorem holds.

2.4 THEOREM. *Let A ⊆ B be discrete abelian groups. If B/A is torsion, and A is coherent, then B is coherent.*

PROOF. We first show that if B/A is torsion, and A is finitely presented, then B is discrete. Given $b \in B$ there is $n \neq 0$ so that $nb \in A$. If $nb \notin nA$ (decidable as A finitely presented), then $b \notin A$. If $nb = na$, for some $a \in A$, then it suffices to decide whether or not $b-a \in A$. But there are only finitely many torsion elements of A.

For the general case, let φ map a finite-rank free abelian group F into B. Then there is $n \neq 0$ such that $\varphi(nF) \subseteq A$. The group $B/\varphi(nF)$ is discrete, by the first paragraph, and $A/\varphi(nF)$ is coherent by (III.2.5); thus the hypotheses of the theorem are inherited by the situation $A/\varphi(nF) \subseteq B/\varphi(nF)$. If $B/\varphi(nF)$ is coherent, then so is B by (III.2.5), so we may assume that $\varphi(nF) = 0$. But F/nF is finite, and B is discrete, so the induced map from F/nF to B has finite kernel, whence the kernel of φ is finitely generated. □

EXERCISES

1. Show that any torsion group is equal to the direct sum its p-primary subgroups.
2. Show that every finitely generated subgroup of $\mathbb{Z}(p^\infty)$ is cyclic,

that the finitely generated subgroups of $\mathbb{Z}(p^\infty)$ form a chain under inclusion, and that for each n there is a finite cyclic subgroup of $\mathbb{Z}(p^\infty)$ of order p^n. Show that any group G with these properties is isomorphic to $\mathbb{Z}(p^\infty)$.

3. Let a be a binary sequence with at most one 1, and let H be the $\mathbb{Q}$-subspace of $\mathbb{Q}$ generated by $\{a_n/n : n = 1,2,\ldots\}$. Show that H is a Brouwerian example of a countable divisible subgroup of $\mathbb{Q}$, in fact a countable direct sum of copies of $\mathbb{Q}$, that is not a summand. Why doesn't (2.1) apply? Show that $G = \mathbb{Q}/H$ is a Brouwerian example of a countable discrete torsion-free divisible group that is not a countable direct sum of copies of $\mathbb{Q}$. Why doesn't (2.2) apply?

4. Find two places where the axiom of dependent choices is used in the proof of (2.2), other than (at one remove) in the appeal to (2.1).

5. Let G be a discrete divisible p-group, and let X be a basis for $\{x \in G : px = 0\}$, viewed as a vector space over the p-element field $\mathbb{Z}/(p)$. For each $x \in X$, construct a sequence $x = y_0, y_1, \ldots$ so that $py_{i+1} = py_i$, and let A_x be the subgroup of G generated by the y's. Show that each A_x is isomorphic to $\mathbb{Z}(p^\infty)$, and that G is the direct sum of the subgroups A_x.

6. Construct a Brouwerian example of a countable abelian group between $\mathbb{Z}$ and $\mathbb{Q}$ that is not detachable from any divisible hull.

3. HEIGHT FUNCTIONS ON p-GROUPS.

Let G be an abelian group and p a prime. We say that G is a **p-group** if for each x in G there is a positive integer n such that $p^n x = 0$. A high point in the theory of abelian groups is the classification of countable p groups by the dimensions of certain vector spaces defined in terms of the notion of *height*. If n is a nonnegative integer, and $x \in G$, then we say that the **height** of x is n if $x \in p^n G$ and $x \notin p^{n+1}G$. We have already seen that there are problems computing heights in torsion-free groups. Moreover, to get the classification theorem for countable p-groups, we must extend the notion of height to include transfinite values.

Let G be an abelian group, p a prime, λ an ordinal and $\lambda_\infty = \lambda \cup \{\infty\}$. A

p-height function on G is a function h from G onto λ_∞ such that

(i) If $hx < \infty$, then $hpx > hx$

(ii) If $(m,p) = 1$, then $hmx = hx$

(iii) If $a < hx$, or $a = hx = \infty$, then there exits y such that $py = x$ and $hy \geq a$.

The ordinal λ is called the **p-length** of G. Note that (i) implies that $h0 = \infty$; if $hx = \infty$ implies $x = 0$, then we say that G is **p-reduced.** If G is a p-group, then (i) and (ii) imply that if q is a prime other than p, then the q-height of each element of G is ∞; in this case we drop the prefix 'p' from 'p-height' and 'p-length'. Such a height function is unique; in fact, both λ and h are isomorphism invariants of G in the following sense.

3.1 THEOREM. *Let G and G' be abelian groups with p-height functions h and h' and p-lengths λ and λ' respectively. Let $\varphi : G \to G'$ be an isomorphism. Then there exists an isomorphism $\rho : \lambda_\infty \to \lambda'_\infty$ such that $h\varphi = \rho h'$.*

PROOF. We say that **ρ is defined at a** if whenever $x,y \in G$ such that $hx = hy = a$, then $h'\varphi x = h'\varphi y$. If ρ is defined at a, we set $\rho a = h'\varphi x$ for any x such that $hx = a$. Let $[0,a] = \{b \in \lambda : b \leq a\}$, and let

$$S = \{a \in \lambda : \rho \text{ is defined at each element of } [0,a], \text{ and } \rho \text{ is an injection on } [0,a]\}.$$

We want to show that $S = \lambda$. Suppose $a \in S$ for all $a < b$. If $hx = b$ we shall show

(i) If $a < b$, then $\rho a < h'\varphi x$.

(ii) If $c < h'\varphi x$, then there is $a < b$ such that $\rho a = c$.

This will show that $b \in S$; so S is hereditary, whence $S = \lambda$.

To show (i) suppose $a < b$. As h is a height function, there is z such that $a \leq hz < b$ and $pz = x$. So $p\varphi z = \varphi x$ and $h'\varphi z = \rho hz \geq \rho a$. Thus $\rho a < h'\varphi x$.

To show (ii), suppose $c < h'\varphi x$. As h' is a height function, there exists z such that $h'z \geq c$ and $pz = \varphi x$. Then $p\varphi^{-1}z = x$, so $h\varphi^{-1}z = d < b = hx$. Therefore $hz = h\varphi\varphi^{-1}z = \rho d$ and ρ is an injection on $[0,d]$. Hence there is $a \leq d < b$ such that $\rho a = c$.

As $S = \lambda$, the function ρ is an injection of λ into λ'. Similarly we get an injection from λ' into λ. Their composition is an injection of λ

(or λ') into itself, hence the identity by (I.6.6). Finally we set $\rho\infty = \infty$. □

To illustrate the kind of structures we shall be dealing with, we consider the simplest example of an abelian p-group with elements of transfinite height other than ∞. This group is constructed by subjecting the generators $x_0, x_1, \ldots$ to the relations $px_0 = 0$, and $p^n x_n = x_0$ for $n > 0$.

3.2 EXAMPLE. Let F be the free abelian group on the discrete set $\{x_n : n \in \mathbb{N}\}$. Let P be the quotient of F by the subgroup R of F generated by the elements px_0, $px_1 - x_0$, $p^2x_2 - x_0$, ... It is a routine exercise to establish that R is detachable from F, so P is discrete, and that each element of P has a **canonical representative** in F of the form $\Sigma\, n_i x_i$ with $0 \leq n_0 < p$ and $0 \leq n_i < p^i$ for $i > 0$. Clearly P is a p-group.

Let λ be the well-ordered set $\{0,1,2,\ldots,\omega\}$. Define a function **h** from P to λ_∞ as follows. Let $\Sigma\, n_i x_i$ be the canonical representative of the element y in P. Set $h(y) = \infty$ if all the n_i are 0; set $h(y) = \omega$ if n_0 is the unique nonzero n_i; set $h(y) = min\{\upsilon_p n_i : i \neq 0\}$ otherwise, where $\upsilon_p m$ is the p-adic value of m, that is, the exponent of p in the prime decomposition of m. It is readily verified that h is a height function on P. □

The next theorem shows that homomorphisms (weakly) increase heights, provided one can make sense of that statement.

3.3 THEOREM. *Let G and H be groups with p-height functions, whose lengths are initial segments of a common ordinal λ. If φ is a homomorphism from G to H, then $h\varphi x \geq hx$ for each x in G.*

PROOF. Let $S = \{a \in \lambda : \text{if } hx = a, \text{ then } h\varphi x \geq a\}$, and suppose $a \in S$ for all $a < b$. If $hx = b$ and $h\varphi x < b$, then $x = py$ for some y such that $hy \geq h\varphi x$. Since $b = hx = hpy > gy$, we have $h\varphi y \geq hy$. But $h\varphi x = hp\varphi y > h\varphi y$, so $h\varphi y > hy$, a contradiction. Thus $S = \lambda$.

Now let $S = \{a \in \lambda : h\varphi x > a \text{ if } hx = \infty\}$, and suppose $a \in S$ for each $a < b$. If $hx = \infty$, then there exists y such that $hy = \infty$ and $py = x$. If $h\varphi y \geq b$, then $h\varphi x > b$, so we have shown $b \in S$. Otherwise $h\varphi y = a < b$, which leads to the contradiction $h\varphi y > a$ as $a \in S$. □

So far the only interaction between a p-height function on a group and the additive structure of the group that we have considered concerns relations between hmx and hx. We turn now to the fundamental relation between height and addition.

3.4 THEOREM. *Let h be a p-height function on a group G. Then*

$$h(x + y) \geq \min(hx, hy),$$

for any pair x,y in G, with equality holding if $hx \neq hy$.

PROOF. By symmetry we may assume that $hx \leq hy$. Let λ be the length of G, and let S be the set of $b \in \lambda$ for which $h(x + y) \geq hx$ whenever $b = hx \leq hy$. We shall show that if $a \in S$ for all $a < b$, then $b \in S$. Suppose, on the contrary, that

$$h(x + y) < hx = b \leq hy.$$

Then $x = pw$ and $y = pz$, where $hw \geq h(x + y)$ and $hz \geq h(x + y)$, so $h(x + y) = h(pw + pz) > h(w + z) \geq \min(hw,hz) \geq h(x + y)$, a contradiction. Thus $S = \lambda$, so the claim is true if $hx \neq \infty$. If $hx = \infty$, then $h(x + y) = \infty$ lest $hy = h((x+y) + (-x)) \neq \infty$. Finally, if $hx < hy$, then $hx = h((x+y) + (-x)) \geq \min(h(x+y),hx)$, so $hx \geq h(x + y)$. □

It follows from (3.4) that $\{x \in G : hx \geq a\}$ is a subgroup of G.

EXERCISES

1. Let p be a prime and G a finite abelian group. Show that G has a p-height function. What is the p-length of G (in terms of the invariants of V.2).

2. Let G be a group with a p-height function h. Show that $h^{-1}(\infty)$ is a p-divisible subgroup of G that contains every p-divisible subgroup of G.

3. Let $G = A \oplus B$ be an abelian group with a p-height function h_G. Show that A has a p-height function which is equal to the restriction of h_G to A, and that the p-length of A is an initial segment of the p-length of G.

4. Give an example of an abelian p-group G, with a p-height function, and a subgroup A of G such that A has a p-height function, but the p-height function on G does not restrict to a p-height function on A.

5. Verify the claims made about the p-group P in Example 3.2.

6. Let a be a binary sequence. Let G be the subgroup of the group P in Example 3.2, generated by the elements $a_n x_n$. Show that G is a countable discrete p-group with a height function. What is the length of G?

4. ULM'S THEOREM

If G is a countable p-group with a height function, then the subgroup $D = h^{-1}(\infty)$ is a countable divisible detachable subgroup of G, and G/D is a p-group, hence coherent. Therefore by (2.1) we can write $G = D \oplus R$, where D is a countable coherent divisible group, hence of known structure (2.2), and R is a countable reduced p-group with a height function. Thus we are led to the study of such groups R, which we call **Ulm groups.** Finite p-groups are clearly Ulm groups. The following theorem shows how to construct lots of big Ulm groups.

4.1 THEOREM. *If λ is an ordinal, then there is a reduced p-group, with a height function, of length λ. If λ is a countable ordinal, then there is an Ulm group of length λ.*

PROOF. Let F be the free abelian group on finite sequences $\sigma = (a_1,\ldots,a_n)$ of elements of λ such that $a_1 < a_2 < \cdots < a_n$ and $n \geq 1$. Let K be the subgroup of F generated by the elements of the form

$$p(a_1), \text{ and}$$
$$p(a_1,a_2,\ldots,a_n) - (a_2,\ldots,a_n), \text{ for } n > 1,$$

and let G be the quotient group F/K. We say that an element of F is **in standard form** if it can be written as $\sum n_i\sigma_i$, where the σ_i are distinct free generators of F, and $0 \leq n_i < p$. Each element of G comes from a unique element of F in standard form: that such an element can be found is clear from the nature of the generators of K; that such an element is unique follows from the fact that each nonzero element $\sum k_i\sigma_i$ of K has a nonzero coordinate k_i that is divisible by p. Define $h : G \to \lambda_\infty$ by setting hx equal to the least element of λ occurring in a sequence that has a nonzero coefficient in the standard form of x, and $hx = \infty$ if $x = 0$. It is readily seen that h is a reduced height function on G under which G has length λ. Clearly G is an Ulm group if λ is a countable ordinal. □

A complete set of invariants for Ulm groups is provided by certain countable discrete vector spaces over the p-element field $\mathbb{Z}/(p)$, called Ulm invariants. It will be convenient to define these invariants in the more general setting of valuated groups.

4.2 DEFINITION. A **valuated p-group** is a p-group H together with an ordinal λ and a function $v : H \to \lambda_\infty$ satisfying

(i) If $vx < \infty$, then $vx < vpx$.

(ii) If $(p,m) = 1$, then $vmx = vx$.

(iii) $v(x + y) \geq min(vx,vy)$.

We say that H (or v) is **reduced** if $v^{-1}(\infty) = 0$.

The model for a valuated p-group is a subgroup H of a p-group with a height function h: the function v is h restricted to H. Every p-group with a height function becomes a valuated group upon setting $v = h$. If λ is an ordinal, and $a < b$ are in λ_∞, then we write $a \ll b$ if $b = \infty$, or if there exists c such that $a < c < b$. Note that if $vx > a$, then $vpx \gg a$.

4.3 DEFINITION. Let H be a valuated p-group with values in λ. For each $\alpha \in \lambda$ we define the αth **Ulm invariant** of H to be the group

$$f_H(\alpha) = \frac{\{x \in H : vx \geq \alpha \text{ and } vpx \gg \alpha\}}{\{x \in H : vx > \alpha\}}.$$

The function f_H is called the **Ulm function** of H. Note that $f_H(\alpha)$ is a discrete vector space over the field $\mathbb{Z}/(p)$.

A countable discrete vector space over a finite field has a countable basis, so an Ulm invariant of an Ulm group is determined by its dimension, which is the cardinality of some detachable subset of $\mathbb{N}$. In the classical case the Ulm invariants may be thought of as elements of $\mathbb{N} \cup \{\infty\}$. One virtue of defining them as vector spaces is that they are then additive functors.

If H is a finite cyclic group of order p^n, with v equal to the height function, then the length of H is $\{0,1,\ldots,n-1\}$, and $dim\ f_H(n-1) = 1$, while $f_H(\alpha) = 0$ for $\alpha < n-1$. The Ulm invariants are clearly additive in the sense that if G is a direct sum of a family of subgroups $H(i)$, then $f_G(\alpha)$ is the direct sum of the vector spaces $f_{H(i)}(\alpha)$. As each finite p-group is a finite direct sum of finite cyclic p-groups, the Ulm invariants provide a complete set of invariants for finite p-groups; more generally, they provide a complete set of invariants for direct sums of finite cyclic p-groups.

We now turn to the proof of Ulm's theorem: an Ulm group is determined, up to isomorphism, by its Ulm invariants. If H is a subgroup of an Ulm group G, and $x \in G$, we say that x is **H-proper** if x is of maximum height among the elements of $x + H$.

4.4 THEOREM (Ulm's theorem). *Let G and G' be Ulm groups of length λ with isomorphic Ulm invariants, and let φ be a height preserving isomorphism from a finite subgroup H of G to a finite subgroup H' of G'. Then φ extends to an isomorphism from G to G'.*

PROOF. Let $x_1, x_2, \ldots$ be an enumeration of G such that px_i is in the subgroup generated by $x_1, \ldots, x_{i-1}$ for each i, and let $x'_1, x'_2, \ldots$ be such an enumeration of G'. We shall construct sequences of finite subgroups $H_1 \subseteq H_2 \subseteq \cdots$ of G, and $H'_1 \subseteq H'_2 \subseteq \cdots$ of G', and height preserving isomorphisms $\varphi_n : H_n \to H'_n$, so that $x_n \in H_{2n-1}$ and $x'_n \in H'_{2n}$, and so that φ_{n+1} extends φ_n. By symmetry, it suffices to show that if φ is a height preserving isomorphism between a finite subgroup H of G and a finite subgroup H' of G', and if $x \in G$ and $px \in H$, then φ can be extended to a height preserving isomorphism of the subgroup generated by H and x.

If $x \in H$ there is nothing to do. Otherwise, by choosing an element of maximum height in the finite set $x + H$, we may assume that x is H-proper. Among such x, pick one that maximizes hpx. Note that $h(x + z) = \min(hx, hz)$ for all $z \in H$ because x is H-proper. Let $hx = \alpha$. We must define φx.

Since $h\varphi px = hpx > \alpha$, we can find $x' \in G'$ such that $hx' \geq \alpha$ and $px' = \varphi px$. If $hx' = \alpha$, and x' is H'-proper (note that these questions are decidable), then we can extend φ by setting $\varphi x = x'$. If $hx' = \alpha$ but x' is not H'-proper, then there exists $z \in H$ such that

$$h(x' + \varphi z) > \alpha,$$

so $h(px' + \varphi pz) \gg \alpha$, and so $h(px + pz) \gg \alpha$. But $h(x + z) \geq \alpha$, since $hz = h\varphi z = \alpha$, so $x + z$ is H-proper. Hence, by our choice of x, we have $hpx \geq h(px + pz)$, so $hpx \gg \alpha$. Finally, if $hx' > \alpha$, then we also have $hpx = hpx' \gg \alpha$.

So we turn our attention to the case when $hpx \gg \alpha$, and hence $x \in f_G(\alpha)$ but, since x is H-proper, $x \notin f_H(\alpha) \subseteq f_G(\alpha)$. We need to find an element $x'_* \in f_{G'}(\alpha)$ that is not in $f_{H'}(\alpha)$. Let σ be an isomorphism from $f_G(\alpha)$ to

$f_{G'}(\alpha)$. Note that the finite set $f_{H'}(\alpha)$ is a detachable subspace of the discrete set $f_{G'}(\alpha)$. If $\sigma x \notin f_{H'}(\alpha)$, then we are done. Otherwise, by repeated application of φ^{-1} and σ we can find a finite subspace V of $f_H(\alpha)$ such that $\sigma x \in \varphi V = \sigma V$, or find our desired x'_* along the way. But if $\sigma x \in \sigma V$, then $x \in V$ which contradicts the fact that $x \notin f_{H'}(\alpha)$. Thus we get our x'_*.

We can choose x'_* so that $px'_* = 0$. Moreover, x'_* is H'-proper, for if $h(x'_* + z') > \alpha$, and hence $hz' = \alpha$, then $hpz' \geq min(hp(x'_*+z'),hpx'_*) \gg \alpha$, so $z' \in f_{H'}(\alpha)$, and $x'_* = -z'$ as elements of $f_{G'}(\alpha)$, a contradiction. Set $\varphi x = x'_* + x'$, where $px' = \varphi px$ and $hx' > \alpha$, noting that $x'_* + x'$ is H'-proper because x'_* is. □

We can use Ulm's theorem to prove Prüfer's theorem characterizing the countable direct sums of finite cyclic p-groups.

4.5 THEOREM (**Prüfer**). *A countable discrete p-group G is a direct sum of finite cyclics if and only if $p^n G$ is detachable for each $n \in \mathbb{N}$, and* $G\backslash\{0\} = \bigcup_{n\in \mathbb{N}} p^n G\backslash p^{n+1}G$.

PROOF. As finite cyclic p-groups G have the specified properties, so do direct sums of them. Conversely, if the specified properties hold, then we can define a height function on G by setting $hx = n$ if $x \in p^n G\backslash p^{n+1}G$, and $h0 = \infty$. We can then construct a countable direct sum of finite cyclic p-groups with the same Ulm invariants as G. By Ulm's theorem, G is isomorphic to this group, hence is a direct sum of cyclics. □

An immediate corollary to (4.5) is that finite p-groups are direct sums of cyclic groups.

EXERCISES

1. A **torsion forest** is a discrete set X with a detachable subset R and a function $\pi : X\backslash R \to X$ such that for each $x \in X$ there exists n such that $\pi^n x \in R$. Define what a height function on a torsion forest is. Let $S(X)$ be the free abelian group on X modulo the relations $px = 0$, if $x \in R$, and $px = \pi x$ if $x \notin R$. Show that $S(X)$ is a discrete p-group, and that if X has a height function then so does $S(F)$. What is X in the proof (4.1)? Groups of the form $S(F)$ are called **simply presented p-groups.** Show that every

finite p-group is simply presented.

2. Let e be the generator of a finite cyclic group H of order p^3, and let υ be the valuation on H such that $\upsilon e = 0$, $\upsilon pe = 2$, and $\upsilon p^2 e = 3$. What are the Ulm invariants of H? Embed H in a group G in such a way that the valuation on H is the restriction of the height function on G, and the Ulm invariants of H and G are isomorphic.

3. What are the Ulm invariants of the group in Example 3.2?

4. Let $\lambda = \{0,1,2,3\}$. What are the Ulm invariants of the Ulm group of length λ constructed in the proof of (4.1)?

5. The classical formulation of (4.5) has $\bigcap_{n\in\mathbb{N}} p^n G = 0$ instead of $G\backslash\{0\} = \bigcup_{n\in\mathbb{N}} p^n G\backslash p^{n+1}G$. Show that this modified version of (4.5) is equivalent to Markov's principle.

6. Construct a Brouwerian example of an Ulm group with length $\lambda \subseteq \{0,1\}$ which is not a direct sum of cyclics. Show that an Ulm group of length λ is a direct sum of cyclics if and only if $\lambda \leq \omega$.

7. A subgroup H of a p-group G is **pure** if $p^n H = H \cap p^n G$ for each $n \in \mathbb{N}$. Use (4.4) to show that a pure subgroup of a finite p-group is a direct summand.

8. Let G be an Ulm group, and $x,y \in G$ elements of order p. Show that $hx = hy$ if and only if there is an automorphism of G taking x to y.

5. CONSTRUCTION OF ULM GROUPS

It simplifies things a great deal to put a mild condition on the length of G. Let λ be a well-ordered set. If $a < b$ are elements of λ, we say that b is the **successor** of a, and write $b = a + 1$, if there is no $c \in \lambda$ such that $a < c < b$. For n a nonnegative integer we define $\mathbf{b = a + n}$ inductively by setting $a + 0 = a$, and $a + (n + 1) = (a + n) + 1$, if the latter exists. A well-ordered set λ is said to **have successors** if whenever there exists $b > a \in \lambda$, then $a + 1$ exists.

Let f be a function that assigns to each element of a countable ordinal with successors, a countable discrete vector space over $\mathbb{Z}/(p)$. When can we construct an Ulm group G such that $f_G = f$? A necessary condition is given in the following theorem.

5.1 THEOREM. *Let λ be a well-ordered set with successors, and G an Ulm group of length λ. If $a \in \lambda$, then we can find a nonnegative integer n such that $a + n$ exists and $f_G(a + n) \neq 0$.*

PROOF. Choose x such that $hx = a$, and let n be the least nonnegative integer such that $hp^{n+1}x = \infty$ or $hp^{n+1}x > hp^n x + 1$. Then $p^n x$ represents a nonzero element of $f_G(a + n)$. □

Let λ be a countable well-ordered set with successors, and f a function that assigns to each element in λ a countable discrete vector space over the p-element field $\mathbb{Z}/(p)$. We say that f is a ***U*-function** if, for each $a \in \lambda$, there is $n \in \mathbb{N}$ such that $a + n$ exists and $f(a + n) \neq 0$. Theorem 5.1 says that an Ulm function of an Ulm group whose length is a well-ordered set with successors, is a U-function. We shall show that, conversely, if f is a U-function on λ, then we can construct an Ulm group G of length λ, such that $f_G(a)$ is isomorphic to $f(a)$ for each $a \in \lambda$. This result is known as **Zippin's theorem.**

Let H be a reduced valuated p-group with values in a countable well-ordered set λ with successors. Let f be a U-function on λ. We say that H is ***f*-admissible** if we can embed $f_H(a)$ as a subspace of $f(a)$ for each $a \in \lambda$. This condition is certainly necessary for H to be embedded as a subgroup of an Ulm group with Ulm function f that extends the valuation on H. It turns out that it is also sufficient if H is finite; the case $H = 0$ is Zippin's theorem. The key construction is to enlarge an f-admissible finite valuated group to an f-admissible finite valuated group in which a given instance of Property (iii) of a height function is satisfied.

5.2 LEMMA. *Let H be a finite reduced valuated p-group with values in a well-ordered set λ with successors. Let f be a U-function on λ, and let $\varphi^c : f_H(c) \to f(c)$ be one-to-one maps for each $c \in \lambda$. If $x \in H$, and $a < vx = b$, then we can construct a finite valuated p-group K containing H, so that*

(i) *the valuation on K extends that on H,*

(ii) *the maps φ^c extend to one-to-one maps from $f_K(c)$ to $f(c)$,*

(iii) *$py = x$ for some $y \in K$ such that $vy \geq a$.*

PROOF. By increasing a, if necessary, we may assume either that $b = a + 1$, or that $f(a) \neq 0$ and there is no element $z \in H$ such that

$vz = \alpha$. We may also assume that x has maximal value in the set $\{x + pz : z \in H \text{ and } hz \geq \alpha\}$, since if $py = x + pz$, then $p(y - z) = x$. Adjoin an element y to H subject only to the relation $py = x$, and set

$$v(z + my) = min(vz, \alpha)$$

for $z \in H$ and $0 < m < p$. It is easily checked that this defines a valuated p-group K containing H, whose valuation extends that on H. We must now extend the maps φ^c.

The embedding $H \subseteq K$ induces one-to-one maps $f_H(c) \to f_K(c)$ for each $c \in \lambda$. Suppose $z + my$ represents an element of $f_K(c)$, where $z \in H$ and $0 < m < p$. Then $c = min(vz, \alpha) \leq \alpha$. As we only care about the subspace generated by $z + my$, we may assume that $m = 1$. If $c < \alpha$, then $z + y = z$ in $f_K(c)$, hence $z + y$ represents an element of $f_H(c)$. If $c = \alpha$, and $b = \alpha + 1$, then $vp(z + y) > b$, so $v(pz + x) > vx$, contrary to the choice of x. So if $c < \alpha$, or $b = \alpha + 1$, then $f_H(c) = f_K(c)$ and there is no problem extending φ^c. There remains the case when $c = \alpha$, there is no element $z \in H$ such that $vz = c$, and $f(c) \neq 0$. Then $f_H(c) = 0$ and $f_K(c)$ is one-dimensional, so we simply embed the latter in $f(c)$ any way we like. □

5.3 THEOREM. *Let λ be a countable well-ordered set with successors, and f a U-function defined on λ. Let H be a finite valuated p-group with values in λ. If H is f-admissible, then H can be embedded in an Ulm group G of length λ so that the height function on G induces the valuation on H, and so that $f_G(\alpha) \cong f(\alpha)$ for each $\alpha \in \lambda$.*

PROOF. If μ is a finite subset of λ, and $\varphi^\alpha : f_H(\alpha) \to f(\alpha)$ is a one-to-one map for each $\alpha \in \lambda$, then by repeated applications of (5.2), we can embed H in a finite valuated p-group $E(H,\mu)$ such that

(i) the valuation on $E(H,\mu)$ extends the valuation on H,

(ii) the maps φ^α extend to one-to-one maps from $f_K(\alpha)$ to $f(\alpha)$,

(iii) if $\alpha \in \mu$ and $x \in H$ with $\alpha < vx$, then there is $y \in E(H,\mu)$ satisfying $py = x$ and $vy \geq \alpha$.

We shall construct a chain of finite valuated groups $H = H_0 \subseteq H_1 \subseteq \cdots$ with a common valuation v, and one-to-one maps

$$\varphi_i^\alpha : f_{H_i}(\alpha) \to f(\alpha)$$

such that φ_{i+1}^α extends φ_i^α for each $i \in \mathbb{N}$ and $\alpha \in \lambda$. Let $x_1, x_2, \ldots$ be an enumeration of the disjoint union of $\{f(\alpha)\}_{\alpha \in \lambda}$, let $\alpha_1, \alpha_2, \ldots$ be an enumeration of λ, and let $\mu_n = \{a_1,\ldots,a_n\}$. If $x_n \in f(b)$, set

$$H_{2n} = E(H_{2n-1},\mu_n).$$
$$H_{2n-1} = H_{2n-2} \text{ if } x_n \in \varphi^b_{2n-2}(\, f_{H_{2n-2}}(b)\,),$$
$$H_{2n-1} = H_{2n-2} \oplus \langle y\rangle \text{ where } py = 0 \text{ and } vy = b \text{ otherwise.}$$

In the last case we extend the map φ^b_{2n-1} by setting $\varphi^b_{2n-2}(y) = x_n$. Then $G = \cup_j H_j$ is clearly a countable valuated p-group taking values in λ. That v is a height function on G follows from the construction $H_{2n} = E(H_{2n-1},\mu_n)$, provided that we can show that v maps G onto λ_∞. This follows from the fact that f is a U-function: If $a \in \lambda$, then there exists $m \in \mathbb{N}$ such that $f(a + m) \neq 0$. By the construction of H_{2n-1} we can find an element y_1 in G such that $vy_1 = a + m$. If $m = 0$ we are done. Otherwise, by the construction of H_{2n} we can find an element y_2 of G such that $a \leq vy_2 < vy_1$, and we are done by induction on m.

The isomorphism from $f_G(a)$ to $f(a)$ is provided by the maps φ^a_j; it is onto because of the construction of H_{2n-1}. □

EXERCISES

1. Give a Brouwerian example of a countable well-ordered set that does not have successors.

2. Show that a well-ordered set has successors if and only if whenever $a < b$, then either $a \ll b$ or $b = a + 1$.

3. Let λ be a countable well-ordered set with successors, and f a U-function on λ. Using arguments similar to the proofs of (5.2) and (5.3), show that there is a simply presented Ulm group with Ulm function isomorphic to f.

4. Let λ be a countable well-ordered set with successors, and f a U-function on λ. If $f(b) \neq 0$, show that there is a U-function g on $[0,b]$, and a U-function f' on λ such that

   (i) $f(a) = g(a) \oplus f'(a)$ for $a \leq b$,

   (ii) $f(a) = f'(a)$ for $a > b$,

   (iii) $\dim g(b) = 1$.

   Conclude that each element of order p and height a of an Ulm group G of length λ is contained in summand H of G of length $[0,a]$ such that $\dim f_H(a) = 1$.

5. Use Exercise 4 to show that if λ is a well-ordered set with successors, and G is an Ulm group of length λ, then G is a direct sum of Ulm groups of lengths $[0,\alpha]$ for various $\alpha \in \lambda$.

6. Let λ be a well-ordered set with successors such that $\omega \leq \lambda$. Show that any Ulm group of length λ contains a summand that is an unbounded direct sum of cyclics. Can you find such a summand in Example 3.2?

NOTES

Our construction of a divisible hull for a countable discrete abelian group is due to Rick Smith (1981) who was working in the context of recursive algebra. In that same paper he shows that the divisible hull of a p-group is unique if and only if pG is detachable from G. The proof of the "only if" part of this theorem provides a dramatic illustration of the difference between recursive and constructive mathematics. To establish the existence of an algorithm that decides for any element of G whether it is in pG, he constructs an algorithm that doesn't quite do the right thing for each element, but cannot be wrong infinitely often. From a recursive point of view such an algorithm is easily modified to yield the desired one: simply change a finite number of answers. From a constructive point of view this algorithm is essentially worthless: not only do we have the wrong answer in a finite number of cases, but we have no idea at what point the answers can be counted upon as being correct.

Smith's treatment of uniqueness of divisible hulls is modeled on the treatment of the uniqueness of algebraic closures in [Metakides & Nerode 1979] where it is shown, for countable discrete fields of characteristic zero, that the algebraic closures is unique if and only if the field is factorial. For more on this issue see the treatment of uniqueness of splitting fields in [Bridges-Richman 1987].

A recursive algebra counterexample to Ulm's theorem is allegedly constructed in [Lin 1981a], but it relies on a faulty proof of Zippin's theorem.

Mal'cev (1971) has a brief discussion of types of rank-one torsion-free abelian groups in the context of recursive algebra.

The material on Ulm's theorem and Zippin's theorem is a simplified version of [Richman 1973].

Rogers (1980) studied p-groups that are not necessarily either discrete or countable. She investigated those p-groups G such that p^nG is a detachable subgroup of G for each n, and proved, for groups with countable dense subsets, that this property is equivalent to having a basic subgroup. Classically every p-group has a basic subgroup; constructively it is already a problem to construct a desnse discrete subset. Until we have some examples in mind, it is idle to speculate about extensions of this result.

Chapter XII. Valuation Theory

1. VALUATIONS

If k is a Heyting field, then a (rank-one) **valuation** on k is a function assigning to each element x in k a nonnegative real number $|x|$ such that

$$|x| \neq 0 \text{ if and only if } x \text{ is a invertible}$$
$$|xy| = |x||y|$$
$$|x + y| \leq |x| + |y|$$

The set $\{|x| : x \in k \text{ is invertible}\}$ forms a group called the **value group** of the valuation. A field with a valuation is called a **valued field.** A valued field is a metric space with metric $|x - y|$.

A **general** (rank-one) valuation on a Heyting field k is a function assigning to each element x in k a nonnegative real number vx such that

(i) $vx \neq 0$ if and only if x is invertible

(ii) $v(xy) = (vx)(vy)$

(iii) There is a constant B such that $v(x + y) \leq B \cdot sup(vx, vy)$ *for all* x *and* y.

The constant B in (iii) is called a **bounding constant** for v. Note that (i) and (ii) imply that $v1 = 1$, so taking $x = 1$ and $y = 0$ in (iii) shows that $B \geq 1$. Note also that a valuation is a general valuation with bounding constant 2.

1.1 THEOREM. *A general valuation is a valuation if and only if it has 2 as a bounding constant.*

PROOF. Suppose v is a general valuation with bounding constant 2. An easy induction show that $v(a_1 + \cdots + a) \leq m \cdot sup_{i=1}^{m} va_i$ if m is a power of 2. As there is always a power of 2 between m and $2m$, we have $v(a_1 + \cdots + a_m) \leq 2m \cdot sup_{i=1}^{m} va_i$ for any m; in particular $v(m) \leq 2m$. Therefore, for each $n > 1$,

$$v(x + y)^n = v\left[\sum_{i=0}^{n} \binom{n}{i} x^i y^{n-i}\right] \leq 2(n+1) sup_{i=0}^{n}\, v(\binom{n}{i} x^i y^{n-i})$$

$$\leq 2(n+1)\sum_{i=0}^{n} 2\binom{n}{i} \upsilon x^i \upsilon y^{n-i} = 4(n+1)(\upsilon x + \upsilon y)^n.$$

Taking n^{th} roots and letting $n \to \infty$ gives $\upsilon(x + y) \leq \upsilon x + \upsilon y$. □

A general valuation is **nontrivial** if $\upsilon x > 1$ for some x in k, **nonarchimedean** if $\upsilon(x + y) \leq \sup(\upsilon x, \upsilon y)$ for all x and y in k, and **archimedean** if $\upsilon n > 1$ for some integer n. We say that two general valuations are **equivalent** if they induce the same uniform structure, that is, υ_1 is equivalent to υ_2 if for each $\epsilon > 0$ there is $\delta > 0$ such that $\upsilon_1 x < \delta$ implies $\upsilon_2 x < \epsilon$, and $\upsilon_2 x < \delta$ implies $\upsilon_1 x < \epsilon$. We say that υ_1 and υ_2 are **inequivalent** if there exists x such that either $\upsilon_1 x < 1$ and $\upsilon_2 x \geq 1$, or $\upsilon_2 x < 1$ and $\upsilon_1 x \geq 1$.

1.2 THEOREM. *Let υ_1 and υ_2 be general valuations and consider the following three conditions.*

(i) *There is a positive real number r such that $\upsilon_1 x = (\upsilon_2 x)^r$ for all invertible x,*

(ii) *υ_1 is equivalent to υ_2,*

(iii) *$\upsilon_1 x < 1$ implies $\upsilon_2 x < 1$ for all x,*

Then (i) implies (ii) implies (iii). Moreover if υ_1 is nontrivial, then (iii) implies (i).

PROOF. If (i) holds and $\epsilon > 0$, choose $\delta = \epsilon^r$. Either $\upsilon_2 x < \epsilon$ or $\upsilon_2 x > 0$, and in the latter case x is invertible so if $\upsilon_1 x < \epsilon^r$, then $\upsilon_2 x < \epsilon$. A similar argument works with the roles of υ_1 and υ_2 interchanged.

Suppose (ii) holds and $\upsilon_1 x < 1$. Choose δ such that if $\upsilon_1 y < \delta$, then $\upsilon_2 y < 1$. Then for some $n > 0$ we have $(\upsilon_1 x)^n < \delta$ so $(\upsilon_2 x)^n < 1$ whereupon $\upsilon_2 x < 1$.

Now suppose (iii) holds and υ_1 is nontrivial. Note that if $\upsilon_1 y > 1$, then $\upsilon_1 y^{-1} < 1$ so $\upsilon_2 y > 1$. As υ_1 is nontrivial we can choose y such that $\upsilon_1 y > 1$ and hence $\upsilon_2 y > 1$. Set $r = (\log \upsilon_1 y)/(\log \upsilon_2 y) > 0$. It suffices to show, for $x \neq 0$, that $\gamma_1 = (\log \upsilon_1 x)/(\log \upsilon_1 y)$ is equal to $\gamma_2 = (\log \upsilon_2 x)/(\log \upsilon_2 y)$. Letting $\sim$ stand for $<$ or $>$, and m and n be integers with n positive, we have $m/n \sim \gamma_1$ implies $\upsilon_1(y^m x^{-n}) \sim 1$ implies $\upsilon_2(y^m x^{-n}) \sim 1$ implies $m/n \sim \gamma_2$. □

1.3 THEOREM. *Let υ_1 and υ_2 be nontrivial general valuations. If υ_1 and υ_2 are inequivalent then there exists x such that $\upsilon_1 x < 1$ and $\upsilon_2 x > 1$.*

If v_1 and v_2 are not inequivalent, then they are equivalent.

PROOF. Suppose v_1 and v_2 are inequivalent. As replacing x by x^{-1} in the conclusion shows that the first claim is symmetric, we may suppose that there exists y such that $v_1y < 1$ and $v_2y \geq 1$. Choose z so that $v_2z > 1$ and set $x = y^nz$. Then $v_2x > 1$, and, for sufficiently large n, we have $v_1x < 1$.

Suppose v_1 and v_2 are not inequivalent. By (1.2) it suffices to show that if $v_1x < 1$, then $v_2x < 1$. Let y be an element such that $v_2y > 1$. Then there is $n \in \mathbb{N}$ such that $v_1(x^ny) < 1$. If $v_2(x^ny) > 1$, then the two valuations would be inequivalent, Therefore $v_2(x^ny) \leq 1$ (see Exercise II.3.1), so $v_2x^n \leq v_2y^{-1} < 1$, whence $v_2x < 1$. □

The nontriviality requirement in Theorem 1.2 is essential even if condition (iii) reads "$v_1x < 1$ if and only if $v_2x < 1$." In fact without the nontriviality condition we can prove neither (iii) implies (ii), nor (ii) implies (i). To see that we cannot prove (iii) implies (ii) let $t \geq 0$ be a real number and define valuations v_1 and v_2 on the rational numbers by setting $v_1x = |x|^t$ and $v_2x = (v_1x)^t$ if $x \neq 0$. Here $|x|$ is the absolute value of x. Then $v_1x < 1$ if and only if $v_2x < 1$, while if we could find $\delta > 0$ such that $v_1x < \delta$ implies $v_2x < 1/2$, then either $t > 0$ or $\delta^t > 3/4$, and in the latter case $t = 0$, for if $t > 0$ we could find x such that $v_1x < \delta$ is arbitrarily close to δ so v_2x would be arbitrarily close to $\delta^t > 3/4$. Thus we would have a procedure for deciding whether $t > 0$ or $t = 0$, which is LPO.

To see that we cannot prove (ii) implies (i) let t be a real number and define v_1 and v_2 on the rational numbers by defining $v_1x = |x|^{|t|}$ for $x \neq 0$ and v_2x to be v_1x if $t > 0$ and v_1x^2 if $t < 0$ and extend to all t by continuity. It is readily verified that v_2 is a general valuation and that v_1 and v_2 are equivalent. But if we could find r such that $v_1x = (v_2x)^r$, then we could determine whether $t \geq 0$ or $t \leq 0$, which is LLPO, as $r < 1$ implies $t \leq 0$ while $r > 1/2$ implies $t \geq 0$.

Note that if $v_1 = v_2^r$, and B is a bounding constant for v_2, then B^r is a bounding constant for v_1. Hence any general valuation is equivalent to a valuation.

1.4 THEOREM. *Let v be a general valuation on k. Let x and y be in k, and m and n integers greater than 1. Then*

(i) $vm \leq \sup(1,vn)^{\log m/\log n}$

(ii) $\sup(1,v2)$ *is a bounding constant for* v.

PROOF. Let $\log = \log_2$ and note that

$$v(x_1 + \cdots + x_j) \leq B^{1 + \log j} \sup_{i=1}^{j} vx_i .$$

To prove part (i), write $m^s = a_0 + a_1 n + \cdots + a_r n^r$ where $0 \leq a_i < n$ and $r \log n \leq s \log m$. Then

$$vm^s \leq B^{1+\log(n-1)(r+1)} \sup(1,vn)^r .$$

Raising both sides to the $1/s$ power, and letting $s \to \infty$ gives the desired result. To prove part (ii), let $q = \sup(vx,vy)$ and consider

$$v(x+y)^s = v\sum_{i=0}^{s}\binom{s}{i}x^i y^{s-i} \leq B^{1+\log(s+1)}\sup_{i=0}^{s} q^s v\binom{s}{i}.$$

By part (i) we have $v\binom{s}{i} \leq \sup(1,v2)^{\log\binom{s}{i}}$. But $\binom{s}{i} \leq \binom{s}{[s/2]}$ so

$$v(x+y) \leq B^{(1+\log(s+1))/s} q\cdot\sup(1,v2)^{(\log\binom{s}{[s/2]})/s}$$

and $(1 + \log(s+1))/s \to 0$ while $\binom{s}{[s/2]}/s \to 1$ (use Stirling's formula), so $v(x+y) \leq q\cdot\sup(1,v2) = \sup(1,v2)\cdot\sup(vx,vy)$. □

As corollaries we have the following characterizations of nonarchimedean and archimedean valuations.

1.5 COROLLARY. *Let* v *be a general valuation on* k. *Then the following are equivalent.*

(i) v *is nonarchimedean,*

(ii) $v(1+x) \leq 1$ *for all* x *such that* $vx \leq 1$,

(iii) $vm \leq 1$ *for all integers* m,

(iv) $v2 \leq 1$,

(v) $vn \leq 1$ *for some integer* $n > 1$.

PROOF. Clearly (i) implies (ii) implies (iii) implies (iv) implies (v). If (v) holds, then $v2 \leq 1$ by (1.4.i), so we can take $B = 1$ by (1.4.ii). □

1.6 COROLLARY. *Let* v *be a general valuation on a field* k. *Then the following are equivalent.*

(i) $v(x+y) > \sup(vx,vy)$ *for some* x *and* y,

(ii) *There is* $q > 0$, *such that if* B *is a bounding constant for* v, *then* $B > 1 + q$,

(iii) $\upsilon 2 > 1$,

(iv) $\upsilon m > 1$ *for some integer* m (υ *is archimedean*),

(v) $\upsilon n > 1$ *for all integers* $n > 1$,

(vi) *For some* x *we have* $\upsilon x \leq 1$ *and* $\upsilon(1+x) > 1$.

PROOF. Clearly (i) implies (ii). That (ii) implies (iii) follows from the fact that $\upsilon 2$ is a bounding constant for υ (1.4.ii). Clearly (iii) implies (iv). That (iv) implies (v) follows from (1.4.i). Clearly (v) implies (vi) upon taking $x = 1$. Finally (vi) implies (i) upon taking $y = 1$. □

It is a consequence of (1.5.iv) and (1.6.iii) that to assert that k is nonarchimedean is the same as to deny that k is archimedean, for to assert $\upsilon 2 \leq 1$ is the same as to deny $\upsilon 2 > 1$.

The following is a Brouwerian example of a nontrivial valuation that is neither archimedean nor nonarchimedean.

1.7 EXAMPLE. Let a be a binary sequence with at most one 1. Let R be $\mathbb{Z}[X]/I$, where I is the (detachable prime) ideal generated by $\{a_n(X - 2^n) : n \in \mathbb{N}\}$. Let k be the field of quotients of R. We define a valuation υ on R by associating with each polynomial $f(X)$ the limit of the Cauchy sequence

$$r_m = |f(2^m)|^{1/m} \quad \text{if } a_n = 0 \text{ for all } n \leq m$$
$$r_m = |f(2^n)|^{1/n} \quad \text{if } a_n = 1 \text{ for some } n \leq m.$$

Here $|\ |$ is absolute value. The valuation υ extends uniquely to k. It is nontrivial because $\upsilon X = 2$. However to settle the archimedean question would be to settle the question of whether $a_n = 1$ for some n or not.

1.8 THEOREM. *Let* υ *be a map from a Heyting field* k *to the nonnegative reals satisfying* (i) *and* (ii) *of the definition of a general valuation. Suppose there is a constant* B *such that if* $\upsilon x \leq 1$, *then* $\upsilon(1 + x) \leq B$. *Then* υ *is a general valuation with bounding constant* B.

PROOF. It suffices to show that we cannot have $\upsilon(x + y) > B \cdot \sup(\upsilon x, \upsilon y)$, so suppose this is the case. Suppose $\upsilon x > 0$. If $\upsilon(y/x) < 1$, then

$$\upsilon(x + y) = \upsilon x \cdot \upsilon(1 + y/x) \leq B \cdot \upsilon x \leq B \cdot \sup(\upsilon x, \upsilon y),$$

a contradiction, so $\upsilon(y/x) \geq 1$, whereupon $\upsilon y > 0$ and $\upsilon(x/y) \leq 1$. But then

$$v(x + y) = vy \cdot v(1 + x/y) \leq B \cdot vy \leq B \cdot \sup(vx, vy),$$

a contradiction; so $vx = 0$ whence $x = 0$ because k is a Heyting field. But then $v(x + y) = vy \leq B \cdot \sup(vx, vy)$. □

EXERCISES

1. Let k be a commutative ring in the definition of a valuation. Show that if k admits a (general) valuation, then k is a local ring. Observe that (1.1) through (1.6) still hold in this more general situation. Show that (1.8) holds if k is a local ring, but construct a (classical) counterexample to (1.8) for general k.

2. **The trivial valuation.** Show that every discrete field admits a valuation v such that $v(x) = 1$ for all nonzero x.

3. **The p-adic valuations on $\mathbb{Q}$.** Let p be a prime. Every nonzero rational number can be written uniquely as $r = \pm p^n a/b$, where a and b are relatively prime positive integers not divisible by p; set $v_p r = p^{-n}$. Show that v_p is a nonarchimedean valuation on $\mathbb{Q}$. Show that every nontrivial nonarchimedean valuation on $\mathbb{Q}$ is equivalent to some v_p.

4. Let v be a nonarchimedean valuation. Show that if $vx < vy$, then $v(x + y) = vy$.

5. Let k be a discrete field. Define v on the rational function field $k(X)$ by setting $v(f) = 2^{-\deg f}$ for $f \in k[X]$, and setting $v(f/g) = v(f)/v(g)$. Show that v is a nonarchimedean valuation on $k(X)$.

6. Let S be the set of nontrivial valuations on a Heyting field k, with equality in S being equivalence of valuations. Show that inequivalence is a tight apartness on S.

2. LOCALLY PRECOMPACT VALUATIONS

A subset B of a metric space X is **bounded** if there exists N such that $d(x,y) \leq N$ for all $x,y \in B$. A metric space is **locally precompact** if we can approximate bounded subsets with finite sets, that is, if for each bounded subset B, and $\epsilon > 0$, there is a finite subset Y such that if $x \in B$ then $d(x,y) < \epsilon$ for some y in Y. A locally precompact space is **locally**

compact if it is complete.

It is readily seen that a valued field k is locally precompact if for each positive integer N and positive number ϵ there is a finite subset Y of k such that if $|x| \leq N$ then $|x - y| < \epsilon$ for some y in Y. Such a subset Y is called an **ϵ-approximation to the N-ball.** Absolute value and the p-adic valuations on $\mathbb{Q}$ have this property.

2.1 THEOREM. *A locally precompact valuation is either archimedean or nonarchimedean.*

PROOF. Choose a finite subset Y of k such that if $|x| \leq 2$, then $|x - y| < 1/3$ for some y in Y. Consider the integers $0,1,\ldots,\text{card } Y$. Viewed in k, either one of them has value greater than 1, so k is archimedean, or they all have value less than 2. In the latter case, two must be within 1/3 of the same $y \in Y$, hence within 2/3 of each other, so their difference has value less than 2/3; so k is nonarchimedean by (1.5.v) as $|1| = 1$. □

Let k be a nonarchimedean valued field. The **residue class field** of k is the set $\bar{k} = \{x \in k : |x| \leq 1\}$ where $x = y$ if $|x - y| < 1$. Note that $x \in \bar{k}$ is invertible if and only if $|x| = 1$, and that $\bar{k}$ is a denial field. The residue class field $\bar{k}$ need not be a local ring as we shall see in the example following Theorem 6.2. However if k has a discrete value group, then $\bar{k}$ is a discrete field. A nontrivial nonarchimedean valuation on k is **discrete** if the value group of k is cyclic. Note that a discrete valuation has a discrete value group and a discrete residue class field.

2.2 THEOREM. *Let k have a nontrivial, nonarchimedean valuation. Then k is locally precompact if and only if the valuation is discrete and the residue class field is finite.*

PROOF. Suppose k is locally precompact. As the valuation is nontrivial we can find z in k such that $0 < |z| < 1$. Choose a $|z|$-approximation $x_1,\ldots,x_n$ to the N-ball for $N \geq 1/|z|$. We will show that the value group is a discrete subset of the positive real numbers. For x in k either $|x^n| \neq 1$, in which case $|x| \neq 1$, or $|z| < |x^n| < 1/|z|$, in which case $|z| < |x^t| < 1/|z|$ for $t = 0,1,\ldots,n$. In the latter case there must be s, t and i, with $s \neq t$, such that $|x^s - x_i| \leq |z|$ and $|x^t - x_i| \leq |z|$. As $|x^s| > |z|$ and $|x^t| > |z|$ and the valuation is nonarchimedean, $|x^s| = |x_i| = |x^t|$, so $|x| = 1$.

Thus the value group is a discrete subset of the positive real numbers. As $|z| < 1 < N$, there is i such that $|z - x_i| \leq |z|$, so $|x_i| \leq \sup(|z-x_i|,|z|) = |z| < 1$. Choose π among $x_1,\ldots,x_n$ of biggest value less than 1. For any invertible y there is an integer m such that $|\pi| < |\pi^{-m}y| \leq 1$. Choose j so that $|x_j - \pi^{-m}y| \leq |z| \leq |\pi|$. Then $|x_j| = |x_j - \pi^{-m}y + \pi^{-m}y| = |\pi^{-m}y|$, so $|\pi^{-m}y|$ cannot be less than 1 by the choice of π. Thus $|\pi^{-m}y| = 1$, so $|y| = |\pi|^m$. This shows that the valuation is discrete. To show that the residue class field is finite, choose a finite 1-approximation Y to the 1-ball. The elements of Y form a system of representatives for the residue class field, which is therefore finite since it is discrete.

Conversely, if the valuation is discrete and the residue class field is finite, let $|\pi| < 1$ generate the value group and let A be a finite system of representatives for the residue class field. Given positive N and ϵ we must find a finite set Y such that if $|x| \leq N$ then $|x - y| < \epsilon$ for some y in Y. Choose m so that $|\pi|^m < \epsilon$ and $|\pi|^{-m} \geq N$ and let

$$Y = \{\sum_{i=-m}^{m} a_i \pi^i : a_i \in A\}.$$

If $|x| \leq N$ then $|\pi^m x| \leq 1$ so

$$\pi^m x = a_{-m} + a_{-m+1}\pi + \cdots + a_m \pi^{2m} + b$$

where $|b| < |\pi|^{2m}$. Thus $x = \sum_{i=-m}^{m} a_i \pi^i + b/\pi^m$, and $|b/\pi^m| < |\pi|^m < \epsilon$. □

Although in the locally precompact case you can distinguish archimedean from nonarchimedean, you cannot necessarily distinguish trivial from nontrivial.

2.3 EXAMPLE. Let k be a finite field and X an indeterminate. Let α be a binary sequence with at most one 1, and $\alpha_0 = \alpha_1 = 0$. Let F be the subfield of the rational function field $k(X)$ generated by k and $\{\alpha_n X : n \in \mathbb{N}\}$. We define a nonarchimedean valuation on F by setting $|X| = n$ if $\alpha_n = 1$. Thus $|f| = n^{\deg f}$ if f is a nonzero polynomial in F. Note that if $\alpha_n = 0$, then for any x in F either $|x| > n$ or $|x| \leq 1$. In the latter case $|x - t| < 1/n$ for some $t \in k$. As $\alpha_0 = \alpha_1 = 0$ the residue class field of F is k. To get an ϵ-approximation to the N-ball choose $n > \sup(N,1/\epsilon)$. If $\alpha_n = 1$, then the valuation is discrete and Theorem 2.2 shows that F is locally precompact. If $\alpha_n = 0$, then k is an ϵ-approximation to the N-ball by the previous argument. □

EXERCISES

1. Show that absolute value and the p-adic valuations on $\mathbb{Q}$ are locally precompact. Show that the trivial valuation on any discrete field is locally compact.
2. Show that the valuation on $\mathbb{Q}(X)$ given by $v(f) = 2^{-\deg f}$ is not locally precompact (see Exercise 1.5).
3. Show that residue class fields are denial fields.
4. Construct a Brouwerian example of a nonarchimedean valuation whose value group is cyclic but not discrete.

3. PSEUDOFACTORIAL FIELDS

A field k with a valuation is **pseudofactorial** if $\inf\{|f(a)| : a \in k\}$ exists for each polynomial f in $k[X]$. The reason for this terminology is that the root test (VII.1.8) says that a discrete field k is factorial if and only if for each polynomial f in $k[X]$, either f has a root in k, or $f(a) \neq 0$ for all $a \in k$. Thus a discrete field k is factorial precisely when k is pseudofactorial under the trivial valuation. An example of a factorial field that is not pseudofactorial is provided by taking k to be $\cup k_n$ where k_n is either $\mathbb{Q}$ or $\mathbb{Q}(\pi i) \subseteq \mathbb{R}$. Then the infinimum of the polynomial $X^2 + 1$ cannot be computed.

If k is a valued field we may form the **completion** $\hat{k}$ of k in the same way we formed $\mathbb{R}$ from $\mathbb{Q}$ using the absolute value valuation on $\mathbb{Q}$. The valuation on k extends naturally to $\hat{k}$, and $\hat{k}$ is a complete Heyting field. The p-adic numbers are constructed in this manner from the p-adic valuation on the rationals. If k is nonarchimedean, and E is a field between k and $\hat{k}$, then k and E have the same the residue class fields and value groups.

3.1 THEOREM. *Let k be a discrete field with a valuation. Let f be a separable polynomial in $k[X]$, and let $\hat{k}$ denote the completion of k. Then there is $\delta > 0$ such that, given a_0 in k for which $|f(a_0)| < \delta$, we can construct a root of f in $\hat{k}$.*

PROOF. If f is constant, then the conclusion is clear, so assume $\deg f > 0$. Choose N so that $|f(a)| \geq 1$ whenever $|a| \geq N$, and thus $|a| \leq N$ whenever $|f(a)| < 1$. Write

$$f(Y+Z) = f(Y) + Zf'(Y) + Z^2g(Y,Z)$$

and choose M so that $|g(a,b)| \le M$ whenever $|a| \le N$ and $|b| \le 1$.

Find s and t in $k[X]$ such that $s(X)f(X) + t(X)f'(X) = 1$. Choose $0 < r \le 1$ so that $|t(a)| \le r^{-1}$ and $|s(a)| \le r^{-1}$ whenever $|a| \le N$. If $|f(a)| < 1$, then $|a| \le N$, so

$$r^{-1}|f'(a)| \ge |t(a)f'(a)| \ge 1 - |s(a)||f(a)| \ge 1 - r^{-1}|f(a)|,$$

whereupon

$$|f'(a)| \ge r - |f(a)|.$$

Choose $\delta \le r/2$ such that if $u \ge 1/\delta$, then

$$2M\left[\frac{1}{ru-1}\right]^2 \le \frac{1}{u}.$$

If $|f(a_0)| < \delta$, then we can use **Newton's method** to construct a root of f as follows. Define a_n inductively by $a_{n+1} = a_n + h_n$, where $h_n = -f(a_n)/f'(a_n)$. We shall show that $|f(a_n)| \le 2^{-n}\delta$. This is true for $n = 0$. If $|f(a_n)| \le 2^{-n}\delta < 1$, then $|a_n| \le N$ and $|f'(a_n)| \ge r - 2^{-n}\delta$ so $|h_n| \le (r2^n/\delta - 1)^{-1} \le 2^{-n}$. But $|f(a_{n+1})| = |f(a_n + h_n)| = |h_n^2||g(a_n,h_n)| \le |h_n|^2M \le 2^{-(n+1)}\delta$ by taking $u = 2^n/\delta$. Thus $\{a_n\}$ is a Cauchy sequence converging to a root of f in $\hat{k}$. □

3.2 COROLLARY. *Let k be a discrete field with a pseudofactorial valuation. If f is a separable polynomial in $k[X]$, then either f has a root in $\hat{k}$, or $|f(a)|$ is bounded away from 0 for all a in k.*

PROOF. Choose δ as in Theorem 3.1. Either $\inf\{|f(a)| : a \in k\} \ge \delta/2$, in which case $|f(a)|$ is bounded away from 0, or $\inf\{|f(a)| : a \in k\} < \delta$ in which case we can find a root of f in $\hat{k}$ by (3.1). □

Using the notion of *winding number*, we can get another proof of the discrete fundamental theorem of algebra.

3.3 COROLLARY. *Let k be a discrete subfield of $\mathbb{C}$, and K the algebraic closure of k in $\mathbb{C}$. Then K is algebraically closed.*

PROOF. Let f be a monic polynomial of degree n with coefficients in K, which is a discrete field by (VI.1.9). Choose $r > 0$ so that if z is a complex number of modulus r, then $|z^n - f(z)| < r^n$. Then the winding number with respect to 0 of the path given by restricting f to the circle of radius r is n. If $\inf\{|f(x)| : |x| \le r\} > 0$, then this winding number would be 0. Hence $\inf\{|f(x)| : |x| \le r\} = 0$. As f is a product of separable polynomials (VI.6.3), we can find a root of f in the complex numbers by Theorem 3.1. By (VI.1.5) such a root is in K, so the corollary is proved. □

Pseudofactorial fields arise as follows.

3.4 THEOREM. *If k is a discrete field with either a locally precompact valuation, or a discrete valuation with a factorial residue class field, then k is pseudofactorial.*

PROOF. Let f be a polynomial in $k[X]$. If k is locally precompact choose N such that $|f(a)| \geq |f(0)|$ whenever $|a| \geq N$, so if $|f(a)| < |f(0)|$, then $|a| \leq N$. Let Y_n be a $1/n$-approximation to the N-ball. Then the sequence $r_n = inf\{|f(y)| : y \in Y_n\}$ is Cauchy and converges to $inf\{|f(a)| : a \in k\}$, so k is pseudofactorial.

Suppose the valuation is discrete with a factorial residue class field. Replacing $f(X)$ by $dc^{deg\ f}f(X/c)$ we may assume that f is monic and the values of the coefficients of f do not exceed 1. Let the value group be generated by $r < 1$, and choose π in k such that $|\pi| = r$. We shall construct a sequence of (possible empty) finite families $B_m \subseteq \{b \in k : |b| \leq 1 \text{ and } |f(b)| \leq r^m\}$, and positive integers $n(m)$ such that if $|f(a)| \leq r^{n(m)}$, then $|a - b| \leq r^m$ for some b in B_m.

Let B_1 consist of one representative of each of the roots of f read as a polynomial over the (factorial) residue class field, and set $n(1) = 1$. If $|f(a)| \leq r$, then $|a| \leq 1$ because f is monic with small coefficients, so a represents a root of f in the residue class field, whereupon $|a - b| < r$ for some b in B_1. To construct B_{m+1} from B_m we proceed as follows. For each b in B_m write

$$f(b + \pi^m X) = \pi^{e(b)}g_b(X)$$

where the maximum value of the coefficients of $g_b \in k[X]$ is 1. As $f(b + \pi^m X) - f(b)$ is divisible by $\pi^m X$, and $|f(b)| \leq r^m$, we have $e(b) \geq m$. Let $c_1^b,\ldots,c_s^b$ be representatives of the roots of g_b read in the residue class field, and let B_{m+1} consist of all elements of the form

$$b + \pi^m c_i^b$$

for b in B_m. Let $n(m+1)$ exceed $n(m)$ and $e(b)$ for each b in B_m. Suppose $|f(a)| \leq r^{n(m+1)}$. Then $|a - b| \leq r^m$ for some b in B_m, so $a = b + \pi^m c$ for some c in k such that $|c| \leq 1$. Since $|f(a)| \leq r^{n(m+1)}$ and $f(a) =$

$\pi^{e(b)}g_b(c)$, we have $|g_b(c)| \le r$ so c represents a root of g_b read in the residue class field. Hence $|c - c_i^b| \le r$ for some i, so $|a - (b + \pi^m c_i^b)| = |\pi^m c - \pi^m c_i^b| \le r^{m+1}$.

If B_1 is empty, then $1 = |f(0)| \le |f(a)|$ for each a in k. Otherwise we compute $\inf\{|f(a)| : a \in k\}$ by taking the limit of the Cauchy sequence $d_n = \inf\{|f(b)| : b \in \cup_{m \le n} B_m\}$. □

3.5 COROLLARY. *If k is a discrete field with a pseudofactorial valuation, then the algebraic (separable) closure of k in its completion $\hat{k}$ is separably factorial.*

PROOF. The algebraic (separable) closure K of k in $\hat{k}$ is discrete (VI.1.9). Moreover K is pseudofactorial since k is dense in K. Hence, if f is a separable polynomial in $K[X]$, then either f has a root in $\hat{K} = \hat{k}$, or $|f(a)|$ is bounded away from 0 for all a in K, by Corollary 3.2. But any root of f in $\hat{k}$ is in K, so K is separably factorial (VII.1.7). □

EXERCISES

1. Let k be a discrete field, and equip the rational function field $K = k(X)$ with the valuation of Exercise 1.5. The completion $\hat{K}$ of K is the **field of formal power series** over k. Show that we may identify $\hat{K}$ with formal sums of the form

$$f = \sum_{i=m}^{\infty} a_i X^i$$

where $m \in \mathbb{Z}$ and $a_i \in k$, and that if $a_m \ne 0$, then $v(f) = m$.

2. The stipulation that f be separable cannot be eliminated from the hypothesis of Theorem 3.1. Let $\{a_n\}$ be a binary sequence of with at most one 1. Let K be the field of formal power series in Y over the two element field $\mathbb{Z}_2$. Let ζ be an element of K that is transcendental over $\mathbb{Z}_2(Y)$. Show that

$$\eta = (1 + \sum a_n Y^{2n+1})\zeta^2$$

is transcendental over $\mathbb{Z}_2(Y)$. Let $k = \mathbb{Z}_2(Y,\eta)$ and let $f(X) = X^2 - \eta$. Show that f has a root in $K = \hat{k}$ if and only if $a_n = 0$ for all n. Show that k is pseudofactorial by Theorem 3.4, so Corollary 3.2 would allow us to decide whether f had a root in K or not.

3. Describe the valuation on the completion $\hat{k}$ of a valued field k. Show that the residue class fields and value groups of k and $\hat{k}$ are the same.

4. Show that the sequences r_n and d_n in the proof of (3.4) are Cauchy.

4. NORMED VECTOR SPACES.

Let k be a valued field. A **normed vector space** over k is a vector space V over k, with a tight inequality, and a map, called a **norm**, taking each $x \in V$ to a nonnegative real number $|x|$ satisfying

$$\text{(i)}\ |x| \neq 0 \text{ if and only if } x \neq 0.$$
$$\text{(ii)}\ |ax| = |a|\,|x|$$
$$\text{(iii)}\ |x + y| \leq |x| + |y|$$

If V has no inequality, and we replace (i) by $|x| = 0$ only if $x = 0$, then we can use (i) to *define* a tight inequality on V. A normed vector space is a metric space with the metric $|x - y|$.

The strong inequality on k^n is given by setting $x \neq y$ if $x_i \neq y_i$ for some i. We turn k^n into a normed vector space by setting $|x| = sup\{|x_i| : i = 1,\ldots,n\}$. If we put the absolute value valuation on $\mathbb{Q}$ and $\mathbb{R}$, then any subfield of $\mathbb{R}$ is a normed vector space over $\mathbb{Q}$.

Let V be a normed vector space over a valued field k. We say that $v_1,\ldots,v_n$ in V are **metrically independent** if for every $\epsilon > 0$ there is $\delta > 0$ so that if $|\Sigma\, a_i v_i| < \delta$, then $|a_i| < \epsilon$ for each i. If, in addition, every element of V can be written as a linear combination of the v_i, then the v_i are said to be a **metric basis** of V. Thus V admits a metric basis if and only if there is a continuous isomorphism from V to k^n, the latter equipped with the supremum norm.

The subfield $\mathbb{Q}(\sqrt{2})$ of $\mathbb{R}$ is a finite-dimensional normed vector space over a discrete field that does not admit a metric basis. On the other hand, as linear transformations of finite-dimensional normed vector spaces are continuous, if a normed vector space over a discrete valued field admits a metric basis, then any basis is a metric basis.

Two norms $|\ |_1$ and $|\ |_2$ are **equivalent** if there exists $\epsilon > 0$ such that $\epsilon|x|_1 \leq |x|_2 \leq \epsilon^{-1}|x|_2$ for each x in V. It is an easy exercise to show that the norms $|\ |_1$ and $|\ |_2$ are equivalent if and only if the identity map from $(V,|\ |_1)$ to $(V,|\ |_2)$ is bicontinuous.

Let V be a normed vector space over k. A set of elements $v_1,\ldots,v_n$ of V is said to be a **basis** for V if the linear transformation φ from k^n to V defined by $\varphi(e_i) = v_i$ is an isomorphism that preserves inequality. We shall see that for locally compact fields k, any basis for V is a metric basis; this is the same as saying that any two norms on k^n are equivalent. Warning: A discrete field is locally compact under the trivial valuation, but not necessarily under others--for example, $\mathbb{Q}$ with absolute value is not complete.

LEMMA 4.1. *Let k be a valued field, and $e_1,\ldots,e_n$ the natural basis for k^n. A norm on k^n is equivalent to the supremum norm if and only if e_i is bounded away from $k^{i-1} = ke_1 + \cdots + ke_{i-1}$ for $i = 2,\ldots,n$.*

PROOF. The 'only if' is clear from the definition of the supremum norm. For the 'if', it suffices to show that the coordinate projections $\pi_i(x_1e_1 + \cdots + x_ne_n) = x_i$ are continuous. Suppose $|e_n - \alpha| \geq \delta > 0$ for all α in k^{n-1}. Then $\delta|x_n| \leq |x_1e_1 + \cdots + x_ne_n|$ so projection of k^n onto its last coordinate, and hence also onto k^{n-1}, is continuous. We are done by induction on n. □

A subset A of a metric space X is **located** if we can find, for each x in X and $\epsilon > 0$, an element a_0 in A such that $d(x,a_0) < d(x,a) + \epsilon$ for all $a \in$ A; thus $d(x,A) = \inf\{d(x,a) : a \in A\}$ exists. If A is a nonempty locally precompact subset, then A is located: Choose N so that the bounded set $B = \{a \in A : d(x,a) < N\}$ is nonempty. Approximate B to within $\epsilon/2$ by a finite set $F \subseteq A$. Choose $a_0 \in F$ so that $d(x,a_0)$ is within $\epsilon/2$ of $\inf\{d(x,a) : a \in F\}$.

THEOREM 4.2. *Let k be a locally compact valued field. Then k^n is a locally compact normed space over k under the supremum norm, and any norm on k^n is equivalent to this one.*

PROOF. Clearly k^n is locally compact under the supremum norm. Suppose that k^n is equipped with another norm. By induction k^{n-1} is locally compact and so is located. It suffices by Theorem 4.1 to show that $d(e_n,k^{n-1}) > 0$. Using the fact that k^{n-1} is located, define a sequence of elements $\theta_i \in k^{n-1}$ so that for each $i > 0$ either

$$|\theta_i - e_n| < 1/i$$

or $\quad d(e_n,k^{n-1}) > 1/i+1$ and $\theta_i = \theta_{i-1}$.

It is easily seen that the θ_i form a Cauchy sequence in k^{n-1}. As k^{n-1} is complete, the θ_i converge to $\theta \in k^{n-1}$. Since e_n is different from each element of k^{n-1} we have $\theta \neq e_n$ so $|\theta - e_n| > 0$ whence $|\theta - e_n| > 1/i$ for some i. As $|\theta - e_n| \leq |\theta_i - e_n|$, we cannot have $|\theta_i - e_n| < 1/i$, whereupon $d(e_n, k^{n-1}) > 1/(i+1)$. □

The condition in Theorem 4.2 that the field k be locally compact cannot be weakened to read that k is complete, as in the classical context; the appeal to the locatedness of k^{n-1} in the proof is essential as the following Brouwerian counterexample shows.

EXAMPLE 4.3. Let $r \geq 0$ be a real number and let k be the completion of the subfield $\mathbb{R}(ri)$ of $\mathbb{C}$. Then k is a closed subfield of $\mathbb{C}$. We shall construct a norm on k^2 such that you cannot bound e_2 away from k^1.

If e_1, e_2 is the natural basis for $\mathbb{C}^2$, then every element of $\mathbb{C}^2$ can be written uniquely in the form $z = \alpha(e_1 - ie_2) + \beta(e_1 + ie_2)$ with α and β in $\mathbb{C}$. Equip $\mathbb{C}^2$ with the seminorm $\|z\| = |\beta| + r|\alpha|$. To see that this defines a norm on k^2, let $Im\ z$ denote the imaginary part of z and note that $|Im\ z| = |\alpha - \overline{\beta}|$. If $z \neq 0$, then either $\alpha \neq 0$ or $\beta \neq 0$. Either $|\alpha - \overline{\beta}| < \sup(|\alpha|, |\beta|)/2$, in which case $|\beta| \neq 0$ so $\|z\| \neq 0$, or $Im\ z \neq 0$ so $r > 0$ whence $\|z\| \neq 0$.

Suppose $d(e_2, k^1) > 0$. If $r > 0$, then $r = \|e_1 - ie_2\| \geq d(e_2, k^1)$. By cotransitivity either $r > 0$ or $r < d(e_2, k^1)$, and in the latter case $r = 0$ because $r > 0$ is impossible. Thus either $r > 0$ or $r = 0$, which is equivalent to LPO. □

EXERCISES

1. The **trivial metric** on a discrete set X is defined by setting $d(x,y) = 0$ if $x = y$, and $d(x,y) = 1$ otherwise. Show that a subset of X is detachable if and only if it is located in the trivial metric.
2. Show that the space $\mathbb{Q}(\sqrt{2})$ is discrete.
3. Define a positive notion of inequivalence of norms on vector spaces, and show that the supremum norm on $\mathbb{Q}^2$ is inequivalent to the norm on $\mathbb{Q}^2$ induced by an isomorphism with $\mathbb{Q}(\sqrt{2})$.

5. REAL AND COMPLEX FIELDS.

All archimedean fields contain a copy of the rational numbers $\mathbb{Q}$, and there is essentially only one archimedean valuation on $\mathbb{Q}$. The following theorem characterizes all nontrivial valuations on $\mathbb{Q}$.

THEOREM 5.1. *Any nontrivial valuation on $\mathbb{Q}$ is equivalent either to absolute value or to the p-adic valuation for some prime p.*

PROOF. If $|n/p| > 1$ for positive integers n and p, then $|n| > |p|$ so cotransitivity implies that either $|n| > 1$ or $|p| < 1$. If $|n| > 1$, then $|m| > 1$ for every integer $m > 1$ by (1.6). Furthermore

$$|m| \leq |n|^{\log m/\log n} \quad \text{and} \quad |n| \leq |m|^{\log n/\log m}$$

by (1.4) so $|m| = |n|^{\log m/\log n} = m^r$ for fixed r; thus the valuation is equivalent to absolute value. If, on the other hand, $|p| < 1$, then the valuation is nonarchimedean (1.5) and we may assume that p is a prime. If $sm + tp = 1$, then

$$1 = |sm + tp| \leq \sup(|sm|,|tp|),$$

but $|tp| = |t||p| < 1$ so $|sm| = 1$ whence $|m| = 1$. Thus the value of an arbitrary positive integer mp^e is $|p|^e$ if $(m,p) = 1$; so the valuation is equivalent to the p-adic valuation. □

Note that every nontrivial valuation on $\mathbb{Q}$ is locally precompact, and that absolute value is the only archimedean valuation on $\mathbb{Q}$ with $|2| = 2$. There are two important kinds of archimedean fields.

DEFINITION. Let k be an archimedean field. If there is $\delta > 0$ such that $|a^2 + 1| \geq \delta$ for all a in k, then k is said to be **real**. If for every $\delta > 0$ there exists a in k such that $|a^2 + 1| < \delta$, then k is said to be **complex**.

Classically every archimedean field is either real or complex. However the field k of Example 4.3 cannot be said to be real or to be complex.

The construction of the valued field $\mathbb{C}$ of complex numbers from the valued field $\mathbb{R}$ of real numbers can be used to construct a unique valued splitting field for $X^2 + 1$ over any real field.

THEOREM 5.2. *If k is a real field, then k can be embedded in a valued field K containing an element i such that $i^2 = -1$ and $K = k(i)$. Moreover if k is embedded in a valued field E containing an element j such that*

$j^2 = -1$, *then* j *is bounded away from* k *and* $k(j)$ *is isomorphic to* K *as a valued field.*

PROOF. Let K be the ring of all formal sums $a + bi$ with $a,b \in k$ and $i^2 = -1$. If we define inequality on K by $a + bi \neq 0$ if $a \neq 0$ or $b \neq 0$, then K is a Heyting field. We can value K by setting $|a + bi| = \sqrt{|a^2 + b^2|}$. To show that this is a general valuation it suffices to find a constant B so that $|(1+a)^2 + b^2| \leq B$ whenever $|a^2 + b^2| \leq 1$. But $|(1+a)^2 + b^2| \leq 1 + |a^2 + b^2| + 2|a|$, so we need only bound $|a|$. If $|a| > 0$, then $|a|^{-2} \geq |1 + b^2/a^2|$ is bounded away from 0. As $|2| \leq 2$, it's a valuation.

To prove the second claim we first show that j is bounded away from k. If $a \in k$, then either $|a| > 2$ or $|a| < 3$. If $|a| > 2$, then $|a - j| \geq |a| - |j| > 1$; if $|a| < 3$, then $|a - j| = |a^2 + 1|/|a + j| > |a^2 + 1|/4$, and the latter is bounded away from 0 because k is real.

If $\varphi : K \to k(j)$ is defined by $\varphi(a + bi) = a + bj$, then φ is a homomorphism of K onto $k(j)$. We must show that φ preserves inequality. If $a + bj \neq 0$, then $a \neq 0$ or $bj \neq 0$ as E is a Heyting field, so $a \neq 0$ or $b \neq 0$, whence $a + bi \neq 0$. Conversely if $a + bi \neq 0$, then $a \neq 0$ or $b \neq 0$, so $0 \neq a^2 + b^2 = (a + bj)(a - bj)$ whence $a + bj \neq 0$. The valuation on $k(j)$ induces, via φ, a valuation on K. Theorem 4.1 says that this valuation is equivalent to the given valuation on K, so one is a positive power of the other (1.2). As they agree on k, they are equal. □

COROLLARY 5.3. *There is only one valuation on the field* $\mathbb{C}$ *that extends absolute value on* $\mathbb{R}$.

PROOF. Immediate from Theorem 5.2. □

THEOREM 5.4 (Gelfand-Tornheim). *If* k *is a normed field over* $\mathbb{C}$, *then* $k = \mathbb{C}$.

PROOF. As $\mathbb{C}$ is complete, it suffices to show that $\mathbb{C}$ is dense in k. Suppose we want an element of $\mathbb{C}$ that is close to a given $a \in k$. By cotransitivity, either 0 is close to a, or $|a| > 0$, so we may assume the latter. Construct a square of width $|8a|$, centered at 0, cut it into $4n^2$ little squares of width $|4a|/n$, and let S be the polygonal path formed by the edges of the little squares that lie on the perimeter of the big square. Let δ be the infimum of $|z - a|$ as z ranges over the midpoints of the sides, and the centers, of the little squares. We will show that

$\delta^3 \le 3|4a|^3/n$, so we can get complex numbers that are arbitrarily close to a by choosing n large. By cotransitivity we may assume $\delta > 0$.

Given a function $f : \mathbb{C} \to k$, and a polygonal path P with successive vertices $z_0,\ldots,z_n$ in $\mathbb{C}$, we define the 'integral' $I_P(f)$ to be $\Sigma_{i=1}^{n}(z_i - z_{i-1})f((z_i + z_{i-1})/2)$. Clearly I_P is linear in f, and if $f(z) = cz + d$, then $I_P(f) = (z_n - z_0)((z_n + z_0)c/2 + d)$. In particular, if P is a closed path, that is, if $z_0 = z_n$, then I_P vanishes on polynomials of degree at most one.

Consider the function $g(z) = 1/(z - a)$ from $\mathbb{C}$ to k. First we get an upper bound on $I_Q(g)$ where Q is a square with sides of length ϵ. Let $\delta > 0$ be at most the infimum of the distances from a to the midpoints of the sides, and to the center, of Q. Our bound will depend only on ϵ and δ, we may assume that 0 is the center of Q. As

$$g(z) = \frac{1}{z - a} = \frac{z^2}{a^2(z - a)} - \frac{a + z}{a^2}$$

and I_Q vanishes on polynomials of degree at most one, we have $|I_Q(g)| \le 4\epsilon(\epsilon/2)^2/\delta^3 = \epsilon^3/\delta^3$.

As $I_S(g)$ is equal to the sum of the integrals $I_Q(g)$, where Q ranges over the $4n^2$ little squares of width $\epsilon = |4a|/n$, we have $|I_S(g)| \le 4|4a|^3/\delta^3 n$. We use this to get an upper bound on $I_S(1/z)$. Now

$$\frac{1}{z} = \frac{1}{z - a} - \frac{a}{z(z - a)}$$

so

$$|I_S(1/z)| \le \frac{4|4a|^3}{\delta^3 n} + \frac{|a|4|8a|}{|4a|(|4a| - |a|)} = \frac{4|4a|^3}{\delta^3 n} + \frac{8}{3}.$$

We now compute a lower bound for $I_S(1/z)$ directly from the definition. Each of the four sides of the square contributes

$$\frac{i}{n}\sum_{k=1}^{n}\left[\frac{1}{1 + i(k - 1/2)/n} + \frac{1}{1 - i(k - 1/2)/n}\right]$$

which, in absolute value, is equal to

$$\sum_{k=1}^{n}\frac{2n}{n^2 + (k - 1/2)^2} \ge \sum_{k=1}^{n}\frac{2n}{2n^2} = 1.$$

Combining this with the upper bound on $|I_S(1/z)|$ we get

$$4 \le \frac{4|4a|^3}{\delta^3 n} + \frac{8}{3}.$$

Thus $\delta^3 \le 3|4a|^3/n$. □

COROLLARY 5.5. *Let k be an archimedean field with $|2| = 2$. Then k is real if only if $\hat{k} \cong \mathbb{R}$ as a valued field.*

PROOF. Clearly any subfield of $\mathbb{R}$ is real. Conversely if k is real, then $\hat{k}$ is real and we can form $\hat{k}(i)$ by Theorem 5.2. Theorem 5.1 allows us to assume that $\mathbb{R} \subseteq \hat{k}$. As $\mathbb{R}(i) \cong \mathbb{C}$ by Theorem 5.2, we have $\hat{k}(i) = \mathbb{R}(i)$ by Theorem 5.4. But i is bounded away from $\hat{k}$ as $\hat{k}$ is real, so $\hat{k} = \mathbb{R}$. □

THEOREM 5.6. *Let k be an archimedean field with $|2| = 2$. Then the following are equivalent.*

(i) *k is complex,*

(ii) *$|a^2 + 1| < 3/4$ for some a in k,*

(iii) *There is i in $\hat{k}$ such that $i^2 = -1$*

(iv) *$\hat{k} \cong \mathbb{C}$.*

PROOF. Clearly (i) implies (ii), and (iv) implies (i). Suppose (ii). If $|a^2 + 1| < 3/4$, then $|a|^2 > 1/4$ so if $b = (a - 1/a)/2$, then $|b^2 + 1| = |(a^2+1)^2/4a^2| \leq |a^2 + 1|^2$. Thus if we define a sequence by $a_0 = a$, and $a_{n+1} = (a_n - 1/a_n)/2$, we have $a_n^2 \to -1$, while $|a_{n+1} - a_n| = |a_n^2 + 1|/2|a_n| < |a_n^2 + 1|$ shows that the sequence is Cauchy.

Now suppose (iii). By (5.1) we may assume $\mathbb{R} \subseteq \hat{k}$, and $\mathbb{R}(i) \cong \mathbb{C}$ by (5.2), so $\hat{k} \cong \mathbb{R}(i)$ by (5.4). □

Using condition (5.6.ii), and cotransitivity, we see immediately that an archimedean field is real if and only if it cannot be complex.

COROLLARY 5.7 (Ostrowski). *If k is a locally precompact valued field with $|2| = 2$, then $\hat{k} \cong \mathbb{R}$ or $\hat{k} \cong \mathbb{C}$.*

PROOF. As k is locally precompact we can calculate the infinimum of $|a^2 + 1|$ as a ranges over k. If this infinimum is less than 3/4, then k is complex by Theorem 5.6. If the infinimum is greater than 0, then k is real. □

As the completion $\hat{k}$ of a discrete field is not discrete in general, it is often convenient to work instead with the separable closure $\tilde{k}$ of k in $\hat{k}$. Because in the nonarchimedean case the field $\tilde{k}$ is intimately tied up with Hensel's lemma, we call $\tilde{k}$ the **Henselization** of k. The following characterizes when a discrete archimedean field k is real or complex in terms of its Henselization.

THEOREM 5.8. *Let k be a discrete archimedean field. Then $\tilde{k}$ is factorial if and only if k is real or k is complex. If k is complex, then $\tilde{k}$ is algebraically closed. If k is real, then the Henselization of $k(i)$ is $\tilde{k}(i)$, which is algebraically closed, and every polynomial over $\tilde{k}$ is a product of irreducible linear and quadratic factors.*

PROOF. If $\tilde{k}$ is factorial, then either $X^2 + 1$ has a root in $\tilde{k}$, or $X^2 + 1$ is irreducible over $\tilde{k}$. In the first case k is complex by (5.6.iii); in the second case k is real because (5.6.ii) cannot hold, so $|a^2 + 1| > 1/2$ for all $a \in k$. If k is complex, then $\hat{k} \cong \mathbb{C}$ so $\tilde{k}$ is algebraically closed, hence factorial. If k is real, then $\hat{k} \cong \mathbb{R}$, so we may assume $\hat{k} = \mathbb{R}$. Let f be a polynomial of positive degree in $\tilde{k}[X]$. Then f has a complex root $a + bi$, so $a - bi$ is also a root of f. Thus a and b are algebraic over $\tilde{k}$ and so are in $\tilde{k}$ (so the Henselization of $k(i)$ is $\tilde{k}(i)$). If $b = 0$, then $X - a$ is a factor of f. If $b \neq 0$, then $X^2 - 2aX + a^2 + b^2$ is an irreducible factor of f in $\tilde{k}[X]$. □

EXERCISES

1. Construct a Brouwerian example of a nonarchimedean valuation on $\mathbb{Q}$ that is neither trivial nor equivalent to a p-adic valuation.
2. Show that $I_S(1/z)$ in the proof of (5.4) approaches $2\pi i$ as $n \to \infty$, using
$$\int_0^1 (1 + x^2)^{-1} dx = \pi/4.$$
3. Show that if every archimedean field can be embedded in $\mathbb{C}$, then the world's simplest axiom of choice holds (use the construction in Exercise VI.3.1).

6. HENSEL'S LEMMA

Let k be a nonarchimedean valued field, and $\bar{k}$ its residue class field. Let f be a monic polynomial in $k[X]$ all of whose coefficients have value at most 1. Then f determines a polynomial $\bar{f}$ in $\bar{k}[X]$. A standard form of Hensel's lemma is that if $\bar{f}$ has a simple root in $\bar{k}[X]$, and k is complete, then f has a root in k. More generally, if $\bar{f}$ factors into strongly relatively prime factors, then this factorization is induced by a factorization of f. Our version of Hensel's lemma concerns how to improve approximate factorizations of f into approximately strongly relatively prime factors.

First we show how to extend a nonarchimedean valuation on k to a valuation on $k(X)$, with any specified positive value for X. The following lemma is needed because we cannot determine which coefficients of a polynomial have maximum value.

6.1 LEMMA. *Let $s_1,\ldots,s_\ell$ be real numbers with supremum σ, and let m be a positive integer. Then for all but at most $(m+1)(\ell-1)$ positive integers n we can construct a finite subset A of $\{1,\ldots,\ell\}$ such that*

$$s_i > \sigma - 2^{-(m+n)} \qquad \text{if } i \in A$$

and

$$s_i < \sigma - 2^{-n} \qquad \text{if } i \notin A$$

PROOF. For each positive integer n construct a finite subset A_n of $\{1,\ldots,\ell\}$ such that

$$s_i > \sigma - 2^{-(n-1)} \qquad \text{if } i \in A_n$$

and

$$s_i < \sigma - 2^{-n} \qquad \text{if } i \notin A_n.$$

Then the A_n form a descending chain of nonempty finite subsets of $\{1,\ldots,\ell\}$. Since $\#A_1 \leq \ell$, there are at most $\ell-1$ values of n for which $A_{n+1} \neq A_n$, and hence at most $(m+1)(\ell-1)$ values of n such that $A_{m+n+1} \neq A_n$. For the remaining values of n we let $A = A_n = A_{m+n+1}$. □

6.2 THEOREM. *Let k be a field with a nonarchimedean valuation, and let λ be a positive real number. For $f(X) = \Sigma\, a_i X^i \in k[X]$ let $|f| = \sup |a_i|\lambda^i$. Then this defines a valuation on $k(X)$.*

PROOF. It suffices to show that we have defined a valuation on $k[X]$. Let $g(X) = \Sigma\, b_i X^i$. The only problem is in showing that $|fg| = |f||g|$. Clearly $|fg| \leq |f||g|$ so it suffices to show that $|fg| + \epsilon > |f||g|$ for each $\epsilon > 0$. If $|f||g| < \epsilon$, then we are done, so we may assume that $|f||g| > 0$, whence $|f| > 0$ and $|g| > 0$. Choose m and n so that

$$2^{-m}\sup(1,|f|+|g|) < \inf(|f|,|g|)$$

and

$$2^{-n}\inf(|f|,|g|) < \epsilon,$$

and use (6.1) to construct finite subsets A and B of the indices of the coefficients of f and g respectively, so that

$$\begin{aligned}
|a_i| &> |f| - 2^{-(n+m)} > 0 && \text{if } i \in A \\
|a_i| &< |f| - 2^{-n} && \text{if } i \notin A \\
|b_j| &> |g| - 2^{-(n+m)} > 0 && \text{if } j \in B \\
|b_j| &< |g| - 2^{-n} && \text{if } j \notin B.
\end{aligned}$$

Let $r = |f||g| - 2^{-n}\inf(|f|,|g|)$. If $i \in A$ and $j \in B$, then

$$\begin{aligned}
|a_i X^i b_j X^j| &> (|f| - 2^{-(n+m)})(|g| - 2^{-(n+m)}) \\
&> |f||g| - 2^{-(n+m)}(|f| + |g|) > r.
\end{aligned}$$

If $j \notin B$, then $|a_i X^i b_j X^j| < |f|(|g| - 2^{-n}) = |f||g| - 2^{-n}|f| \le r$. Similarly for $i \notin A$. Let $f = f_1 + f_2$ and $g = g_1 + g_2$, where $f_1 = \Sigma_{i\in A} a_i X^i$ and $g_1 = \Sigma_{j\in B} b_j X^j$. Then $fg = f_1g_1 + f_1g_2 + f_2g_1 + f_2g_2$ and $|f_1g_2 + f_2g_1 + f_2g_2| < r$. But considering the monomial of highest degree in f_1g_1 shows that $|f_1g_1| > r$. Hence $|fg| = |f_1g_1| > r > |f||g| - \epsilon$. □

We can now give an example of a residue class field that is not a local ring. Let k be a field with a nonarchimedean valuation, and let α and β be positive real numbers such that $\sup(\alpha,\beta) = 1$. Applying Theorem 6.2 twice we get a valuation on the rational function field $k(X,Y)$ such that $|X| = \alpha$ and $|Y| = \beta$ and $|X + Y| = \sup(\alpha,\beta) = 1$. In the residue class field $\bar{k}$ we have $X + Y$ is invertible but we cannot assert that either X or Y is invertible (see Exercise II.3.5).

The following theorem bounds the value of the remainder polynomial in the division algorithm.

6.3 THEOREM. *Let k be a field with a nonarchimedean valuation and let*

$$g(X) = b_0 + b_1X + \cdots + b_mX^m$$

and

$$\varphi(X) = a_0 + a_1X + \cdots + a_nX^n.$$

be elements of $k[X]$ such that a_n is a unit. Let $s = |\varphi(X)|/|a_nX^n|$. and $M = \max(m-n+1,0)$. Then there exist $q,r \in k[X]$ such that

$$g(X) = q(X)\varphi(X) + r(X)$$

$\deg r(X) < \deg \varphi(X)$ and $|r(X)| \le s^M|g(X)|$.

PROOF. We induct on m, which we may assume is at least n. Write

$$g_1(X) = g(X) - \varphi(X)X^{m-n}b_m/a_n.$$

Then $|\varphi(X)X^{m-n}b_m/a_n| = s|b_mX^m| \le s|g(X)|$, so $|g_1(X)| \le s|g(X)|$. By induction, $g_1(X) = q_1(X)\varphi(X) + r(X)$ where $\deg r(X) < \deg \varphi(X)$ and $|r(X)| \le$

$s^{M-1}|g_1(X)| \leq s^M|g(X)|$. Set $q(X) = q_1(X) + X^{m-n}b_m/a_n$. □

The setting for Hensel's lemma is an approximate factorization of a polynomial into approximately strongly relatively prime factors.

6.4 DEFINITION. Let k be a field with a nonarchimedean valuation, extended to $k[X]$ via (6.2). An **Henselian context** is

$$f,\varphi,\psi,h,A,B,C \in k[X],$$
$$d \in k, \quad L,M \in \mathbb{N}, \quad \epsilon \in \mathbb{Q}$$

such that

(i) $f = \varphi\psi + h$,

(ii) $A\varphi + B\psi = d + C$,

(iii) $|\psi| \leq 1$ and $|B| \leq 1$,

(iv) $\varphi = \Sigma_{i=0}^{n} a_i X^i$ and $|a_n| \neq 0$,

(v) $deg\ f$, $deg\ h$, $n \leq L$,

(vi) $deg\ A \leq M - L - 1$ and $deg\ B \leq M - 2(L-n) - 1$,

(vii) $0 < |d| \leq 1$ and $\epsilon < 1$,

(viii) $s^M|C/d| \leq \epsilon$ and $s^{2M}|h/d^2\varphi| \leq \epsilon$ where $s = |\varphi|/|a_nX^n|$.

Condition (i) says that $\varphi\psi$ approximates f to within h, and (viii) says that h is small compared to φ. Condition (viii) also says that $|C|$ is small compared to d, so (ii) says that φ and ψ are approximately strongly relatively prime.

6.5 LEMMA. *Let $f,\varphi,\psi,h,A,B,C,d,L,M,\epsilon$ be an Henselian context. Then*

(i) $|A\varphi| \leq 1$,

(ii) $deg\ \psi \leq L - n$,

(iii) $deg\ C \leq M$,

(iv) $deg\ Ch$, $deg\ Bh \leq M + n - 1$.

PROOF. From (6.4.ii) we get

$$|A\varphi| \leq \sup\{|B\psi|,|d|,|C|\} \leq 1,$$

using the bounds on $|B|,|\psi|,|d|$, and $|C|$ in (6.4). As $deg\ \varphi\psi = deg\ f-h \leq L$, and the leading coefficient a_n of φ is a unit, we have $deg\ \psi \leq L - n$. Therefore

$$deg\ C \leq \sup(deg\ A + n,\ deg\ B + L - n) \leq M - L - 1 + n$$

so $deg\ Ch \leq M + n - 1$. Finally

$$\deg Bh \le M - (L-n) + n - 1 \le M + n - 1. \quad \square$$

The heart of Hensel's lemma is the refinement of Henselian contexts.

6.6 THEOREM. *There is a function from Henselian contexts to Henselian contexts, taking*

$$f, \varphi, \psi, h, A, B, C, d, L, M, \epsilon$$

to

$$f, \varphi^*, \psi^*, h^*, A, B, C^*, d, L, M, \epsilon$$

such that

(i) $|h^*| \le \epsilon |h|$,

(ii) φ^* *has the same degree and leading coefficient as* φ,

(iii) $|\varphi^* - \varphi| \le s^M |h/d| \le s^{-M}\epsilon |d\varphi|$, so $|\varphi^*| = |\varphi|$,

(iv) $|\psi^* - \psi| \le s^M |h/d\varphi| \le s^{-M}\epsilon |d|$.

PROOF. We shall construct $\varphi^* = \varphi + \beta$ and $\psi^* = \psi + \alpha$ satisfying (ii), (iii) and (iv), and define

$$h^* = f - \varphi^*\psi^* \quad \text{and} \quad C^* = A\varphi^* + B\psi^* - d.$$

We must then show (i), that $\deg h^* \le L$, and $s^M|C^*/d| \le \epsilon$. That $|\psi^*| \le 1$ follows from (iv).

In view of (6.5.iv) we can apply (6.3) with $g = Bh/d$ and with $g = Ch/d$ obtaining

$$Bh/d = q\varphi + \beta$$
$$Ch/d = p\varphi + r$$

where $\deg \beta < n$ and $\deg r < n$ and

$$|\beta| \le s^M |Bh/d| \le s^M |h/d|$$
$$|r| \le s^M |Ch/d| \le \epsilon |h|.$$

Setting $\varphi^* = \varphi + \beta$ we have (ii) and (iii). Let

$$\alpha = Ah/d + q\psi - p.$$

Then multiplying (6.4.ii) by h/d we get

$$\alpha\varphi + \beta\psi = h + r$$

so $\deg \alpha\varphi \le L$, because $\deg \beta\psi \le n + (L-n)$ by (6.5.ii), whence $\deg \alpha \le L - n$. Also

$$|\alpha\varphi| \le \sup\{|\beta\psi|, |h|, |r|\}$$
$$\le \sup\{|\beta|, |h|, |r|\}$$
$$\le \sup\{s^M|h/d|, |h|, \epsilon|h|\} = s^M|h/d|$$

so

$$|\alpha| \leq s^M |h/d\varphi|.$$

Thus (iv) holds for $\psi^* = \psi + \alpha$.

Let $h^* = f - \varphi^*\psi^*$. Then

$$h^* = h - (h + r) - \alpha\beta = - r - \alpha\beta$$

so

$$deg\ h^* \leq sup(deg\ r,\ deg\ \alpha\beta) \leq L.$$

Also

$$\begin{aligned}|h^*| &\leq sup(|r|, |\alpha\beta|)\\ &\leq sup\{s^M|Ch/d|, (s^M|h|/|d\varphi|)(s^M|h/d|)\}\\ &\leq \epsilon|h|,\end{aligned}$$

which is (i). Define C^* by

$$A\varphi^* + B\psi^* = d + C^*$$

Then the only thing left to check is the bound (6.4.viii) on $|C^*|$. We have

$$C^* = C + A\beta + B\alpha$$

so

$$s^M|C^*| \leq s^M sup\{|C|, |A\beta|, |B\alpha|\} = s^M sup\{|C|, |A\varphi||\beta/\varphi|, |\alpha|\} = sup\{s^M|C|, s^{2M}|h|/|d\varphi|, s^{2M}|h|/|d\varphi|\} \leq \epsilon|d|. \quad \square$$

6.7 DEFINITION. Let k be a field with a nonarchimedean valuation, extended to $k[X]$ via (6.2). We say that k is **Henselian** if whenever $f, \varphi, \psi, h, A, B, C, d, L, M, \epsilon$ is an Henselian context, then there is an Henselian context $f, \hat{\varphi}, \hat{\psi}, 0, A, B, \hat{C}, d, L, M, \epsilon$ such that $\hat{\varphi}$ has the same degree and leading coefficient as φ.

Note that this definition depends on the value $\lambda = |X|$ chosen for the extension of the valuation on k to $k[X]$ via Theorem 6.2. In fact this dependence is only apparent, at least for discrete fields, as we shall show in Theorem 6.11.

6.8 HENSEL'S LEMMA. *If k is a complete nonarchimedean field, then k is Henselian.*

PROOF. Let $f, \varphi, \psi, h, A, B, C, d, L, M, \epsilon$ be an Henselian context. From (6.6) we can construct sequences $\{\varphi_i\}$, $\{\psi_i\}$, $\{h_i\}$ and $\{C_i\}$ so that $f, \varphi_i, \psi_i, h_i, A, B, C_i, d, L, M, \epsilon$ is a Henselian context, $\varphi_0 = \varphi$, $\psi_0 = \psi$, $h_0 = h$, $C_0 = C$ and

(i) $|h_{i+1}| \leq \epsilon|h_i|$,

(ii) φ_i has the same degree and leading coefficient as φ,

(iii) $|\varphi_{i+1} - \varphi_i| \leq s^M|h/d| \leq s^{-M}\epsilon|d\varphi|$,

(iv) $|\psi_{i+1} - \psi_i| \leq s^M|h/d\varphi| \leq s^{-M}\epsilon|d|$.

From (i), (iii) and (iv) we see $h_i \to 0$, and that the other sequences are Cauchy. As all degrees are bounded, and k is complete, we have $\varphi_i \to \hat{\varphi}$, $\psi_i \to \hat{\psi}$, and $C_i \to \hat{C}$, with $\hat{\varphi},\hat{\psi},\hat{C} \in k[X]$. As all the conditions on $\hat{\varphi},\hat{\psi}$ and $\hat{C}$ are equations and weak inequalities satisfied by φ_i,ψ_i and C_i, the conclusion easily follows. □

Note that if k is a discrete field, then the algebraic closure of k in $\hat{k}$ is Henselian by (VII.1.6), and discrete by (VI.1.9). Thus we have an ample supply of discrete Henselian fields: for example, the algebraic closure of the rational numbers in the p-adic numbers.

6.9 COROLLARY. *Let $f(X) = a_rX^r + \ldots + a_1X + a_0$ be a polynomial over an Henselian field k. Suppose there is $n < r$ such that $a_n \neq 0$ and*

$$|(a_jX^j/a_nX^n)^{2(2(r-n)+1)}a_mX^m| < \sup\{|a_iX^i| : 0 \leq i \leq n\}$$

whenever $j \leq n < m \leq r$. Then f has a factor of degree n in $k[X]$.

PROOF. We set up an Henselian context. Let $\varphi = a_nX^n + \cdots + a_0$ and $\psi = 1$, so $h = f - \varphi\psi = a_rX^r + \cdots + a_{n+1}X^{n+1}$. Let $A = C = 0$ and $B = d = 1$ and $L = r$ and $M = 2(r-n) + 1$. The hypothesis says that there is $\epsilon < 1$ so that $s^{2M}|h| \leq \epsilon|\varphi|$, so $f,\varphi,\psi,h,A,B,C,d,L,M,\epsilon$ is an Henselian context. As k is Henselian, we get the desired factor. □

6.10 COROLLARY. *Let $f = a_rX^r + \cdots + a_1X + a_0$ be a polynomial over an Henselian field k. If*

$$0 < |a_rX^r| < |f| = \sup\{|a_iX^i| : 0 < i < r\},$$

then f is reducible over k.

PROOF. By (6.1) we can construct a finite subset A of $\{0,\ldots,r\}$ so that

$$|a_iX^i| > |f| - 2^{-(m+t)} \qquad \text{if } i \in A$$

and

$$|a_iX^i| < |f| - 2^{-t} \qquad \text{if } i \notin A.$$

Furthermore we can choose m and t so that $N = 4r + 2 < 2^m$ and $2^{-t} < |f|$. Then

$$(|f| - 2^{-(m+t)})^N > (|f| - 2^{-t}/N)^N$$
$$> |f|^N - 2^{-t}|f|^{N-1} = |f|^{N-1}(|f| - 2^{-t})$$

so

$$\left[\frac{|f|}{|f|-2^{-(m+t)}}\right]^N (|f| - 2^{-t}) < |f|.$$

We can also require $|a_rX^r| + 2^{-(m+t)} < |f|$, so $r \notin A$. Let n be the biggest integer in A. As $|f| = \sup\{|a_iX^i| : 0 < i < r\}$ we have $n > 0$. The hypotheses of Corollary 6.9 are met, so f has a factor of degree n in $k[X]$. □

6.11 THEOREM. *Let k be a discrete Henselian field with a trivial or nontrivial valuation, and f a separable polynomial in $k[X]$. Then there is $\delta > 0$ such that if $|f - \varphi\psi| < \delta$, and $\deg f = \deg \varphi\psi$, then f has a factor in $k[X]$ of the same degree as φ.*

PROOF. The trivial case is trivial so we may assume that there exists $e \in k$ such that $0 < |e| < 1$. For any nonzero polynomial p in $k[X]$, let $L(p)$ be the value of the leading monomial in p. As the map taking the polynomial $F(X)$ to $F(X/e^m)$ is an automorphism of $k[X]$ that is uniformly bicontinuous on polynomials of bounded degree, we may assume that $|f| = L(f)$. Using the Euclidean algorithm, and multiplying by an appropriate power of e, we can find a and b in $k[X]$ so that $|a| \leq 1$ and $|b| \leq 1$ and

$$0 \neq af + bf' = d \in k.$$

with $|d| \leq 1$, and $|af| < |e|^2$ and $|bf| < |e|^2|X|$. Let

$$\delta = |d|^2 \inf(|X|,|f|)$$

Now suppose $|f - \varphi\psi| < \delta$, and $\deg f = \deg \varphi\psi$. As $\delta < |f| = L(f)$ we have $|f| = |\varphi\psi|$ and $L(f) = L(\varphi\psi)$, so $|\varphi| = L(\varphi)$, that is, $s = 1$ in (6.4). Set $h = f - \varphi\psi$. By multiplying ψ by an appropriate power of e, and multiplying φ by the same power of e^{-1}, we may assume that $|e|^2 < |\psi| < 1$, so $|f| = |\varphi\psi| < |\varphi| < |f|/|e|^2$. Then $|a\varphi| < |af|/|e|^2 < 1$ and $|b\varphi| < |bf|/|e|^2 < |X|$. There is $\epsilon < 1$ so that $|h| \leq \epsilon\delta \leq \epsilon|d^2|\inf(|X|,f|) \leq \epsilon|d^2\varphi|$. Note that if $F \in k[X]$, then $|F'| \leq |F|/|X|$, so $|f' - \varphi'\psi - \varphi\psi'| \leq \epsilon|d|^2$ and $|b\varphi'| < 1$. Thus

$$\epsilon|d|^2 \geq |ah + b(f' - \varphi'\psi - \varphi\psi')| = |d - [(a\varphi + b\varphi')\psi + b\psi'\varphi]|,$$

so setting

$$A = b\psi',$$
$$B = a\varphi + b\varphi',$$
$$C = A\varphi + B\psi - d,$$
$$L = max(deg\ f,\ deg\ h,\ n),$$

and taking M as large as necessary ($s = 1$), we get an Henselian context. As k is Henselian, the desired factor can be constructed. □

Is this theorem true without the hypothesis that the valuation on k is trivial or nontrivial?

Recall that if k is a discrete valued field, then we define the **Henselization** $\tilde{k}$ of k to be the separable closure of k in $\hat{k}$. This terminology is justified by the fact that a discrete field with a nonarchimedean valuation is Henselian exactly when it is separably closed in its completion.

6.12 LEMMA. *Let $\varphi(X) = a_nX^n + \cdots + a_0$ be a polynomial over a nonarchimedean valued field k. Suppose $|a_n| \neq 0$ and set $s = |\varphi|/|a_nX^n|$. Let $C \in k[X]$ and $d \in k$ be such that $s^M|C| < |d|$ and $\deg C \leq M$. Then φ and $d + C$ are relatively prime.*

PROOF. Let $\theta = \theta_0 + \theta_1X + \cdots + \theta_rX^r$ be a common factor of φ and $d + C$. Let $d + C = \theta\sigma$ where $\sigma = \sigma_0 + \sigma_1X + \cdots + \sigma_tX^t$. As $|C| < |d|$ we have $|\theta_0\sigma_0| = |d| = |d + C| = |\theta||\sigma|$, so $|\theta| = |\theta_0|$ and we may assume that $\theta_0 = 1$. Suppose, by way of contradiction, that $r > 0$ and $\theta_r \neq 0$. Because θ is a factor of φ, and $|\varphi/\theta| = |\varphi|$, we have $|\theta_rX^r| \geq |a_nX^n|/|\varphi| = 1/s$. Thus $|\sigma_tX^t| \leq s|C|$. Similarly $|\sigma_{t-1}X^{t-1}| \leq s^2|C|$ and so on until $|\sigma_0| \leq s^{t+1}|C|$. Thus $t + 1 > M \geq \deg C$ whence $\sigma_t = 0$ and we are done by induction on t, getting a contradiction when $t = 0$. □

6.13 THEOREM. *Let k be a discrete field with a nonarchimedean valuation. Then k is Henselian if and only if k is separably closed in its completion.*

PROOF. Suppose k is separably closed in its completion $\hat{k}$. Since $\hat{k}$ is Henselian we can construct the required $\hat{\varphi}$ and $\hat{\psi}$ with coefficients in $\hat{k}$. The leading coefficient of $\hat{\varphi}$, and hence the leading coefficient of $\hat{\psi}$, is in k. By (VI.1.6) all the coefficients of $\hat{\varphi}$ and $\hat{\psi}$ are algebraic over k. Lemma 6.12 says that $\hat{\varphi}$ and $\hat{\psi}$ are relatively prime, hence strongly relatively prime as their coefficients lie in a discrete subfield. As k

is separably closed in $\hat{k}$, the coefficients of $\hat{\varphi}$ and $\hat{\psi}$ in fact lie in k by (VII.1.7).

Now suppose k is Henselian and θ in $\hat{k}$ is separable over k. Then θ satisfies a separable polynomial f in $k[X]$. If $deg\ f = 1$, then $\theta \in k$ and we are done. Otherwise we can find α in k so that $|f(\alpha)|$ is as small as we please. So either f has a root in k, and we get a polynomial of smaller degree satisfied by θ, or the valuation on k is nontrivial. But $f(\alpha) = f - (X - \alpha)\psi$ for some ψ in $k[X]$, so by Theorem 6.11 we can find a linear factor of f in $k[X]$, and so get a polynomial of smaller degree satisfied by θ. □

As remarked earlier, this theorem shows that the definition of Henselian for discrete fields is independent of the choice of $|X|$.

EXERCISES

1. Let $f \in \mathbb{Z}[X]$ and let $\bar{f}$ be the image of f in $\mathbb{Z}_p[X]$. Show that if $\bar{f}$ has a simple root in $\mathbb{Z}_p$, then there exists x in the p-adic completion of $\mathbb{Q}$ such that $f(x) = 0$.
2. **Hensel's lemma.** Let k be an Henselian field and f a monic polynomial in $k[X]$ all of whose coefficients have value at most 1. If the image $\bar{f}$ of f in $\bar{k}[X]$ is a product of strongly relatively prime monic polynomials λ and μ, then f is a product of monic polynomials φ and ψ such that $\bar{\varphi} = \lambda$ and $\bar{\psi} = \mu$.

7. EXTENSIONS OF VALUATIONS

We turn to the question of extending valuations from k to $k(\theta)$ where θ is algebraic over k. In general we must restrict ourselves to the case when $k(\theta)$ is finite dimensional over k, that is, when θ satisfies an irreducible polynomial (VI.1.13). The following example illustrates this for both the archimedean and the nonarchimedean cases.

EXAMPLE. Equip the field of rational numbers $\mathbb{Q}$ with either absolute value or the 7-adic valuation, and let $\sqrt{2}$ be a fixed root of 2 in the completion $\hat{\mathbb{Q}}$ of $\mathbb{Q}$. Let a be a binary sequence with at most one 1. The field

$$k = \cup\, \mathbb{Q}(a_n\sqrt{2}) \subseteq \hat{\mathbb{Q}}$$

is discrete, countable, and locally precompact. Define a field $E = k(\theta)$ with θ algebraic over k by

$$E = \{s + t\theta : s,t \in k\}$$

where $\theta^2 = 2$ and equality is defined by setting $s + t\theta = 0$ if $s = t = 0$, or if $a_m = 1$ and $s = (-1)^m t\sqrt{2}$. Note that E is a discrete field. But if we could extend the valuation to E, then in $\hat{E}$ either $|\theta - \sqrt{2}| \neq 0$ or $|\theta + \sqrt{2}| \neq 0$, so either $a_m = 0$ for all odd m, or $a_m = 0$ for all even m, giving us LLPO. □

7.1 LEMMA. *Let K be a nonarchimedean valued field and k a discrete Henselian subfield of K on which the valuation is trivial or nontrivial. Let $\theta \in K$ satisfy a separable irreducible polynomial of degree n over k. Then there is $\epsilon > 0$ such that $|g(\theta)| \geq \epsilon|g|$ whenever $g \in k[X]$ is of degree less than n. Thus the elements $1,\theta,\ldots,\theta^{n-1}$ form a metric basis for $k(\theta)$ over k.*

PROOF. We construct a sequence $1 = \epsilon_0 \geq \epsilon_1 \geq \cdots \geq \epsilon_{n-1} > 0$ so that $|g(\theta)| \geq \epsilon_m|g|$ if $\deg g = m$. Let $g(X) = g_0 + g_1X + \cdots + g_mX^m$. As $|g| \leq \sup(1,|X|^n)\sup\{|g_i| : 0 \leq i \leq m\}$, we may assume that $|X| = 1$. We may also assume that g is monic, so $|g| \geq 1$.

Let f be the minimal polynomial of θ and suppose we have constructed ϵ_{m-1}. Set $\mu = \sup(1,|\theta|^n)$ and set

$$\epsilon_m = \delta(\epsilon_{m-1}/2\mu)^{n-m+2}|f|^{-1}$$

where $\delta \leq 1$ is gotten from Theorem 6.11. Write $f = qg + r$ where $\deg r < m$. By (6.3) we have $|r| \leq |g|^{n-m+1}|f|$ so

$$|q| = |f - r|/|g| \leq |f||g|^{n-m}$$

whence

$$|q(\theta)| \leq |q|\mu \leq |f||g|^{n-m}\mu.$$

Either $|g| > \mu/\epsilon_{m-1}$ or $|g| < 2\mu/\epsilon_{m-1}$. If $|g| > \mu/\epsilon_{m-1} \geq 1$, then, by induction,

$$|g(\theta) - \theta^m| \geq \epsilon_{m-1}|g(X) - X^m| = \epsilon_{m-1}|g| > \mu \geq |\theta^m|.$$

As the valuation is nonarchimedean we have

$$|g(\theta)| = |g(\theta) - \theta^m| \geq \epsilon_{m-1}|g| \geq \epsilon_m|g|.$$

Otherwise $|g| < 2\mu/\epsilon_{m-1}$ and

$$|g(\theta)| = |r(\theta)|/|q(\theta)| \geq \epsilon_{m-1}|r|/|f||g|^{n-m}\mu$$

since $deg\ r < m$. As f is irreducible, Theorem 6.11 says $|r| \geq \delta$ so

$$\epsilon_{m-1}|r|/|f||g|^{n-m}\mu > \epsilon_{m-1}\delta|g|/|f|(2\mu/\epsilon_{m-1})^{n-m+1}\mu = 2\epsilon_m|g| > \epsilon_m|g|. \quad \square$$

The hypothesis that the valuation is trivial or nontrivial in this lemma was needed in order to apply (6.11). Is the lemma true without that hypothesis?

7.2 LEMMA. *Let k be a discrete Henselian field and K a finite-dimensional separable extension of k. Then any two valuations on K extending the valuation on k are equal.*

PROOF. Let $K = k(\theta)$ where θ is separable (VI.4.3) of degree n and let f be the minimal polynomial of θ. Let $|\ |_1$ and $|\ |_2$ be valuations on K extending the one on k, and suppose $|b|_1 > |b|_2$ for some b in K. As $f(\theta) = 0$, if we take $|X| = 1$, then $|\theta|_i \leq |f(X)|$. Each element a in K can be written uniquely as $a = g_a(\theta)$ where $g \in k[X]$ has degree less than n. If the valuation on k is trivial or nontrivial, then by Lemma 7.1, for some $\epsilon > 0$ we have

$$|f(X)|^n|g_a(X)| \geq |a|_i \geq \epsilon|g_a(X)|.$$

But there is m such that $|b^m|_1 > (|f(X)|^n/\epsilon)|b^m|_2$, which is impossible. So if $|b|_1 > |b|_2$ then it is impossible that the valuation be trivial or nontrivial; but a valuation is trivial if and only if it is not nontrivial, so $|b|_1 > |b|_2|$ is impossible, whence $|b|_1 \leq |b|_2$ for each b in K. $\square$

Let E be a finite-dimensional extension of a discrete field k. The **field norm $N_{E/k}$ from E to k** is defined on elements x of E by setting $N_{E/k}(x)$ equal to the determinant of the linear transformation of E induced by multiplication by x.

7.3 THEOREM. *Let k be a discrete Henselian field and E a finite-dimensional extension of k. Then the valuation on k extends uniquely to E, and $V_E^\ell \subseteq V_k$ for some positive integer ℓ. Thus if the value group of k is discrete, so is the value group of E.*

PROOF. Let F be the separable closure of k in E. Since $E^m \subseteq F$ for some positive integer m (VI.4.3), any valuation on F extends uniquely to E so we may assume that E is separable. Lemma 7.2 takes care of uniqueness. To construct a valuation on E, for θ in E set

$$|\theta| = |N(\theta)|^{1/n}$$

where N is the field norm from E to k and n is the dimension of E over k. The only problem in showing that this defines a (general) valuation, is verifying that $|N(1 + \theta)| \leq 1$ if $|N(\theta)| \leq 1$. Let

$$f(X) = X^m + a_{m-1}X^{m-1} + \ldots + a_0$$

be the minimal polynomial of θ. Since $N(\theta)$ is a power of $|N_{k(\theta)/k}(\theta)|$, we have $|a_0| = |N_{k(\theta)/k}(\theta)| \leq 1$. By Corollary 6.10 we have $|a_i| \leq 1$ for all i, as $\sup |a_i| > 1$ implies f is reducible, which is impossible. But $f(X-1)$ is the minimal polynomial of $1 + \theta$ and its constant term is $\pm(1 + \Sigma \pm a_i)$. As the valuation is nonarchimedean $|1 + \Sigma \pm a_i| \leq 1$. Since $N(1 + \theta)$ is a power of this term, we have $|N(1 + \theta)| \leq 1$.

Let $\ell = mn$. □

7.4 THEOREM. *Let k be a discrete Henselian field with a trivial or nontrivial valuation, and E a finite dimensional separable extension of k. Then E is Henselian.*

PROOF. By Theorem 6.13 it suffices to show that E is separably closed in its completion. Let $E = k(\theta)$. As k is separably closed in $\hat{k}$, it follows by (VIII.2.2) that E is separably closed in $\hat{k}(\theta)$. Lemma 7.1 shows that the norm on $k(\theta)$ given by the valuation is equivalent to the supremum norm on k^n, so $\hat{k}(\theta) = \hat{E}$. □

By Corollary 3.5, the Henselization of a discrete field with a pseudofactorial valuation is separably factorial. Thus there are plenty of situations in which the following theorem applies.

7.5 THEOREM. *Let k be a nontrivial discrete valued field that is either real, complex, or nonarchimedean. Let $\tilde{k}$ denote the Henselization of k, and let $E = k(\theta)$ be a finite-dimensional separable extension of k. If the minimal polynomial of θ over k is a product $g_1 g_2 \cdots g_s$ of irreducible factors in $\tilde{k}[X]$, then there exist distinct valuations $|\ |_1, \ldots, |\ |_s$ on E, extending the valuation $|\ |$ on k, such that if $\tilde{E}_i$ is the Henselization of E under $|\ |_i$, then*

(i) *g_i is the minimal polynomial of θ over $\tilde{k} \subseteq \tilde{E}_i$.*

(ii) *Every valuation on E extending $|\ |$ is equal to some $|\ |_i$.*

(iii) *$n_i = [\tilde{E}_i : \tilde{k}]$ is the degree of g_i, and $\Sigma n_i = n$.*

PROOF. Let α be a root of g_i in a discrete extension field of $\tilde{k}$

(possible since g_i is irreducible over $\tilde{k}$). By (7.3), in the nonarchimedean case, and by (5.2) and (5.8) in the archimedean case, we can extend $|\ |$ uniquely to $\tilde{k}(\alpha)$. As α satisfies the minimal polynomial of θ over k, there is a monomorphism $\varphi : E \to \tilde{k}(\alpha)$, that is the identity on k and takes θ to α. For $x \in E$, set $|x|_i = |\varphi(x)|$. As $\tilde{k}(\alpha)$ is the Henselization of $k(\alpha)$ by (7.4) and (5.8), we can identify $\tilde{E}_i$ with $\tilde{k}(\alpha)$, so (i) and (iii) are clear.

To prove (ii), suppose we have a valuation on E extending the valuation on k, and let $\tilde{E}$ be the Henselization of E with respect to that valuation. Then there is i such that $g_i(\theta) = 0$, so the valuation on $\tilde{E}$ must be $|\ |_i$ by (7.2) and (5.2).

We want to show that $|\ |_i \neq |\ |_j$ if $i \neq j$. As $g_i g_j$ is separable, we can find polynomials s and t such that $sg_i + tg_j = 1$. Then $g_i(\theta) = 0$ in $\tilde{E}_i$ and $g_i(\theta) \neq 0$ in $\tilde{E}_j$. By choosing a sufficiently close approximation $g_i^* \in k[X]$ to g_i, we can make $|g_i^*(\theta)|_i$ as small as we wish and keep $|g_i^*(\theta)|_j$ bounded away from 0, so $|\ |_i \neq |\ |_j$. □

EXERCISES

1. Show that if θ is in a field K with a nonarchimedean valuation, and $f \in K[X]$ has θ as a root, then $|\theta| \leq |f|$, where we take $|X| = 1$.
2. Find all archimedean valuations on the following fields

   (i) $\mathbb{Q}[X]/(X^2 - 2)$

   (ii) $\mathbb{Q}[X]/(X^2 + 1)$

   (iii) $\mathbb{Q}[X]/(X^3 - 2)$

   Find all extensions of the 5-adic valuation on $\mathbb{Q}$ to the same fields.
3. Let K be a finite-dimensional extension field of $\mathbb{Q}$. Show that an element x of K is integral over $\mathbb{Z}$ if and only if $|x| \leq 1$ for every nonarchimedean valuation on K.

8. e AND f

Let k be a discrete field with a discrete valuation, and E a finite dimensional separable extension of k with a valuation extending that of k. If the quotient V_E/V_k of the value groups of E and k is finite, define the **ramification index** $e = e(E/k)$ to be the number of elements in V_E/V_k.

Similarly if the residue class field $\overline{E}$ is finite dimensional over $\overline{k}$, define the **residue class degree** to be the dimension $f = f(E/k)$ of $\overline{E}$ over $\overline{k}$. As passing to the Henselizations $\tilde{k} \subseteq \tilde{E}$ leaves the value groups and residue class fields unchanged, we may assume that k is Henselian.

The classical method for constructing e is to choose π in k such that $|\pi| < 1$ generates the value group V_k of k. Then (Theorem 7.3) V_E is a subgroup of the cyclic group generated by $|\pi|^{1/\ell}$, hence is cyclic and contains V_k as a subgroup of finite index e. As we cannot establish that a nontrivial subgroup of a cyclic group is cyclic, our construction of e must be more elaborate. The problem with computing f is that finitely generated field extensions need not be finite dimensional. We need to impose Seidenberg's condition P on the residue class field.

8.1 THEOREM. *Let k be a discrete Henselian field with a discrete valuation. Let E be an n-dimensional separable extension of k, and suppose the residue class field $\overline{k}$ satisfies Seidenberg's condition P. Then the quotient group V_E/V_k has finite cardinality e, the field $\overline{E}$ has finite dimension f over $\overline{k}$, and $n = ef$. In particular, the valuation on E is discrete, and $\overline{E}$ satisfies condition P.*

PROOF. By Theorem 7.3 the group V_E is discrete so $\overline{E}$ is discrete. Also V_E/V_k is discrete because V_k is cyclic and $V_E^{\ell} \subseteq V_k$.

We shall construct a finite subset S of E that maps one-to-one onto V_E/V_k, and a finite subset W of $\{w \in E : |w| = 1\}$ that maps one-to-one onto a basis for $\overline{E}$ over $\overline{k}$, so that $SW = \{sw : s \in S \text{ and } w \in W\}$ is a basis for E over k. We build these sets up inductively starting with $S = W = \{1\}$.

At each stage of our construction we will have a finite set S of nonzero elements of E containing 1 and mapping one-to-one into V_E/V_k, and a finite subset W of $\{w \in E : |w| = 1\}$ that maps to a $\overline{k}$-linearly independent family $\overline{W}$ in $\overline{E}$. Moreover we will require that $\overline{kW}$, the $\overline{k}$-vector subspace of $\overline{E}$ spanned by $\overline{W}$, be a field.

If $b_w \in k$ for $w \in W$, then $|\Sigma_{w \in W}\, b_w w| = max\, |b_w|$ for if $|\Sigma\, b_w w| < |b_{w'}| = max\, |b_w|$, then, dividing by $b_{w'}$, we get a dependence relation among $\overline{W}$. So if $a_{sw} \in k$, for $s \in S$ and $w \in W$, then

$$|\Sigma_{S\times W}\, a_{sw} sw| = max_W\, |\Sigma_S\, a_{sw} s| = max_{S\times W}\, |a_{sw} s|$$

as the values of the nonzero $a_{sw}s$ are distinct for fixed w. So SW is linearly independent over k. Let kSW be the k-vector subspace of E spanned by SW. If $|a_{sw}s| = 1$, then $s = 1$, so if $x \in k\text{SW}$ and $|x| \leq 1$, then the image $\bar{x}$ of x in $\bar{E}$ is in $\overline{k\text{W}}$.

If $E = k\text{SW}$, then we are done as S maps onto V_E/V_k and $\bar{E} = \overline{k\text{W}}$. Otherwise there is α in $E \backslash k\text{SW}$. As k is Henselian, E has a metric basis over k by Theorem 7.1, so $\{\alpha\} \cup \text{SW}$ is metrically independent, whence α is bounded away from kSW by some positive distance. Let π in k be such that $|\pi| < 1$ generates V_k. We may assume that if $s \in S\backslash\{1\}$, then $|\pi| < |s| < 1$. Let r be the maximum of the values of elements of $S\backslash\{1\}$, with $r = |\pi|$ if $S = \{1\}$. Given b in kSW, we shall construct one of the following:

(i) $b' \in k\text{SW}$ such that $|\alpha - b'| \leq r|\alpha - b|$.

(ii) $\beta \in E$ so that $|\beta|/|s|$ is not in V_k for any s in S.

(iii) A field extension $K \subseteq E$ of k of dimension $m > 1$ such that $\bar{K}$ is m-dimensional over $\bar{k}$ (so $V_K = V_k$).

(iv) A proper finite-dimensional field extension of $\overline{k\text{W}}$ in $\bar{E}$.

As V_E/V_k is discrete, either $\beta = \alpha - b$ satisfies (ii) or we can find t in k and s in S such that $\theta = (\alpha - b)/st$ has value 1. In the latter case we shall show that either $\bar{\theta} \in \overline{k\text{W}}$ or (iii) holds or (iv) holds. First let's see what to do if $\bar{\theta} \in \overline{k\text{W}}$. If $\bar{\theta} \in \overline{k\text{W}}$, then there is u in kW such that $\bar{\theta} = \bar{u}$ so $|\theta - u| < 1$. Either $|\theta - u| \leq r$ or $\beta = \theta - u$ satisfies (ii). If $|\theta - u| \leq r$ we set $b' = b + stu$, and (i) holds since $\alpha - b = st\theta$ and $|\theta| = 1$.

We return to the problem of whether $\bar{\theta} \in \overline{k\text{W}}$. As V_k is discrete, and θ is algebraic over k, we can divide the minimal polynomial of θ by its largest coefficient to show that $\bar{\theta}$ is algebraic over $\bar{k}$. By (VI.6.3) there is a monic polynomial g with coefficients in k, each of value at most 1, so that $\bar{g}$ is separable and $\bar{g}(\bar{\theta}^q) = 0$ where q is 1 or a power of the finite characteristic of $\bar{k}$. Since E is Henselian (Theorem 7.4), there is ω in E such that $g(\omega) = 0$ and $\bar{\omega} = \bar{\theta}^q$. The minimal polynomial of ω, which exists as E is finite dimensional over k, divides g and, by Gauss's lemma, all its coefficients have value at most 1. Thus we may assume that g is irreducible. As E is Henselian and $\bar{g}$ is separable, $\bar{g}$ is irreducible. Thus $\bar{k}(\bar{\omega})$ is finite dimensional over $\bar{k}$. If $\deg g > 1$, set $K = k(\omega)$ and we get (iii). If $\deg g = 1$, then $\bar{\theta}$ is purely inseparable over $\bar{k}$, and hence over $\overline{k\text{W}}$. By condition P we then have that $\overline{k\text{W}}(\bar{\theta})$ is finite dimensional over $\overline{k\text{W}}$. If this dimension is 1, then $\bar{\theta} \in \overline{k\text{W}}$. Otherwise $\overline{k\text{W}}(\bar{\theta})$ is the field needed in (iv).

If (ii) occurs we can increase S; if (iv) occurs we can increase W. As α is bounded away from kSW, situation (i) can occur only finitely many times. Finally, suppose (iii) occurs. Then $V_K = V_k$ and K is Henselian (Theorem 7.4). Moreover $\overline{K}$ satisfies condition P since $\overline{K}/\overline{k}$ is finite dimensional (VII.3.1.ii). We are done by induction on n. □

The requirement in Theorem 8.1 that $\overline{k}$ satisfy condition P cannot be removed, at least not entirely. To show this we first prove the following theorem.

8.2 THEOREM. *Let K be a discrete field of characteristic p. Then there exists a discrete Henselian field k with a discrete valuation and residue class field $\overline{k} = K$ such that for each y in $\overline{k}$ there is a finite-dimensional separable extension E of k with $y \in \overline{E}^p$.*

PROOF. If k is a field of characteristic p, then by $k^{1/p}$ we mean a field containing k such that $(k^{1/p})^p = k$. That such a field exists is obvious upon contemplating the isomorphism $k \cong k^p \subseteq k$. Let k be the Henselization of the rational function field $K(T)$ under the T-adic valuation. If $y = 0$, let $E = k$. If $y \neq 0$, consider the separable polynomial $f(X) = X^p + TX + y$ in $k[X]$. That $f(X)$ is irreducible (over $k^{1/p}$) is easily seen by substituting $X = Z - \lambda$, where λ is the pth root of y in $k^{1/p}$, and applying Eisenstein's criterion. Thus we can construct $E = k(\theta)$ where $f(\theta) = 0$. Clearly $|\theta| = |y|^{1/p} = 1$ and $y = \overline{\theta}^p$. □

A discrete field $\overline{k}$ of characteristic p satisfies condition P if and only if K^p is detachable from K for any finite-dimensional extension field K of $\overline{k}$ (VII.3.1.ii), that is, for each y in K either $y \in K^p$ or $y \notin K^p$. Thus the following corollary shows that the $\overline{k}$ in Theorem 8.1 must at least satisfy a weak form of condition P.

8.3 COROLLARY. *Let $\overline{k}$ be a discrete field. Suppose that whenever k is a discrete Henselian field with residue class field $\overline{k}$, and E is a finite-dimensional separable extension field of k, that the residue class field $\overline{E}$ is finite dimensional over $\overline{k}$. Then $\overline{k}^p$ is detachable from $\overline{k}$.*

PROOF. Let $y \in \overline{k}$ and apply Theorem 8.2 to construct k and E with $y \in \overline{E}^p$. Then $\overline{E}$ is finite dimensional over $\overline{k}$ so $\overline{E}^p$ is finite dimensional over $\overline{k}^p$. Thus $y \in \overline{E}^p$ is either in $\overline{k}^p$ or it is not. □

The requirement in Theorem 8.1 that the extension be separable cannot be removed. Referring to Example 3.6 let $k = \mathbb{Z}_2(Y^2,\eta^2)$ and $E = \mathbb{Z}_2(Y^2,\eta)$ with the valuation inherited from the field of formal power series in Y. Then the value group of E equals the value group of k if and only if $a_n = 0$ for all n. Henselizing doesn't change the situation.

The p-adic valuations on $\mathbb{Q}$ satisfy the hypotheses of the following theorem.

8.4 THEOREM. *Let k be a discrete field with a valuation such that*

(i) *The valuation is discrete,*

(ii) *The Henselization is separably factorial (for example, if k is pseudofactorial),*

(iii) *The residue class field satisfies condition P.*

If E is a finite separable extension of k, then any valuation on E extending the one on k satisfies (i), (ii) *and* (iii).

PROOF. Let $\tilde{E}$ be the Henselization of E, and $\tilde{k}$ the Henselization of k. We may assume that $\tilde{k} \subseteq \tilde{E}$. Choose α in E such that $E = k(\alpha)$. As $\tilde{k}$ is separably factorial, $\tilde{k}(\alpha)/\tilde{k}$ is finite dimensional so $\tilde{k}(\alpha)$ is Henselian by Theorem 7.4. Hence $\tilde{k}(\alpha) = \tilde{E}$. Thus $\tilde{E}$ is separably factorial by (VII.2.3), so (ii) holds. As the value groups and residue class fields of E and $\tilde{E}$ are the same, (i) and (iii) hold by Theorem 8.1. □

8.5 THEOREM. *Let k be a field with a valuation satisfying the hypotheses of Theorem* 8.4. *Let E be an n-dimensional separable extension of k. Then there are a finite number s of valuations on E extending the valuation on k. Let $E_1,\ldots,E_s$ denote the field E equipped with these valuations. Then $e(E_i/k)$ and $f(E_i/k)$ are defined and*

$$n = \Sigma\, e(E_i/k)f(E_i/k).$$

PROOF. Theorem 7.5 constructs the finite number of valuations on E. We have $e(\tilde{E}_i/\tilde{k})f(\tilde{E}_i/\tilde{k}) = n_i = [\tilde{E}_i:\tilde{k}]$ by Theorem 8.1, and $n = \Sigma\, n_i$ by Theorem 7.5. But Henselization does not change the value group or the residue class field. □

EXERCISES

1. Calculate e and f for each extension of the 5-adic valuation on $\mathbb{Q}$ to $\mathbb{Q}[X]/(X^3 - 2)$. Verify Theorem 8.5 for this case.

NOTES

For a more complete treatement of the constructive theory of real numbers and metric spaces the reader is referred to [Bishop 1967] and [Bishop-Bridges 1985].

The theory of valuations plays a central role in the development of the theory of fields from a constructive point of view. The study of fields with valuations provides an interesting mix of discrete, purely algebraic constructions combined with analytic constructions such as the completion.

Although it is commonly felt that algebraic number theory is essentially constructive in its classical form, even those authors who pay particular attention to the constructive aspects of the theory employ highly nonconstructive techniques which nullify their efforts. In [Borevich-Schafarevich 1966], for example, it is assumed that every polynomial can be factored into a product of irreducible polynomials (every field is factorial) and that given a nonempty subset of the positive integers you can find its least element.

The definition of a general rank-one valuation on a field was adopted by Staples (1971) in his constructive treatment of valuations, and the proof of Theorem 1.6 is in that paper. The problem here is that we cannot decide whether $vx \leq vy$ or $vy \leq vx$, or whether either is zero. Our proof of Theorem 1.1 is taken from [Weiss 1963, Proposition 1-1-8]. The last two sentences of our proof of Theorem 1.2 follow [Weiss 1963, Theorem 1-1-4]

The notion of a pseudofactorial field is a purely constructive one having no classical counterpart (all fields with valuations are pseudofactorial from a classical point of view). However it provides just the information we need either to construct a root of a polynomial, or bound the polynomial away from zero, in the completion of the field (Corollary 3.2). The proof of Corollary 3.3 using winding number is from [Brouwer-de Loor 1924]. Theorem 3.4 shows that there are lots of pseudofactorial fields. Corollary 3.5 shows how to factor separable polynomials over the algebraic or separable closure of a pseudofactorial field in its completion. So, for example, we can factor polynomials over the algebraic p-adic numbers.

Our proof of the Gelfand-Tornheim theorem follows [Artin 1967, page 24]

The somewhat elaborate Hensel's lemma of Section 4 is a modification of

[Artin 1967]. The hypotheses of the lemma are relaxed slightly by not requiring the leading term of φ to have the same value as φ (not requiring $s = 1$), to cover situations where we cannot determine which coefficients of a polynomial have maximal value, and we must worry about some constructive peculiarities. The definition of *Henselian*, when $s = 1$, is the conclusion of Hensel's lemma as given in [Artin 1967, Theorem 5]. The main theorem is Theorem 6.13 that a discrete field with a nonarchimedean valuation satisfies the conclusion of Hensel's lemma precisely when it is separably closed in its completion.

The construction of the valuation in (7.3) follows [O'Meara 1963, Theorem 14:1].

Chapter XIII. Dedekind Domains

1. DEDEKIND SETS OF VALUATIONS.

If S is a set of valuations on a Heyting field k, then we will denote a member of S by $|\ |$ with a subscript. Instead of writing $|\ |_p \in S$, we will often write $p \in S$.

For each prime number p we get a discrete valuation on the field $\mathbb{Q}$ of rational numbers by defining $|p|_p = 1/p$, and $|q|_p = 1$ if q is a prime different from p, and extending multiplicatively. This family of valuations forms a Dedekind set in the sense of the following definition.

1.1 DEFINITION. A nonempty discrete set S of nontrivial discrete valuations on a Heyting field k is a **Dedekind set** if

(i) For each $x \in k$ there is a finite subset T of S so that $|x|_p \leq 1$ for each $p \in S \backslash T$.

(ii) If q and q' are distinct valuations of S, and $\epsilon > 0$, then there exists $x \in k$ with $|x|_p \leq 1$ for each $p \in S$, such that $|x - 1|_q < \epsilon$ and $|x|_{q'} < \epsilon$. Hence distinct valuations are inequivalent.

Note that a nonempty detachable subset of a Dedekind set is a Dedekind set.

Let S be a Dedekind set of valuations on a Heyting field k. If $p \in S$, then, because p is nonarchimedean, the set $R(p) = \{x \in k : |x|_p \leq 1\}$ is a ring, which is local as p is discrete. We call $R(p)$ the **local ring at p.** The elements of the ring $\cap_{p \in S} R(p)$ are called the **integers at S.** A ring a **Dedekind domain** if it is the ring of integers at a Dedekind set of valuations on a Heyting field.

1.2 THEOREM. *The ring of integers at a Dedekind set S of valuations on a Heyting field k, is a detachable subset of k.*

PROOF. Let $x \in k$. There is a finite subset T of S so that $|x|_p \leq 1$ for each $p \in S \backslash T$. Then x is an integer at S if and only if $|x|_p \leq 1$ for each p in the finite set T of discrete valuations. □

Given inequivalent valuations $|\ |_1,\ldots,|\ |_n$ on k, and elements $x_1,\ldots,x_n$ in k, we will be interested in constructing an element of k that simultaneously approximates each x_i with respect to the valuation $|\ |_i$. As a first step, we have the following lemma.

1.3 LEMMA. *Let* $|\ |_1,\ldots,|\ |_n$ *be inequivalent nontrivial valuations on a Heyting field* k. *Then there is* $x \in k$ *such that* $|x|_1 > 1$, *and* $|x|_i < 1$ *for* $i \neq 1$.

PROOF. If $n = 1$, then, as $|\ |_1$ is nontrivial, there is $x \in k$ such that of $|x|_1 > 1$. If $n = 2$, then since $|\ |_1$ and $|\ |_2$ are nontrivial inequivalent valuations, there exists $x \in k$ with $|x|_1 < 1$ and $|x|_2 > 1$. For $n > 2$ we proceed by induction, so we may assume we have $y \in k$ such that $|y|_1 > 1$, and $|y|_i < 1$ for $2 \leq i < n$. Choose $z \in k$, using the case $n = 2$, with $|z|_1 > 1$ and $|z|_n < 1$, and choose m so large that $|y^m|_i|z|_i < 1$ for $2 \leq i < n$. As $|z|_n^{-1} > 1$, it follows that either $|y^m|_n > 1$ or $|y^m|_n < |z|_n^{-1}$. In the latter case, $|y^m z|_n < 1$ so $x = y^m z$ is the desired element. If, on the other hand, $|y^m|_n > 1$, let $x_m = zy^m/(1+y^m)$. Then the sequence $|x_m|_i$ converges to $|z|_i$ if $i = 1$ or n, and converges to 0 otherwise. Set $x = x_m$ where m is large enough so that $|x_m|_1 > 1$ and $|x_m|_i < 1$ for $2 \leq i \leq n$. □

If we let k_p denote the field k with the metric given by the valuation $|\ |_p$, then the next theorem says that the diagonal is dense in $\Pi_p k_p$.

1.4 THEOREM (weak approximation). *Let* $|\ |_1,\ldots,|\ |_n$ *be inequivalent nontrivial valuations on a Heyting field* k, *let* $x_1,\ldots,x_n$ *be elements of* k, *and let* $\epsilon > 0$. *Then there exists* $x \in k$ *such that* $|x - x_i|_i < \epsilon$ *for each* i.

PROOF. For each i use Lemma 1.3 to construct $y_i \in k$ such that $|y_i|_i > 1$, and $|y_i|_j < 1$ if $j \neq i$. For each $m \in \mathbb{N}$ define

$$z_{im} = y_i^m x_i/(1 + y_i^m).$$

Then $\lim_{m\to\infty}|z_{im} - x_i|_i = 0$, and $\lim_{m\to\infty}|z_{im}|_j = 0$ if $j \neq i$. Let $z_m - \Sigma_{i=1}^n z_{im}$. If m is large enough, $|z_m - x_i|_i < \epsilon$ for all i. □

The strong approximation theorem allows to choose integer approximations.

1.5 THEOREM (strong approximation). *Let T be a finite subset of a Dedekind set S of valutions on a Heyting field k. Let $\epsilon > 0$, and for each $p \in T$ let $x_p \in k$. Then there exists $y \in k$ such that*

(i) $|y - x_p|_p < \epsilon$ *for each* $p \in T$,

(ii) $|y|_q \leq 1$ *for each* $q \in S\setminus T$.

PROOF. We may assume $\epsilon \leq 1$. For each $p \in T$, there are only finitely many $q \in S$ with $|x_p|_q > 1$. For each p in T and q in $S\setminus T$ such that $|x_p|_q > 1$, add q to T and define $x_q = 0$. Thus we may assume that $|x_p|_q \leq 1$ for each $p \in T$ and $q \in S\setminus T$.

Fix $p \in T$. For each $q \in T\setminus\{p\}$, construct, by Definition 1.1.ii, an element of k that is an integer at S, that is close to 1 at p, and that is close to 0 at q. Let y_p be the product of these elements (the empty product is 1). The elements y_p are close to 1 at p and close to 0 at all valuations q in $T\setminus\{p\}$. Finally let $y = \Sigma_{p\in T} y_p x_p$, Then $|y|_q \leq 1$ for each $q \in S\setminus T$, and y is close to x_p at p. □

The strong approximation theorem allows us to write elements of k as quotients of integers.

1.6 THEOREM. *Let S be a Dedekind set of valuations on a Heyting field k, and let R be the set of integers at S. Then each element of k is a quotient of elements of R, and R is integrally closed in k.*

PROOF. Let $x \in k$; we shall write x as a quotient of elements of R. By Definition 1.1.i there is a finite subset T of S so that $|x|_q \leq 1$ for each $q \in S\setminus T$. If $|x|_p \leq 1$ for each p in the finite set T, then $x \in R$, so we may assume that $x \neq 0$. As each valuation in the finite set T is discrete, we may assume that $|x|_p > 1$ for each $p \in T$.

By Theorem 1.5 there is $y \in k$ with

$$|y - x^{-1}|_p < min\{|x^{-1}|_p : p \in T\} < 1$$

for each p in T, and $|y|_q \leq 1$ for each $q \in S\setminus T$. As $|y - x^{-1}|_p < |x^{-1}|_p$ for each $p \in T$, and $|\ |_p$ is nonarchimedean, it follows that $|y|_p = |x^{-1}|_p < 1$ for each p in T. In particular, $y \in R$. If $p \in T$, then $|xy|_p = |x|_p|y|_p = |x|_p|x^{-1}|_p = 1$, and if $q \in S\setminus T$, then $|xy|_q \leq 1$. Therefore $xy \in R$, so $x = (xy)y^{-1}$ is a quotient of elements of R.

As each valuation in S is nonarchimedean, it follows that any element of k, satisfying a monic polynomial over R, is in R. □

EXERCISES

1. Let S the set of all p-adic valuations on $\mathbb{Q}$. Show that the ring of integers at S is $\mathbb{Z}$. What is the ring of integers if S = {3}?
2. Verify that if $|\ |$ is a discrete valuation, then $\{x : |x| \leq 1\}$ is a local ring.
3. Complete $\mathbb{Q}$ in the p-adic valuation to show that a Dedekind domain need not be discrete.
4. **Chinese remainder theorem.** Use the strong approximation theorem to prove that if a_1 and a_2 in $\mathbb{Z}$ are relatively prime, and b_1 and b_2 are in $\mathbb{Z}$, then there exists x in $\mathbb{Z}$ congruent to b_i modulo a_i for $i = 1,2$.

2. IDEAL THEORY.

Throughout this section k will denote a Heyting field, S a Dedekind set of valuations on k, and R the associated Dedekind domain. A **fractional ideal** is a nonzero R-submodule A of k such that $|A|_p = max\{|x|_p : x \in A\}$ exists for each $p \in S$, and is 1 for all but finitely many $p \in S$. That is, for each $p \in S$, there exists $x(p) \in A$ such that $|y|_p \leq |x(p)|_p$ for each $y \in A$, and $|x(p)| = 1$ for p outside of a finite subset of S.

2.1 THEOREM. *Any nonzero finitely generated R-submodule of k is a fractional ideal.*

PROOF. Let A be a nonzero R-submodule of k generated by $a_1,\ldots,a_n$. As k is a Heyting field, one of the a_i is nonzero, and if $a_i \neq 0$, then either $a_j \neq 0$ or $a_i + a_j \neq 0$; thus we may assume that the a_i are all nonzero. Suppose $a = \Sigma_{i=1}^{n} r_i a_i$ is an element of A, and $p \in S$. As $|\ |_p$ is nonarchimedean, $|a|_p = |\Sigma_{i=1}^{n} r_i a_i|_p \leq max\ |r_i a_i|_p \leq max\ |a_i|_p$. As each $|\ |_p$ is discrete, we have $|A|_p = max\{|a_i|_p : i = 1,\ldots,n\}$ exists. We can choose a finite subset T of S so that $|a_i|_p = 1$ for $p \in S\backslash T$ and $i = 1,\ldots,n$. Then $|A|_p = 1$ if $p \in S\backslash T$. □

We shall show later that, conversely, every nonzero fractional ideal is generated by two elements as an R-module.

Let A and B be fractional ideals. Define the **sum** of A and B by $A + B = \{a + b : a \in A \text{ and } b \in B\}$, and the **product** of A and B to be set AB of finite sums of elements of the form ab, with $a \in A$ and $b \in B$.

2.2 THEOREM. *Let* A *and* B *be fractional ideals. Then* A + B *and* AB *and* A ∩ B *are fractional ideals such that*

$$\text{(i) } |A + B|_p \le \max(|A|_p, |B|_p)$$
$$\text{(ii) } |AB|_p = |A|_p |B|_p$$
$$\text{(iii) } |A \cap B|_p = \min(|A|_p, |B|_p)$$

for each $p \in S$.

PROOF. Choose $a \in A$ and $b \in B$ so that $|A|_p = |a|_p$ and $|B|_p = |b|_p$. Then $|x|_p \le |a + b|_p \le \max(|A|_p, |B|_p)$ for each $x \in A + B$, so $|A + B|_p = |a + b|_p$ exists and (i) holds. If $|A|_p = |B|_p = 1$, then clearly $|A + B|_p = 1$, so A + B is a fractional ideal.

For x ∈ AB we have $|x|_p \le |ab| = |A|_p|B|_p$, because $|\ |_p$ is multiplicative and nonarchimedean, so (ii) holds and AB is a fractional ideal.

If $x \in A \cap B$, then clearly $|x|_p \le \min(|A|_p, |B|_p)$, so it suffices to construct an element x of A ∩ B such that $|x|_p = \min(|A|_p, |B|_p)$. As $|\ |_p$ is discrete, the values $|A|_p$ and $|B|_p$ are comparable, so we may assume that $|A|_p \le |B|_p$. Now

$$T = \{p\} \cup \{q \in S : |a/b|_q < 1\}$$

is finite, so, by Theorem 1.5, there exists $y \in k$ with $|y|_p = |a/b|_p \le 1$, and $|y|_q \le \min(1, |a/b|_q)$ for all $q \in S$. The latter implies that $yb/a \in R$, so $yb \in A$, and also that $y \in R$, so $yb \in B$. But $\min(|A|_p, |B|_p) = |A|_p = |a|_p = |yb|_p$, so $x = yb$ is the desired element. □

A fractional ideal A is completely determined by its values $|A|_p$.

2.3 THEOREM. *Let* A *and* B *be fractional ideals. Then*

$$\text{(i) } A = \{x \in k : |x|_p \le |A|_p \text{ for each } p \in S\}.$$
$$\text{(ii) } A \subseteq B \text{ if and only if } |A|_p \le |B|_p \text{ for each } p \in S.$$

PROOF. To prove (i) let $x \in k$ be such that $|x|_p \le |A|_p$ for each $p \in S$. Choose a nonzero element $z \in A$ and consider xz^{-1}. As $|xz^{-1}|_p \le 1$ outside a finite set, and $|\ |_p$ is discrete, either $xz^{-1} \in R$, whence $x = (xz^{-1})z \in A$, or $xz^{-1} \ne 0$, whence $x \ne 0$. Thus we may assume that $x \ne 0$.

Replacing A by x^{-1}A, it suffices to show that $1 \in A$ whenever A is a fractional ideal such that $|A|_p \ge 1$ for each $p \in S$. Replacing A by $R \cap A$, and using (2.2), this is the same as showing that $1 \in A$ whenever A is a fractional ideal contained in R such $|A|_p = 1$ for each $p \in S$.

Let z be a nonzero element of A. By Definition 1.1.i there is a finite subset T of S so that $|z^{-1}|_q \leq 1$ for $q \in S\backslash T$. As $z \in A \subseteq R$ we have $|z|_q = 1$ for each $q \in S\backslash T$. If $|z|_p = 1$ for each $p \in T$, then $z^{-1} \in R$, so $1 \in A$. Therefore, as T is finite and each $|\ |_p$ is discrete, we may assume that T is nonempty and that $|z|_p < 1$ for each $p \in T$.

For each $p \in T$ choose $x_p \in A$ with $|x_p|_p = 1$. As $|z|_q = 1$ outside of a finite set, Theorem 1.5 says that there exists $y_p \in k$ such that $|y_p - x_p^{-1}|_p \leq |z|_p < 1$, and $|y_p|_q \leq |z|_q \leq 1$ for each $q \in S\backslash\{p\}$. As $|x_p^{-1}|_p = 1$, and $|\ |_p$ is nonarchimedean, it follows that $y_p \in R$. If $q \in T$, then

$$\begin{aligned} |1 - \Sigma_{p\in T}x_py_p|_q &= |(1 - x_qy_q) - \Sigma_{p\in T\backslash\{q\}}x_py_p|_q \\ &\leq max(|1 - x_qy_q|_q, max\{|y_p|_q : p \in T\backslash\{q\}\}) \\ &\leq max(|x_q(x_q^{-1} - y_q)|_q, |z|_q) = |z|_q. \end{aligned}$$

and if $q \in S\backslash T$, then

$$|1 - \Sigma_{p\in T}x_py_p|_q \leq 1 = |z|_q.$$

Therefore $(1 - \Sigma_{p\in T}x_py_p)z^{-1} \in R$, so $1 - \Sigma_{p\in T}x_py_p \in zR \subseteq A$. But $x_py_p \in A$, so $1 \in A$.

Statement (ii) follows immediately from (i). □

2.4 COROLLARY. *If $A \subseteq B$ are fractional ideals, then A is a detachable subset of B. Hence discrete Dedekind domains have detachable ideals.*

PROOF. If $x \in B$, then (2.3) says that $x \in A$ if and only if $|x|_p \leq |A|_p$ for each $p \in S$. There is a finite subset T of S such that $|A|_p = |B|_p = 1$ for $p \in T\backslash S$. Then $x \in A$ if and only if $|x|_p \leq |A|_p$ for each $p \in T$, which is decidable. Take $B = R$ to establish the second claim. □

We can now show that fractional ideals are finitely generated.

2.5 THEOREM. *For each $p \in S$ let d_p be an element of the value group of $|\ |_p$, such that $d_p = 1$ outside of a finite subset of S. Let $A = \{x \in k : |x|_p \leq d_p$ for each $p \in S\}$. Then $A \neq 0$, and if y is any nonzero element of A, there is nonzero $x \in A$* such that A is generated by *x and y* as an R-module. *Morever, $|A|_p = d_p$ for each $p \in S$.*

PROOF. By Theorem 1.5, there is a nonzero element $y \in A$. Let T be a finite subset of S so that $|y|_p = d_p = 1$ for each $p \in S\backslash T$. By Theorem 1.5 there exists $x \in k$ such that

(i) $|x|_p = d_p$ for each $p \in T$

(ii) $|x|_q \leq 1$ for each $q \in S \setminus T$.

Let B be the submodule of A generated by x and y. Then

$$|B|_p = max(|x|_p, |y|_p) = d_p \text{ for each } p \in S$$

so $|A|_p = |B|_p$ for each $p \in S$, whence $A = B$ by Theorem 2.3. □

Let I be the set of functions $\alpha : S \to \mathbb{Z}$ such that $\alpha(p) = 0$ outside a finite set (depending on α). Then I is the free abelian group on the set S. Let g_p be the generator of the value group of $|\ |_p$ that is less than 1. Let φ map the monoid of fractional ideals under multiplication to I, by taking A to the function α such that $|A|_p = g_p^{\alpha(p)}$. Theorem 2.2.ii says that φ is a homomorphism, Theorem 2.3 says that it is one-to-one, and Theorem 2.4 says that it is onto. Moreover, if $\varphi(A) \neq \varphi(B)$, then $A \neq B$ as subsets of k; that is, either there exists an element $x \in A$ such that $x \neq y$ for all $y \in B$, or there exists an element $y \in B$ such that $y \neq x$ for all $x \in A$.

The natural basis of the free group I on S consists of the functions δ_p such that $\delta_p(p) = 1$, and $\delta_p(q) = 0$ for $q \neq p$. The corresponding fractional ideals M_p are a maximal ideals in R, for if $x \in R$, and $|x|_p < 1$, then $x \in M_p$, while if $|x|_p = 1$, then the ideal A of R generated by x and M_p has $|A|_q = 1$ for each $q \in S$, so is equal to R.

EXERCISES

1. Show that any fractional ideal is detachable from k. Construct a Brouwerian example of a finitely generated submodule of the p-adic completion of $\mathbb{Q}$ that is not detachable.
2. Show that a discrete Dedekind domain is an integrally closed Lasker-Noether ring in which every nonzero proper prime ideal is maximal.

3. FINITE EXTENSIONS

We shall be interested in discrete valuations p on a discrete field k that satisfy the classically trivial condition:

(3.1) The Henselization $\tilde{k}_p$ is separably factorial, and the residue class field $\overline{k}_p$ satisfies condition P.

Let p be a discrete valuation on k satisfying (3.1), and let E be a finite separable extension of k. Then XII.8.4 and XII.8.5 say that the each valuation on E that extends p is discrete, and that the set of such valuations is finite.

The p-adic valuations on $\mathbb{Q}$ satisfy (3.1). The fields of algebraic number theory are finite-dimensional extensions of $\mathbb{Q}$, and they acquire Dedekind sets of valuations as follows.

3.2 THEOREM. *Let k be a discrete field and S a Dedekind set of valuations on k, each of which satisfies (3.1). Let E be a finite dimensional separable extension of k, and let S' be the set of all valuations on L that extend a valuation on S. Then S' is a Dedekind set of valuations.*

PROOF. We first show that S' is discrete. Let P and Q be in S'. If P and Q are extensions of distinct valuations in S, then they are distinct. So suppose that P and Q both extend $p \in S$. By XII.7.5 the set of valuations of S' that extend p is finite, so either $P = Q$ or $P \neq Q$.

That each valuation in S' is discrete follows from XII.8.4.

Given $x \in E$, we must construct a finite subset T' of S' such that $|x|_{P'} \leq 1$ for each $P \in S'\backslash T'$. As E is a finite dimensional extension of k, then by VI.1.13 there exists an irreducible polynomial $f(X) = X^n + a_1X^{n-1} + \cdots + a_n \in k[X]$ such that $f(x) = 0$. Let T be a finite subset of S such that $|a_i|_p \leq 1$ for each i and each $p \in S\backslash T$. Let T' be the finite subset of S' consisting of those valuations on E that extend elements of T. If $P \in S'\backslash T'$, then $|a_i|_P \leq 1$, so $|x|_P \leq 1$ as P is nonarchimedean.

Finally we show that if P and P' are distinct valuations in S', and $\epsilon > 0$, then there exists $x \in E$ such that $|1 - x|_P$ and $|x|_{P'}$ are less than ϵ, and $|x|_Q \leq 1$ for each $Q \in S'\backslash\{P,P'\}$. Let P and P' be extensions of p and p' in S respectively. If $p \neq p'$, then, since S is a Dedekind set, there is $x \in k$ satisfying the required properties; so we may assume that $p = p'$. We may also assume that $\epsilon \leq 1$. Let S'_p be the finite set of valuations on E that extend p. By Theorem 1.4 there exists $y \in E$ such that $|y|_Q < \epsilon$ for each $Q \in S'_p\backslash\{P\}$, and $|y - 1|_P < \epsilon$; consequently, $|y|_Q \leq 1$ for each $Q \in S'_p$. By the preceding paragraph, there is a finite set T' of S', disjoint from S'_p, such that $|y|_Q \leq 1$ for each $Q \in S'\backslash T'$. Let T be the finite subset of S that have extensions in T'. Then $p \notin T$, and, by Theorem 1.5, there is $x \in k$ so that $|x|_q \leq |y|_Q^{-1}$ for each Q extending $q \in T$, and $|1 - x|_p < \epsilon$, and $|x|_q \leq 1$ for each $q \in S\backslash T$. Then $|xy|_Q = |x|_Q|y|_Q \leq 1$ for each $Q \in S'$, and $|1 - xy|_P = |1 - y + (1-x)y|_P < \epsilon$, and $|xy|_{P'} < \epsilon$. □

Suppose R is a discrete Dedekind domain with field of quotients k, and that E is a finite dimensional separable extension of k. If the valuations that determine R satisfy (3.1), then Theorem 3.2 shows how to construct a Dedekind domain with field of quotients E. We now give a purely algebraic characterization of that Dedekind domain.

3.3 THEOREM. *Let E be a finite dimensional separable extension of a discrete field k. Let S be a Dedekind set of valuations on k so that each $p \in S$ satisfies (3.1). Let S' be the Dedekind set on E consisting of those valuations that extend a valuation in S. Then the following are equivalent for an element $x \in E$.*

(i) *x is an integer at S',*

(ii) *x satisfies a monic irreducible polynomial with coefficients in R,*

(iii) *x is integral over R.*

PROOF. Clearly (ii) implies (iii), while (iii) implies (i) because each valuation in S' is nonarchimedean and extends a valuation on S. To prove that (i) implies (ii), suppose $x \in E$ is an integer at S'. As E is finite dimensional, (VI.1.13) says that x satisfies an irreducible polynomial $f(X) = X^n + a_1X^{n-1} + \cdots + a_n \in k[X]$. Thus $k[\alpha]$ is a finite dimensional extension of k, so we may assume that $k[\alpha] = E$. It suffices to show that $|a_i|_p \leq 1$ for each $p \in S$ and each i. Let $p \in S$, and let $\tilde{k}_p$ be the Henselization of k at p. As E is separable, so is f. As p satisfies (3.1), the field $\tilde{k}_p$ is separably factorial, so (VII.2.4) says that there is a splitting field K_p for f over $\tilde{k}_p$. Let $r_1,\ldots,r_n$ be the roots of f in K_p, and let P be the unique extension of the valuation p to K_p guaranteed by XII.7.3. Define monomorphisms $\sigma_i : E \to K_p$, for $i = 1,\ldots,n$, by setting $\sigma_i x = r_i$. By (XII.7.5), each valuation $|\ |$ on E which extends p is of the form $|y| = |\sigma_i y|_P$ for some i. As $|x|_Q \leq 1$ for each $Q \in S'$, it follows that $|r_i|_P \leq 1$ for $i = 1,\ldots,n$. The coefficients of $f(X)$ are symmetric polynomials in the roots r_j, and since P is nonarchimedean, $|r_i|_P \leq 1$ for each i. But $r_i \in k$, so $|r_i|_p = |r_i|_P \leq 1$. □

Bibliography

Artin, E.

1967 *Algebraic numbers and algebraic functions*, Gordon & Breach.

Baumslag, G., F. B. Cannonito and C. F. Miller, III

1981 Computable algebra and group embeddings, *J. Algebra* **69**, 186–212.

Bishop, E.

1967 *Foundations of constructive analysis*, McGraw-Hill.

1973 *Schizophrenia in contemporary mathematics*, AMS Colloquium Lectures, Missoula, Montana.

Bishop, E. and D. S. Bridges

1985 *Constructive analysis*, Grundlehren der mathematischen Wissenschaften 279, Springer-Verlag

Borevich, Z. I. and I. R. Schaferevich

1966 *Number theory*, Academic Press, New York.

Bourbaki, N.

1961 *Algèbre commutative: I. Modules plats*, Hermann, Paris.

Bridges, D. S.

1979 *Constructive functional analysis*, Research notes in mathematics 28, Pitman, London.

Bridges, D. S. and F. Richman

1987 *Varieties of constructive mathematics*, London Math. Soc. Lecture Notes 97, Cambridge Univ. Press.

Brouwer, L. E. J.

1981 *Cambridge lectures on intuitionism*, D. van Dalen editor, Cambridge Univ. Press.

Brouwer, L. E. J. and B. de Loor

1924 Intuitionistischer Beweis des Fundamentalsatzes der Algebra, *Proc. Acad. Amsterdam* **27**, 186–188.

Diaconescu, R

1975 Axiom of choice and complementation, *Proc. Amer. Math. Soc.* **51**, 176–178.

Feferman, S.

1975 Impredicativity of the existence of the largest divisible subgroup of an abelian p-group, *Model theory and algebra*, Springer Lecture Notes 498, 117–130.

Fourman, M. P. and A. Scedrov
1982 The "world's simplest axiom of choice" fails, *Manuscripta Math.* **38**, 325–332.

Greenleaf, N.
1981 Liberal constructive set theory, *Constructive mathematics*, Springer Lecture Notes 873, 213–240.

Heyting, A.
1941 Untersuchungen über intuitionistische algebra, *Verhandelingen Akad. Amsterdam*, 1. sectie **18**.
1971 *Intuitionism, an introduction*, North-Holland.

Julian, W., R. Mines and F. Richman
1978 Algebraic numbers, a constructive development, *Pac. J. Math.* **74**, 91–102.
1983 Alexander duality, *Pac. J. Math.* **106**, 115–127.

Kaplansky, I.
1949 Elementary divisors and modules, *Trans. Amer. Math. Soc.* **66**, 478–479.
1969 *Infinite abelian groups*, Univ. of Mich. Press.

Kronecker, L.
1882 Grundzüge einer arithmetischen Theorie der algebraischen Grossen (section 4), *Journal für die reine und angewandte Mathematik* **92**, 1–122.

Lin, C.
1981 Recursively presented abelian groups: effective p-group theory I, *JSL* **46**, 617–624.
1981a The effective content of Ulm's theorem, *Aspects of effective algebra*, J. Crossley (ed), Upside down A book company, Yarra Glen, Victoria, Australia

Magnus, W., A. Karrass and D. Solitar
1966 *Combinatorial group theory*, Interscience, New York

Mal'cev, I. A.
1971 On recursive abelian groups, *The metamathematics of algebraic systems*, North-Holland, Amsterdam.

Metakides, G., and A. Nerode
1979 Effective content of field theory, *Annals of Math. Logic* **17**, 289–320.

Mines, R. and F. Richman
1981 Dedekind domains, *Constructive mathematics*, Springer Lecture Notes 873, 16–30.
1982 Separability and factoring polynomials, *Rocky Mtn. J. Math.* **12**, 91–102.
1984 Valuation theory: a constructive view, *J. Number Theory* **19**, 40–62.
1986 Archimedean valuations, *J. London Math. Soc.* **34**, 403–410.

Myhill, J. and N. D. Goodman
1978 Choice implies excluded middle, *Zeit. Math. Log.* **23**, 461

Nagata, M.
1962 *Local rings*, Interscience, New York

Olson, P. L.
1977 *Difference relations and algebra: a constructive study*, Univ. of Texas at Austin dissertation.

O'Meara, O. T.
1963 *Introduction to quadratic forms*, Springer-Verlag, Berlin.

Richman, F.
1973 The constructive theory of countable abelian p-groups, *Pac. J. Math.* **45**, 621-637.
1974 Constructive aspects of Noetherian rings, *Proc. Amer. Math. Soc.* **44**, 436-441.
1975 The constructive theory of KT-modules, *Pac. J. Math.* **61**, 263-274
1977 A guide to valuated groups, *Abelian group theory*, Springer Lecture Notes 616, 73-86.
1977a Computing heights in Tor, *Houston J. Math.* **3**, 267-270.
1981 Seidenberg's condition *P*, *Constructive mathematics*, Springer Lecture Notes 873, 1-11.
1982 Finite dimensional algebras over discrete fields, *L.E.J. Brouwer centenary symposium*, A.S. Troelstra and D. van Dalen (editors), North-Holland Pub. Co., 397-411.
1988 Nontrivial uses of trivial rings, *Proc. Amer. Math. Soc.*

Rogers, L.
1980 Basic subgroups from a constructive viewpoint, *Communications in algebra* **8**, 1903-1925.

Rootselaar, B. van
1960 On intuitionistic difference relations, *Indag. math.* 22, 316-322. Corrections: *Indag. math.* 25, 132-133.

Rudin, W.
1985 Unique right inverses are two-sided, *Amer. Math. Monthly* **92**, 489-490.

Scott, D.
1979 Identity and existence in intuitionistic logic, Springer Lecture Notes 753, 660-696.

Scedrov, Andre
1986 Diagonalization of continuous matrices as a representation of intuitionistic reals, *Ann. Pure Appl. Logic* **30**, 201-206.

Seidenberg, A.
1970 Construction of the integral closure of a finite integral domain, *Rend. Sem. Mat. Fis. Milano* **40**, 100-120.
1971 On the length of a Hilbert ascending chain, *Proc. Amer. Math. Soc.* **29**, 443-450.

1972 Constructive proof of Hilbert's theorem on ascending chains, *Trans. Amer. Math. Soc.* 174, 305–312.
1973 On the impossibility of some constructions in polynomial rings, *Proc. Int. Cong. Geom*, Milano 1971, 77–85
1974 Constructions in algebra, *Trans. Amer. Math. Soc.* **197**, 273–313.
1974a What is Noetherian?, *Rend. Sem. Mat. e Fis. Milano*, **44**, 55–61.
1975 Construction of the integral closure of a finite integral domain. II, *Proc. Amer. Math. Soc.* 52, 368–372.
1978 Constructions in a polynomial ring over the ring of integers, *Amer. J. Math.* **100**, 685–703.
1984 On the Lasker-Noether decomposition theorem, *Amer. J. Math.* **106**, 611–638.
1985 Survey of constructions in Noetherian rings, *Proc. Symp. Pure Math.* **42**, 377–385.

Smith, H. J. S.
1861 On systems of linear indeterminate equations and congruences, *Phil. Trans.* **151**, 293–326, in the collected mathematical papers of H. J. S. Smith, J. W. L. Glaisher (ed) Chelsea, NY 1965.

Smith, Rick L.
1981 Two theorems on autostability in *p*-groups, *Logic year* 1979–1980, Springer Lecture Notes 859, 302–311.

Soublin, J-P.
1970 Anneaux et modules cohérents, *J. Algebra* **15**, 455–472.

Staples, J.
1971 On constructive fields, *Proc. London Math. Soc.* (3) **23**, 753–768.

Stoltzenberg, G.
1968 Constructive normalization of an algebraic variety, *Bull. Amer. Math. Soc.* 74, 595–599.

Uspenskii and Semenov
1981 Springer Lecture Notes in Computer Science 122

Waerden, B. L. van der
1930 Eine Bemerkung über die Unzerlegbarkeit von Polynomen, *Math. Annalen* **102**, 738–739.
1953 *Modern Algebra*, Ungar, New York.

Wang, H.
1974 *From mathematics to philosophy*, Routledge & Kegan Paul, London.

Weiss, E.
1963 *Algebraic number theory*, McGraw-Hill, New York.

Index